1001 COMPUTER TELEPHONY TIPS, SECRETS AND SHORTCUTS

Covering Service Planning, Ordering Telco Services, Testing, Installation, Fraud Control, Platform Selection, Buying Questions, Selling Strategies, Application Hints, Script Design and Component Gotchas.

by Edwin Margulies

A Telecom Library, Inc. Book
Published by Flatiron Publishing, Inc.

ISBN 0-936648-86-4

Manufactured in the United States of America

First Edition, May, 1996
Cover Designed by Mara Leonardi and Saul Roldan
Printed at BookCrafters, Chelsea, MI.

Acknowledgments

This document was made possible through the efforts of the top consultants, developers, installers, sales people and reporters in the computer telephony industry. Literally hundreds of tips, white papers, presentations and articles were submitted by our "tipster" friends. Get to know them -- they are stars of this industry.

Special thanks go to 4-Sight, L.C., ABConsultants, Access Graphics, Inc., ACS Wireless, Active Voice, Adtran, Inc., Alliance Systems, Inc., Applied Language Technologies, Inc., ART, Inc., Aspect Telecommunications, Brite Voice Systems, Brooktrout Technology, Inc., Calo Software, Inc., Center-Core, Inc., Cintech Tele-Management Systems, Computer Telephony Magazine, Crystal Group, Inc., CT Division Dialogic Corporation, Cybernetics Systems International, Davidson Consulting, Davox International, Dialogic Corporation, Diamond Head Software, Inc., Diamond Multimedia Systems, Inc. -- Supra Communications Division, DIgby 4 Group, Inc., Digital Systems International, ENSONIQ, Enterprise Integration Group, Excel, Inc., Expert Systems, Inc., GammaLink Division of Dialogic Corporation, Genoa Technology, GM Productions, Hammer Technologies, Inc., Harris Digital Telephone Systems, Ibex Technologies, Inc., I-Bus PC Technologies, Intel Corporation, JABRA Corporation, MediaSoft Telecom, Micom, Inc., New Pueblo Communications, Novell, Nuntius Corporation, Octel Communications Corporation, Optus Software, Panamax, Parity Software, PIKA Technologies Inc., Plantronics, PureSpeech, QNX Software Systems Ltd., RAAC Technologies, Inc., Results Technologies, Rhetorex, Inc., Robins Press, Rockwell International Switching Systems Division, Siemens ROLM Communications, Inc., Soft-Com, SoftLinx, Inc., TALX Corporation, Technology Solutions Company, Teknekron Infoswitch Corporation,

TELECONNECT Magazine, Telephone Response Technologies, Inc., The Kaufmann Group, TKM Communications, Veritel Corporation of America, Voice Information Associates, Inc., Voysys Corporation, WTS Bureau Systems, Inc. and Xircom Systems Division.

All the contact information for editorial contributors is in Appendix B. It's a treasure of the greatest "tipsters" in the world. Enjoy their knowledge.

Ed Margulies
Spring, 1996
EdMargulies@mcimail.com

Who Should Read This Book and Why

Finally! A book crammed with the best advice from the top experts in telephones, computers and communications. This is a must read if you are new to computer telephony or if you want a treasure trove of pointers from the pioneers of this industry. Written in short, no-nonsense "sound bite" style, this book delivers dollar and sanity-saving tips on:

- Configuring your PC for Computer Telephony
- Installing Analog and Digital circuits
- Creating The Best Application flow
- User Interface Design
- Service planning for Call Centers and Clients
- How to Choose and buy the Right Components
- Scripting, Prompts and Operating System Hints
- Mistakes to Avoid in Design and The Best Topologies

This book is a compendium of the best tips from the best minds. I can't think of a better book to read if you want to be successful in this field. Read it if you are a system designer — you'll learn both what to do and what not to do. Read it if you are in sales — you'll discover the best approach in pleasing your customers. Read it if you are about to buy something — you'll know what questions to ask. Read it to learn the best strategies for your call center or help desk if you want to improve customer service and increase productivity. I guarantee you'll pass this book around to your best friends.

Harry Newton
Spring, 1996
harrynewton@mcimail.com

or (snail mail)
12 West 21st Street
New, York, NY 10010

Who Should Read This Book and Why

Table of Contents

Chapter 1

Service Planning 1 - 1

Application Planning Tips	1 - 2
Application Planning Time	1 - 2
Applications Requirements List	1 - 4
Planning And Gathering Ideas	1 - 5
Objectives For A Proposed IVR System	1 - 7
Applications From "Discovery" Input	1 - 8
Planning - Questions Checklist	1 - 8
Pre-Launch Education and Promotion	1 - 10
Wrong Reasons Don't Work	1 - 10
ASR Arguments	1 - 12
Automated Attendant Considerations	1 - 13
Caller ID - Upgrade	1 - 14
Caller ID And Why To Use It	1 - 14
Caller ID Planning Process	1 - 18
When To Use Consultants	1 - 20
Credit Cards And Applications	1 - 21
Fax Planning Tips	1 - 22
Fax Servers - Why To Use Them	1 - 24
Fax Server - LAN Advantages	1 - 25
Fax Reception Truths	1 - 26
Fax Server Productivity	1 - 29
Fax-On-Demand Benefits	1 - 31
Fax-On-Demand Success Tips	1 - 35
Fax On Demand With WWW	1 - 36
Fax-On-Demand & WWW Planning	1 - 38
Faxing Quality Tips	1 - 40

Fax Machines 1 - 41
Fax Broadcasting Guidelines 1 - 42
Fax Statements 1 - 42
Fax Bureau Considerations 1 - 43
Integration of Data, Voice and Fax 1 - 45
Debit Calling Service Planning 1 - 46
International Call-Back Planning 1 - 46
Digit Assignment 1 - 50
Disaster Prevention 1 - 50
Failure Planning - Networks 1 - 54
International Service Planning Lessons 1 - 54
Intranet Planning 1 - 56
Middleware And CTI 1 - 57
On "Hold" Buying 1 - 60
Peer To Peer Vs. Internet 1 - 61
Reseller Or Distributor CT Tips 1 - 64
Traffic And Port Sizes 1 - 67
Traffic Thumbnail Sketch 1 - 68
Thumbnail Traffic Analysis 1 - 69
Voice And Fax Impact On LAN 1 - 70

Chapter 2

Ordering Telco and Carrier Services2 - 1

Caller ID And Centrex 2 - 2
Dedicated 800 vs. Virtual Lines 2 - 2
Detecting Loop Current Reversal 2 - 3
DID Or DNIS Digits 2 - 3
DNIS - DID Digit Passing 2 - 3
Document Service Changes 2 - 4
Fax On Demand Services 2 - 4
Installation Bills 2 - 5

Installation On Time 2 - 6
ISDN Is Real 2 - 6
Why Use ISDN 2 - 7
ISDN Standards 2 - 8
ISDN vs. Switched 56 2 - 8
ISDN FAQ 2 - 9
ISDN - BRI Versus PRI 2 - 11
ISDN - Ordering BRI 2 - 13
ISDN Vs. T-1 2 - 18
PRI Vs. T-1 Speed 2 - 20
PRI Vs. T-1 Reliability 2 - 20
PRI Vs. T-1 Cost 2 - 20
Loop Vs. Ground Start 2 - 21
T-1 Carrier Lines Ordering 2 - 21
CSU Companies 2 - 25
T-1 Line Installation 2 - 25
Localized T-1 2 - 26
Touch-Tone Charges 2 - 26
Traffic Tips 2 - 26
Virtual Nets Cheap 2 - 28

Chapter 3

Application Development And OS Tips 3 - 1

ADSI Information Sources 3 - 2
Answering Consistency 3 - 2
Application Development 3 - 3
Use AppsGens For Simple IVR 3 - 5
Application Menu Design 3 - 7
Area Code Look-Ups 3 - 9
ASR Application Design Tips 3 - 10
ASR Vocabulary Break-Down 3 - 10

Auto Attendant Questions	3 - 11
Backup With XCOPY	3 - 11
Caller ID Problems	3 - 12
Caller Input Quality	3 - 12
Channel Buffer Size for OS/2	3 - 15
Conventional Memory & Event Queue	3 - 16
Credit Card Software For IVR	3 - 17
Dialing International Numbers	3 - 19
Document Imaging For Fax	3 - 19
Error Handling	3 - 21
Expanded Memory Use	3 - 22
Extended Memory In DOS	3 - 23
Blaster Software Features	3 - 24
Fax-on-Demand Document Changes	3 - 25
Fax Server Integration	3 - 26
File Handles - Opening 20+	3 - 26
Forms-Based Development Tools	3 - 28
Forms - Text Mode Saves	3 - 32
ISDN DOS Driver Configuration	3 - 32
IVR Programming Tips	3 - 32
Japanese Disconnect Supervision	3 - 34
LAN Survival	3 - 35
List Management	3 - 35
Logic Flow Considerations	3 - 38
Mail Merge With Powerfax	3 - 40
Memory Terms	3 - 43
Memory Saving Tips	3 - 45
Modem Hang-Up	3 - 46
Operating System Selection	3 - 46
OS For Fax Servers	3 - 55
Outbound Calling Considerations	3 - 56
Pager Tone Detection	3 - 57
PBX Integration With D/42-SX;SL	3 - 58
Prompt For Zero Messages	3 - 59
Prompt Silence	3 - 59
Prompt Succinctly	3 - 60
Prompt To Touch	3 - 62

Prompts - Human Interface 3 - 63
Prompt Recording Tipfest 3 - 64
Rotary Pulse Conversion Snags 3 - 70
Scheduling Development 3 - 71
Script-Based Development 3 - 74
Scripts and Recording Tips 3 - 79
Synchronous Callback Mode 3 - 80
Talk-Off And How To Beat It 3 - 81
TSAPI-Based Applications 3 - 81
Type-Ahead 3 - 82
UNIX Kernal Tuning 3 - 83
UNIX SRL Parameters 3 - 84
Upper Memory, D40DRV And TSRs 3 - 85
Upper Memory DOS Device Drivers 3 - 87
Upper Memory And QEMM-386 3 - 87
V&H Routing 3 - 88
Voice Files From RAM Disk 3 - 88
Voice Processing Menus 3 - 89
Voice Storage 3 - 89
Voice Won't Stac 3 - 89
Warning Messages 3 - 90
Watchdog For PCs 3 - 90
Windows 95 And EMM386 3 - 92
Windows CT Performance Tips 3 - 92
Win 3.1 And NT Server 3 - 94
Windows INI Files 3 - 94
Windows Memory 3 - 94
Win NT - Compiling Visual C++ 3 - 95
Windows NT A Good Choice 3 - 96
Windows Programming Preamble 3 - 98

Chapter 4

Successful Call Center Strategies4 - 1

ACDs And What They Do	4 - 2
ACDs And Web Callback	4 - 4
ANI Considerations	4 - 7
Advanced Auto Attendant Features	4 - 10
Customer Relationship Sensitive Service	4 - 11
DMI - Call Center Desktop Management	4 - 12
DMI In Action	4 - 18
DSVD For Technical Support	4 - 19
Ergonomics For The Call Center	4 - 23
Ergonomic Workplace Creation	4 - 24
Headset Considerations	4 - 26
Help Desk - Handling Support Calls	4 - 27
Help Desk Support Roles	4 - 33
Home Agent Planning	4 - 35
On-Line Call Center Strategies	4 - 37
Screen Pop Test	4 - 38
Agent Scripting	4 - 39
Staffing Levels	4 - 40
Wireless Phone Considerations	4 - 41
Workforce Management	4 - 43

Chapter 5

Platforms For Computer Telephony 5 - 1

Buying Industrial PCs 5 - 2
Mission Critical Platforms 5 - 9
Mission Critical Benefits 5 - 10
PC Buses, Fans And Slots 5 - 15
PC Switching Configurations 5 - 16

Chapter 6

Component Magic Tricks 6 - 1

AEB Signaling on D/41E 6 - 2
AMI WinBios Configuration 6 - 3
B8ZS Line Encoding Detection 6 - 4
D/41E-SC - Non-SCbus Releases 6 - 6
Detecting Dial Tone With D/42-NS 6 - 7
DOS Communications Software 6 - 7
DSVD Setup 6 - 7
DTMF Dialing Levels 6 - 10
Acoustically Coupled DTMF Generators 6 - 11
Fax On Demand Parts 6 - 13
Flash ROM Updates 6 - 14
Hard Disks Must Be Fast 6 - 14
Interrupt Dueling 6 - 15

LAN Card Priority 6 - 15
Memory Conflicts 6 - 15
Modems And Call Waiting 6 - 15
NEC Mark II PBX Integration 6 - 16
Norstar Busy Signals 6 - 16
Norstar And Caller ID 6 - 17
PEB Configuration - HD Series 6 - 18
R2MF High Rate Signaling 6 - 18
Reboot Switch For Computer 6 - 19
SC2000 Internal Registers 6 - 21
Serial Port Overruns 6 - 21
Shared Memory - ISA on Pentium PCs 6 - 27
Shared RAM Voice Card Exclude 6 - 28
Software Interrupt 6D 6 - 28

Chapter 7

Testing And Installation Tips 7 - 1

BRI ISDN Installation Tips 7 - 2
Cables Are Always Bad 7 - 5
Cool Your Phone System 7 - 5
DID Mistake to Avoid 7 - 5
T-1 Card Installation Tips 7 - 6
European Tip & Ring 7 - 9
Fax Archive Consideration 7 - 9
FAX On Demand Installation Warnings 7 - 9
Fax Server IRQ Problems 7 - 10
Fax Server Compatibility Issues 7 - 11
Fax Testing 7 - 12
Grounding Problems 7 - 13
Wireless Considerations 7 - 13
Key System Alert 7 - 14

LAN Segmentation Strategy 7 - 14
Load Testing IVR Apps 7 - 20
Meridian 1 PBX T-1 Conditions 7 - 20
Mistakes To Avoid - Large Systems 7 - 21
PBX Disconnect Supervision Tones 7 - 24
Power Protection On Phone Lines 7 - 26
Power Protection On AC 7 - 26
Power Failure Transfer 7 - 26
Power Outlets 7 - 27
Power Problem Solvers 7 - 27
Protection For ISDN 7 - 31
Reboot Your Phone System 7 - 32
PBX Routing Tables 7 - 32
Simulators For Testing 7 - 32
T-1 Diagnostic Gear 7 - 32
T-1 Troubleshooting Gear Vendors 7 - 33
T-1 Trouble - Lock-Out 7 - 34
Test Everywhere 7 - 34
Test Platform Economics 7 - 35
Testing The Entire Lifecycle 7 - 36
Thunderstorms 7 - 40
Troubleshooting for T-1 7 - 40
Troubleshooting PC Boards 7 - 41
Trunks - Check For Bad Ones 7 - 42
Water Alert 7 - 42

Chapter 8

Buying and Selling Tips 8 - 1

Alarm Subsystems 8 - 2
Buying Application Tools 8 - 6
Application Generator Buying Tips 8 - 18

Application Generators For UNIX 8 - 19
Application Tool Software Choices 8 - 20
ASR Application Opportunities 8 - 21
ASR Benefits 8 - 21
ASR Challenges 8 - 24
ASR Considerations 8 - 28
ASR Integrator Guidelines 8 - 28
ASR System Evaluation 8 - 30
ASR Technology Evaluation 8 - 30
Caller ID Buying Tips 8 - 32
Caller ID - Things To Look For 8 - 32
Choosing The Right Voice Card 8 - 33
Consultants and Honesty 8 - 38
Debit Calling Service Tips 8 - 39
Fax Application Generator Questions 8 - 41
Fax-on-Demand And Who Needs It 8 - 43
Fax Cards And What To Look For 8 - 43
Fax Modem Co-Processors 8 - 45
Fax Server Pricing 8 - 45
Fax Server Retries 8 - 46
Fax Server Vendor Questions 8 - 46
Fax Server Vendor Segmentation 8 - 50
Fax Service Bureau Questions 8 - 51
Fax-On-Demand Bureau Selection 8 - 52
Fax-On-Demand Via Fax Server 8 - 57
Fax On Demand Software Features 8 - 57
Fax On Demand System Pricing 8 - 60
Fax-on-Demand System Sizing; Pricing 8 - 62
Getting Started Successfully 8 - 63
Headsets - The Wireless Kind 8 - 67
IVR Buying Reasons 8 - 68
IVR Selling Approach 8 - 69
IVR Cost Justification 8 - 71
OS Buying Tips - Fault Tolerant Platforms 8 - 72
PBX or ACD Buying Tips 8 - 73
PC Component Selection 8 - 75
Answering Machine Detection 8 - 78

Predictive Dialer Buying Tips 8 - 79
Professional Services Fees 8 - 83
Selling Ideas In An Organization 8 - 84
Selling Fax On Demand 8 - 84
Selling Fax Servers 8 - 90
Selling Customers Is Knowing Them 8 - 91
Selling With Simulators 8 - 96
SOHO-Perfect PC Phone 8 - 96
Switching System Selection 8 - 99
Text-To-Speech Evaluation 8 - 101
Universal Messaging 8 - 102
Used Equipment Makes Sense 8 - 105
Videoconferencing Buying Tips 8 - 107
Cheap Video Conferencing System 8 - 108
Voice Mail System - Keep It Simple 8 - 109
Voice Production Outsourcing 8 - 109
Voice Talent Agency Selection 8 - 111

Chapter 9

Fraud and Money Saving Tips9 - 1

Airplane Phone Warnings 9 - 1
Auto Attendant Security 9 - 2
Block Calls To 900, Etc. 9 - 2
Caller ID Curbs LD Abuse 9 - 2
Dialback Security 9 - 3
DISA And Voice Mail Fraud 9 - 3
DISA - PBX Maintenance Ports 9 - 3
Fax And Its Hidden Cost 9 - 4
Fax Boards And Wasteful Speed 9 - 4
Fraud Buster Top Tips 9 - 4
IVR Logic Flow Security 9 - 12

PBX Fraud Stoppers 9 - 12
Shoulder Surfing Fraud 9 - 13
Switchroom Security 9 - 13
Telex, Fax And Directory Scams 9 - 14
Toll Abuse Responsibility 9 - 14
Toll Restrict These Numbers Instantly 9 - 14
Voice Authentication Alert 9 - 14

Appendix A

Resources and Bibliography A - 1

Appendix B

Tipster Contact Information B - 1

Chapter 1

Service Planning

Before you launch into a development effort and spend loads of cash, read this chapter for the best advice on goal-setting, "specsmanship" and sensitivity to the discovery processes. There's no sense in putting a computer telephony system on line unless you are solving the right problems. Perhaps the most important aspect of service planning is to get everyone involved. Ask your customers what they want. Ask your colleagues what they want. Take a look at how your competitors are handling their information flow.

Above all, you should pretend to be a customer for a few days and call in to your own company. Are you directed to the correct telephone extensions when you ask for a particular department? Can you get your hands on obvious things like the corporate fax number, Web address and directions? How easy is it to get a brochure faxed to you? How long does it take to find out about the status of an order?

What you find out may surprise you. If you think you already know what type of switch, IVR system or fax server you are going to install – put it off long enough to read this advice from the top planners in our industry.

Application Planning Tips

A sad but true fact is that many voice systems are developed with woefully inadequate planning before the application itself is constructed. A number of "tried and true" techniques exist to aid the development of computer telephony applications. The quality of voice systems that follow these guidelines should show a significant improvement over those developed without them. Your customers will notice the difference and you will be rewarded with far fewer support calls. For every hour spent on planning, it will save you more than double that time over the duration of the project.

Even if you are developing systems for in-house use, the steps discussed in this chapter are quite important to follow. It is not advisable to "jump right in" and start developing without the proper planning. This makes for disappointment and a lot of bugs.

(Telephone Response Technologies, Inc.)

Application Planning Time

Give yourself time for planning. It's the most important thing when putting in a new business telephone system. For a system with 250 telephones, start a year in advance, longer if new space is being built. Get everyone's input, particularly the people who will daily be working with the system such as the system administrator, switchboard attendant and those responsible for covering the telephones and screening calls.

Multiple areas of expertise required. Implementation of computer telephony requires expertise in computer hardware, software and telephony. Don't expect one person or one company to be knowledgeable in all three. Estimate the time required by each participant to get the application up and running and then double the estimate! This is based upon experience.

Documentation. Create accurate documentation as your computer telephony application is put together. Without it, system changes will be more trouble, will take longer and will cost more.

Good documentation covers the telephone and computer hardware and software and the outside telephone lines. Designate a person to be responsible for keeping the information current.

ANI Availability. If your Computer Telephony will rely on Automatic Number Identification (ANI), you may first want to find out what percentage of your calls actually arrive with this information and how well it matches up with your database. To do this, set up a Call Accounting System to track incoming calls and the ANI (also called Caller ID) they carry.

Expansion. If your Computer Telephony application is successful, you will expand it. When setting up the system, find out the growth capabilities and what it will take in terms of time, effort and cost to expand the system. This includes telephone and computer hardware and software and outside lines.

Links to other software. Determine if the information collected by your Computer Telephony applications will feed information to another system. Software interfaces may need to be written and the cost for this factored into the budget. For example, some call centers want information about the representatives performance fed into a payroll system. Beware of the terms *flat ACSII file* and *no problem* used in the same sentence.

Functions -- all on the keyboard. If your Computer Telephony application uses both a telephone and a computer on the desk, make sure the telephone functions can be controlled by the computer keyboard (answering, transferring, ending a call, etc.) Don't create confusion for the representative by having some functions performed on the telephone and others on the keyboard.

Computer Processing Time. If your call center is using screen pops with forms to be filled out, set up your application so that the computer does not wait for the end of the call to process the information. Otherwise, you may create a delay in receiving the next call while the representative waits for the computer to process the information.

(DIgby 4 Group, Inc.)

Applications Requirements List

Usually, a written list of desired features or a Request-For-Proposal (RFP) has been generated by the "customer" (or head of the company or department where the voice system will be installed). You should be clear on what the CT system should do and how it should do it. Everyone who will be involved in the application or system development should thoroughly read this list of requirements. If one does not exist, ask for one to be generated. The following is a simplified example of a list of application requirements and objectives that you might get from a typical voice system customer:

XYZ Corporation Order Hotline Requirements. XYZ Corporation would like to have a voice system developed that will be installed at our corporate headquarters behind our phone system. This system is to have the features listed in figure 1.1.

Figure 1.1 Typical Feature/Function List

Feature	Requirement Description
1	Answer all incoming calls within one ring and play a brief "Welcome to XYZ Corporation" greeting to callers.
2	A main menu will allow callers to select from one of several operations.
3	From the main menu, allow callers to transfer to any department phone extension using an audio directory of departments. If the party is not in or the extension is busy, allow callers to leave a recorded voice message.
4	Allow callers to press "0" to get connected to a live operator at any voice menu.
5	From the main menu, allow callers to place automated product orders 24-hours a day. Assume that customers who use this service will have a password and a list of product stock numbers in front of them when they call.
6	From the main menu, allow callers to hear product information messages from a list of products they can hear.
7	Maintain a database of orders that will be transferred to the host computer once per day.
8	Allow callers to hang up at any time without losing their entered order information.
9	Handle any caller Touch-tone errors in a friendly manner, encouraging callers to try again.

Since the customer is not the CT "expert" (you will be soon), look for areas in the RFP that need clarification. For example, in the above example requirements list, item #7 indicates that the company wants to maintain a database of orders and upload it to the host once per day. Many additional details would be required before you could estimate how long it would take to implement that feature: the type of host computer, method of transfer, database format, and so on need to be specified in detail. Keep notes on these open areas and prepare a list of questions.

The Initial Planning Meeting. Whether conducted on the phone or in person, this first planning meeting is very important. One of its purposes is to get, with as much detail as possible, answers to all the questions you have about their list of requirements. You also need to get a good idea from the customer as to when they need to have the system installed and running. Often the timing of "going online" is critical. One of the exercises you will be performing after this meeting is determining whether or not all the features they want can be developed within the desired time frame.

(Telephone Response Technologies, Inc.)

Application Planning And Gathering Ideas

The first step in identifying IVR application is to understand the way the company's business is run. The primary focus is on telephone calls, with a secondary focus on the processes that are spawned by these calls. The goal is to find ways to help the organization improve its delivery of information over the telephone.

Talk to the employees and observe what they do. Find out:

1. **Who Calls?** Is it customers, potential customers, dealers, suppliers, external employees, or some combination? Frequent callers are good candidates for IVR, because they generally do the same transactions over and over.

2. **Why are they calling?** To place orders, to get pricing, to get general information, to ask about the company's processes and procedures (e.g., how to file a claim)? Generally, calls that involve inquiry only and do not generate revenue for the company are the best candidates for automation.

3. **How does the call currently get handled?** Are there processes in place for handling the various requests? How much of the process is procedural vs. decision-based? The process must be entirely procedural in order to be automated.

4. **Where does the information reside?** Is it on-line an kept up-to-date? Or can it be put on-line? Remember, an IVR system is only an information delivery system only; it is not responsible for the integrity of the data.

5. **Do many people call with the same questions?** Or do many people call with many different questions? Or do a few people call? The more repetition that is involved (i.e. answering the same questions over and over), the better the payback for automating.

6. **What hours of the day can people call?** Does the company handle calls differently after normal business (e.g. via an answering service) or does it run multiple shifts? Being able to provide IVR after hours can extend a company's hours without incurring additional staff expense.

7. **What other actions result from the telephone calls?** Is information faxed or mailed, is a record entered or updated in a database? Often, the IVR can do some of these tasks, and eliminate the need for others. For example, an IVR system can send a fax confirmation of a telephone inquiry, or update a database directly after a transaction.

8. **Does the company use other telephone-based automation?** Do the employees have voice mail, automated attendant, ACD queuing, predictive dialing, etc.? IVR must seamlessly integrate with existing technologies.

9. **How many inbound calls are taken?** Do most callers get to someone right away? Do many have to hold? How long do they hold?

Is there a way to prioritize incoming calls? Companies can improve service by giving the caller a choice of how their inbound call is handled. Callers can choose to wait to be answered live, or they can select the faster "self-service" IVR option. Companies save money because hold times are reduced and staff take fewer calls. Staff can then be re-deployed to handle more value-added assignments.

10. **Does the company do outbound calling** (telemarketing, collections, reminders, etc.)?

In summary, the best candidates for IVR applications have these characteristics:

Many people requesting the same information.

The call is non-revenue-generating.

The process to deliver the information is simple and fully procedural.

The data being accessed is on-line and up-to-date.

(Voysys Corporation)

Application Objectives For A Proposed IVR System

When identifying applications within an organization, be sure to indicate which objectives they will meet. Be able to define concrete and measurable criteria for success -- for example:

Handled 20% more calls during the day

Took 25 calls after hours

Processed 5% more orders

Cut processing time by 10%

You will want to establish these criteria up front, based on information gathered during discovery. They have a way to measure them after the application is implemented.

(Voysys Corporation)

Applications From "Discovery" Input

You must find applications that will benefit the company and their callers by improving service and perhaps reducing costs. Remember, automating is easy. Automating well is a challenge.

You now know how the company handles its telephone calls. And you know how from experience what other IVR applications are out there. It is time to match the two. Start with an inquiry application. They are the simplest to implement. And if you have several to choose from, pick the one with the quickest payback (i.e., it will save the most money or make the most noticeable improvement in service). The faster the company sees results from the first application, the more likely they are to implement others. Make sure that the process to be automated is currently being done easily by a person. You do not want to implement a new process and a new system at the same time.

(Voysys Corporation)

Applications Planning -- Questions Checklist

Plenty can go wrong if you don't plan for a new switch or computer telephony system properly. There's organizational and business issues, processes to consider, the impact of a new system on job performance and technology acquisition. Here's a checklist of important questions to ask yourself during the planning process.

Do I have a measure of success? You should define your goals and decide on how you're going to quantify them. Once way of doing this is to choose the right reports you wish to use in measuring success.

In some cases, poorly chosen reports do not identify "hidden states" of agent or trunk activity. It's important to go through the entire process of mapping-out each type of call you make and take and define a report for each one.

Have I given employees the tools they need? You must define the right productivity tools right up front. If, for example, 90% of your transaction require interaction with multiple hosts – then by all means use products that give you instant and simultaneous access to these databases. Ask your staff to give you a wish list individually. If you get a wish list handed to you by a manager, you may be overlooking critical needs. Solicit everyone individually.

Do I have the essential features? Concentrate on the word essential. You should try to tackle only those items that are directly relate to your goals. Ask yourself upon the review of each feature: "Do I need this to solve my problem?" Too many features and too many options require extensive training and refresher training. There are many costs and productivity impacts associated with each feature.

Do I know my capacity requirements? Too much capacity can mean the difference between a $50,000 solution and a $200,000 solution. Don't buy too much. Take into consideration the time-of-day you are most busy and how much capacity you need for these times. Discuss contingency plans for handling busy hour traffic and alternatives to a major upgrade.

Strategy varies with time-of-day, as does staffing, the quality of your agents and the ability to manage these resources. Find out what the actual agent talk times are. Don't use average measures. Consider how much "hidden time" there is in each type of call and on each type of busy hour. You may find that you don't need an upgrade at all but better management of your staff.

Also, there is no use in having plenty of telephone lines if your "back end capacity" for data handling is weak. There is a direct corollary between telephone and data traffic. Consider one along with the other at all times.

What are the Outside Influences? If a programmer required to change scripts, then make sure you have access to these resources and can make changes quickly. Make sure you have a procedure for your marketing, legal and finance departments to approve and help create the scripts before you commit them to the system.

(Digital Systems International)

Applications Pre-Launch Education and Promotion

Promotion-Educate your callers. Let them know what your system will do. Tell them about it in advance. Give them graphics to help them understand how to use it. Callers may be reluctant to try even a simple system unless they are shown how easy it can be. Complex systems demand clear diagrams and charts to facilitate easy usage.

Educate your employees. Ideally, they should be able to call in and use the system themselves before the customers start. The system is not there to replace them, but to handle repetitive and unrewarding calls more efficiently. The employees must understand the system so they can answer questions that callers have and so they can recommend the voice response system when appropriate.

(Enterprise Integration Group)

Applications -- Wrong Reasons Don't Work

IVR may not always be the answer. Some companies have adequate staff to cover their calls, or simply have a philosophy of answering every call live. Customers may need or demand to talk with a person. Management may be against it for their own reasons. And many times, the process of handling the caller's request simply can't be performed by a computer.

Here are some examples of times that IVR won't work:

1. **The employee makes judgment calls during the process.** Judgment calls can not be programmed into a computer. Decisions made based on intangibles such as how "important" a certain relationship is, or whether something is "in the company's best interest" cannot be automated. Think carefully before attempting to change a process to remove those decision points.

2. **The employee adds value or generates revenue from the call.** Up-selling is a common example. The caller orders *Widget A* and the seller convinces them to also purchase a *Widget A Accessory Pack.* An IVR system can't persuade. Can you imagine a system that said "To order the *Widget A Deluxe Accessory Pack*, Press 1"?

3. **The information being delivered is not clear.** The most annoying thing to a caller is to be stuck listening to a machine trying to "guess" why the caller is calling ("If you are calling for A, press 1, if you are calling for B, press 2 ... if you are calling for XX press 99").

The caller should be able to dial a telephone number and immediately begin a specific transaction. If you don't know why callers are calling, or if there is not a clear reason for the call, you have nothing to automate.

4. **The data is not on-line or out-of-date.** IVR systems can get data from many sources, but those sources do not include people's heads and little pieces of paper. The IVR is an information delivery vehicle only; it is not responsible for the integrity of the data. If the data is wrong, it will be given out wrong. If the data can't be kept updated, don't use IVR to deliver it.

Remember IVR works best when it delivers better service than what it replaced. It can't, don't force it. A bad implementation is much worse than none.

(Voysys Corporation)

ASR Arguments

Touch-tones (DTMF) are a very inhuman interface to IVR and the rest of computer telephony voice- and call-processing. You've got these little buttons with three groups of letters to a button, and no "Q" or "Z" letters where they belong, and they make funny dual-tone sounds that blurt right into your ear, etc. Still, most IVR systems use them to get input from callers.

At the opposite end of the spectrum is speech recognition, which allows you to talk to the IVR system almost as you would the human agent the system replaces.

Figure 1.2 - Growth Of Speech Recognition

Telephony Speech Recognition Growth

Worldwide unit sales (000)

	1995	1996	1997	1998	1999	2000
Network	63	338	845	1,916	3,660	7,586
Call centers	20	45	218	457	956	2,035
Medium-to-small	90	208	596	1,639	4,052	10,279
SOHO and home	21	275	1,258	4,047	11,807	37,538

TMA Associates, *The Advanced Speech Technology Market*, 1996.

People love to yak on the phone, even if it's to a computer. It's faster, easier and more natural than "enter your *American Airlines AAdvantage Card ID* (if you know it?) and press the *star* key" DTMF commands. Other advantages:

Speech-Rec works from anywhere. More than half of the world's phones are rotary and don't produce DTMF. How do you get these people to communicate with an IVR system?

Speech-Rec can be handsfree. Car phones. Speakerphones. Conference phones. Computer-based headset phones. Talking while you type.

Speech-Rec is faster and easier to use than DTMF. No more layers of menus. You just utter the magic word or ID number instead of having to fumble and punch it in.

Speaker-dependent speech-Rec (a.k.a. verification) is secure. Anybody can steal a password. Heck, a hacker doesn't even need a password. But they can't steal your voice. We tried to fool the ingenious system of Veritel (Mount Prospect, IL - 708-670-1780) -- we tried it and we couldn't fool it, not even with DAT.

Think about banking applications, any kind of personal account control, medical, and thousands of other critical apps that can benefit from password-less security.

The first ASR systems had to be laboriously trained by those who would use them and could only identify a few words. Today, all that has changed. Digital Signal Processing (DSP) and other new technologies have resulted in reliable, speaker-independent, speech-recognition systems. If it proliferates, IVR systems will actually become more voice-oriented than DTMF or data oriented.

(Computer Telephony Magazine)

Automated Attendant Considerations

Cut down the time your own employees spend on hold when they call in from the field. By adding an automated answering device to your phone system (known as an "automated attendant"), they won't have to wait for your receptionist to answer and connect them.

Since you are paying for their calls, holding time is expensive. Your receptionist will also have time for other work. An automated attendant simply allows callers to push-button in the extension they want and to transfer themselves. Thus the name "automated attendant." A most useful device. Good for employees. Not so good for customers who reach it unexpectedly.

(Teleconnect Magazine)

Caller ID -- Upgrade For Your Business

You can save time and money for your business by using Caller ID. It's not hard to improve customer service, get more business and makes the telephone a profit center rather than a time waster !

What Is Caller ID? Caller ID is used for the transmission of the calling party's phone number and name over an analog phone line. "ANI" is the term used when this information is sent over digital (high-density) phone lines.

How Is The Information Conveyed? Caller ID is transmitted via 1200 baud modem tone (bell 202 signaling). In the U.K. and other European countries, FSK-based signaling is used. You can also get Caller ID with ADSI (Analog Display Services Interface) – it uses a combination of modem tones and DTMF strings.

Where Is Caller ID Available? You can get Caller ID in North America, South America (Chile), Europe & the Middle East.

(Pika Technologies, Inc.)

Caller ID And Why To Use It

Caller ID is a service you get from your local telco on ordinary analog phone lines. They send the data down between the first and second rings. It's a very short burst at 1200 baud.

Unless you have a fancy ADSI-enabled version of the service, the phone must be on hook for the transfer to work. To capture it, you stick some type of Caller ID receiver device on your phone line (or lines) and it collects the number and perhaps the name of the person calling (again depending on the level of service from your local phone company).

Once the device has collected the data and informed you in some way, you can then pick up the phone and begin a conversation or, conversely, not answer if you see on an LCD or PC screen that it's, say, someone you want to avoid.

Until now, Caller ID has only been a spotty locally available service. Anyone calling from outside your local area was reported "Out of Area." Fortunately, the FCC has had enough. Beginning December, 1995, local phone companies and long distance carriers were commanded to work together and deliver the information coast to coast. This includes those in California, where the FCC deemed that state's restrictive ID regulations as not cost effective and an unnecessary burden on the rest of the nation. It's truly great news for consumers and developers alike.

The problem in the U.S. was money. The long-distance carriers wanted to get paid for their part in the delivery of Caller ID information. They figured since the Local Exchange Carrier (LEC) was allowed to charge for Caller ID they should get a piece of the coast-to-coast income too.

Finally the FCC just said "no" to the LD companies, leaving the service revenues solely to the local telcos (as high as $12 a month per line for businesses and somewhere between $4.50 and $2.50 a line for consumers -- regulated by local Public Utility Commissions) and ordering the carriers to stop blocking their SS7 Caller ID signals when they pass calls down to LECs' local loops.

This is huge CT news. Yes, local Caller ID was good enough for the corner pizza parlors or courier companies, but it lost all of its appeal for all the other companies who deal with nationwide callers (but aren't big enough for T-1 or ISDN PRI).

The only way they can reap the benefits (and profits) from the service was through nationwide coverage. Now they have it.

The invasion of privacy has been addressed. First, Caller ID Blocking must be allowed on a per call basis through the *67 command. All telcos must enable this. Don't want your number passed? Punch *67 before you touch-tone in your destination number.

The FCC order also includes a mandate that PBX makers must pass the *67 Caller ID Blocking command. Most currently can.

There are also hardware devices on the market that outpulse *67 as soon as a phone goes off hook. These are selling to those who forever want to keep their phone partners in the dark. It should be noted they can't simply call up their telco and say "turn off my Caller ID signals." Big CO makers Northern Telecom, AT&T and Ericsson are on record saying the blocking schemes to pull this off are impractical and would take years to complete.

Meantime there are some FCC restrictions on the use of collected Caller ID data as well. A company cannot market this information outside their own company.

Best news: it doesn't restrict in any manner the way a company can use Caller ID for more effective handling of inbound phone calls. That makes it a perfect fit for computer telephony, which is all about adding intelligence.

What To Use CALLER ID For. You can also use the Caller ID information to spring a computer telephony application:

1. **Pop a screen.** Know who's calling instantly and bring up their buying records. Previously, with reliable ID only available through digital T-1 / ISDN ANI (Automatic Number Identification), this was usually only for big call centers.

2. **No more crank calls.** Simple Caller ID opens this app to the masses, including mom and pop businesses with a couple of analog lines.

(Side-benefit: It also cuts down repeated crank calls. It also verifies that people are calling from where they're calling. Harry's local pizza store in Manhattan, NY uses it to verify where it's delivering.)

3. **Third-party call control.** It can be as complicated as skills-based routing (knowing who in your company to send a particular caller based on a database search of available call takers) to as simple as literally zeroing in on your callers for faster service.

As for the latter, taxi companies, for one, are using Caller ID to shorten the time it takes to pick up people. They keep databases of payphone numbers all around their area. When someone calls, they know instantly where to go to grab their passengers and which hack (the closest available) to scramble over.

4. **Use it to automate first-party call control.** Voice-mail systems can store the Caller ID information with each call. Then, instead of listening to a message 14 times to scribble the person's always rapid-fire slurred phone number down on an envelope, you can simply punch a touch-tone command or, better, click a GUI icon for automated callbacks.

Caller ID can also come in handy for making call-taking decisions and, once you've agreed to speak to a particular person, call handling in general. For example, the phone rings. Your CT integrated desktop (from a single-line SOHO communicating PC to a networked unified messaging station) pops a card file telling you who wants to talk to you. If you're not interested, route them to a messaging app or another extension. If you do decide to take the call, spring a PIM, review notes on previous conversations, update your records, etc.

5. **Do call-accounting on "inbound" calls.** Lawyers and other professionals who bill for phone calls can use a Caller ID device to log time on specific inbound client calls automatically (no touch-toning in messy client-matter codes). Makes it infinitely easier to track and bill those outrageous charges of course.

Not that it's just for ambulance chasers. Ask any telecom manager. Call-accounting nirvana is getting reports on both outbound and inbound calls. After all, they'll tell you, an employee's time is just as valuable as money (if not more). So even if a call isn't on a company's nickel, people who yack away on personal inbound calls all day affect the bottom line as well.

The possibilities are endless. And now that the service is available (keep your fingers crossed) nationwide, the time is especially right to integrate it into your CT solution.

(Computer Telephony Magazine)

Caller ID Planning Process

Before you order Caller ID, you need to establish how you want to use the incoming information. Then you need to determine the level of Caller ID service you need before mapping out an implementation plan.

Defining your needs. Let's say you are a manufacturer with many outlets – and you want to offer prospects more control of which outlet they are routed to when they call. You would need to maintain a database of all incoming calls in order to determine buying patterns and habits first.

There are several levels of service available depending on the telephone company:

Basic Service: caller's phone number, time & date.

CLASS: adds the caller's name and the phone number called.

ADSI: adds the ability to receive Caller ID while on the phone.

ADSI stands for Analog Display Services Interface. ADSI is a Bellcore standard defining a protocol on the flow of information between something (a CO, a switch, an IVR server, a voice mail system, a service bureau) and a subscriber's telephone, PC, data terminal or other communicating device with a screen.

Most of ADSI's benefits revolve around the visual, i.e. it's a standard signaling format to send visual instructions to users of CT applications over analog phone lines (the ones available around the world). So instead of having to remember confusing prompt instructions on their own ("press one for this, two for that, three for this, four for that..." etc.), you get a visual screen of your choices in the words you're hearing. It sort of makes dealing with IVR and other CT voice-processing applications like dealing with a voice-enabled ATM.

ADSI's signaling is DTMF and Bell 202 modem blurts, which include Caller ID information (thus it's our third level). And while it hasn't really taken off (chicken and egg problem -- ADSI-compatible phones aren't selling because there are no service apps and there are no service apps because there are cheap ADSI no phones) like it was meant to, it does come in handy in straight Caller ID apps. Forget the visual stuff for a moment.

The problem is standard and enhanced Caller ID won't work with call waiting. In other words, when you flash over to your second call, Caller ID information on that call will be blocked. ADSI is a fix for this. The reason is the spec was built with special "muting" tones. These were created so users wouldn't hear modem signaling every time ADSI services sent their "visual" information.

With call-waiting, these same tones can be used to mute one call so you can pick up the second waiting call and get the Caller ID. You'll need two things to pull this off: first the service, check with your telco; second, an ADSI compliant unit. Put them together and you can collect Caller ID information from a second caller while you're already on the phone.

Caller ID Implementation Strategies. You can put CallerID to work in a very simple fashion – or you can emulate what the big call centers do with T-1 based ANI. Here's a look at the three ways of moving ahead:

1. **Basic.** For additions to existing IVR systems, you can use Caller ID to facilitate screen pops for your agents. You can use Visual Basic tools to do database look-ups and provide hook-flash transfers from Caller ID compliant voice boards over to agent telephones.

You can also provide enhanced routing services the same way by transferring callers back into the network (depends on your phone company tariff). In addition, you can go the SOHO route and enable each PC with its own Caller ID set-up for simple first party call control.

2. **Intermediate.** Software add-ons to programmable switches is a good intermediate step. Programmable switches count on outboard software state control and registry based routing. Caller ID can be used to trigger the routing table in your switch with a PC-based application. This is useful for multi-zone call centers. You can also use links to databases for call tracking.

3. **Advanced.** Network level routing of calls (using a CT server on your PBX) can be achieved at this level. You can provide messaging services with Caller ID routing, third party call control, selective audio logging and automatic paging.

Caller ID Effectiveness Issues. Avoid relying exclusively on Caller ID. It's a great productivity enhancer, but your whole business should not fall apart if it is not working. Make sure you have a back-up plan. Also, remember that Caller ID does not pass with every telephone call, so be prepared to answer and route the phone "the old fashioned way" for as many as 25% of your calls. This goes for "Caller ID Blocked" lines, calls from cell phones, payphones and from within a company (behind a PBX). Always offer the customer the option to choose what they want. Remember the idea is to offer the appropriate information and no more. Respect privacy issues.

(PIKA Technologies Inc.)

Consultants And When To Use Them

The mark of any professional computer telephony application developer is knowing when to bring in outside help. How do you know when to do it? One reason might be that the deadline imposed by the customer is impossible with your current level of help. Another is that one or more sections of the application are outside your area of expertise. In either case, finding the right consultant can save your project and make you look very good to your customer.

Where can you find qualified consultants? Start by calling your voice hardware or software firms and asking them for referrals. Usually, they will have a fairly good list of qualified individuals who can work for you on a project-by-project basis.

You should submit your Application Specification to two or three of them, with the tasks you want performed highlighted, and ask for a time and cost bid on the project. After discussing the technical details with each of them and reviewing their bids, the decision should be easy.

(Telephone Response Technologies, Inc.)

Credit Cards And CT Applications

You can spend lot of time trying to figure out the tangled web of electronic fund transfers (EFT). Here are some things to consider:

1. **Hardly anyone connects directly to credit card agencies.** You normally connect to a credit card clearinghouse who has the connections to the credit card companies.

These clearinghouses normally charge a per transaction handling fee as well as the percentage that goes to the credit card company. The per transaction fee ranges from 5 to 15 cents. The credit card percentage ranges from 2% to 15% and also varies depending on the average ticket charges, but is typically 2% to 5%.

2. **Getting your application to talk** to a credit card clearinghouse is only one problem; getting a merchant account is a bigger problem.

Many credit card clearinghouses won't accept credit card charges where the card and a signature is not present. The risk for fraud is a large one and charge-backs can appear anytime within 12 months from the date of the original charge.

3. **There are several generic protocols**, but almost every provider has its own enhancement that makes it unique. The last thing you probably want to do is write your own protocol handler for credit card processing. This often costs as much as $15,000 to get certified with a single clearinghouse. All of this has to be done before it ever sends the first actual credit card call.

Even if you write your own, that would only certify that your system works with a single provider and all of your customers would have to work with that same provider for it to be worthwhile. Once you get one clearinghouse working, you have about 1,400 others to do (in the US).

4. **Be careful when selecting a clearinghouse.** They should have coverage in the areas you are intending to sell your products and they should offer reasonable rates. Most clearinghouses charge the merchant a per transaction fee in addition to the percentage charged by the credit card company. These rates are usually around ten to 20 cents per call.

Many credit card clearinghouses don't understand applications that don't involve a storefront with a cash register. Secondly, they don't want to have to directly work with a large number of developers to connect to their network. They don't have the staff to support the problems of lots of developers trying to figure out all the protocols from each clearinghouse.

(Computer Telephony Magazine)

Fax Planning Tips

Fax Server Rationale. Here's 15 reasons why you should be championing a fax server at your company:

1. Saves time by sending faxes directly from your desktop.

2. Eliminates need to wait for a fax machine that's already sending or receiving.

3. Substantially improves fax image resolution and clarity.

4. Improves fax transmission control, accuracy and reliability (no hard copy to get misplaced or removed from fax machine).

5. Enhances fax reporting using computer-controlled accounting.

6. Routes incoming faxes without manual intervention.

7. Provides access to fax anytime (server always available even if your PC is powered off), anyplace (fax mailboxes can forward inbound faxes anywhere, such as your hotel or home fax).

8. Enables secure receipt of faxes (without hardcopy) to individual users: "for your eyes only faxing."

9. Keeps archive of received and transmitted faxes.

10. Centralizes and shares fax modems to reduce costs and administration complexity.

11. You can use either digital or analog office phone lines. Traditional PC fax cards won't work on digital lines.

12. Network fax software can be integrated into other enterprise-wide server applications.

13. Works with industrial strength, high performance, scaleable and expandable server platforms.

14. The average user takes 10 minutes (?) sending manual faxes, including printing, walking, transmitting. Network fax takes one minute. Send-time is reduced by 90%.

15. Average single fax machine telecom costs a Fortune 500 company $7,000 a year. Network fax can schedule faxing at night. Cut telecom costs by 80% and more. Faxing at night is best for junk fax.

(SoftLinx, Inc.)

Fax Servers -- Why To Use Them

Fax servers offer many benefits. Versus using fax machines, the major reason is to save time by sending from one's desktop PC rather than walking to the fax machine.

Versus single-user fax/data modems, fax servers offer 1. cost-justification for use of intelligent (microprocessor-based) fax boards, which send faxes more quickly due to use of MR and MMR fax compression along with faster handshaking, which means reduced phone charges; 2. the off-loading of file-conversion processing to the server so it doesn't take time on the end user PC, and 3. the ability to keep fax servers on 24-hours-a-day, which means its ability to receive faxes is constant (as opposed to what happens when an individual user turns off (or undocks) their PC with a fax modem in it).

What are fax servers? Fax servers are shared fax resources installed on LAN and multi-user computer networks. They typically are installed on a gateway PC or file server. They enable network users to:

1. **Have computer files transmitted** as faxes to any fax machine or device in the world.

2. **Receive faxes from any fax machine** or device in the world at the fax server (where the fax phone call terminates) either for automatic print-out or to be routed via some mechanism to fax or universal-message mailboxes associated with each individual end user (or department or workgroup) on the LAN. Faxes are received as faxes -- image files! -- not as computer-editable alphanumerics

What is the difference between fax servers, fax gateways & LAN fax? One can make distinctions, but mostly all those terms are used loosely to mean the same thing: some kind of fax node or capability on a network which allows network users to fax out computer files and sometimes to receive faxes to their workstations.

The term *fax server* is often used two ways: 1. as an overall LAN fax system and 2. just as the single computer-and-software node that acts as the fax gateway for the rest of the nodes on the network.

What does a fax server consist of? A computer platform, fax server software, one or more fax modems, boards or fax machines, and end user fax software (including management software for the fax server administrator).

(Davidson Consulting)

Fax Server Advantages On The LAN

The fax server captures the document electronically and queues the transmission along with other fax jobs. All handling from that point on is automatic: If the number is busy or does not answer, it will try again. If the line is answered by a person or modem not capable of receiving a fax, the call will be terminated and the fax job suspended. In either case, the sender can be alerted to problem jobs' status.

Look at the steps and time saved:

No printer queue, printing, or print job pickup time required;

Manual cover sheet creation eliminated;

Feeding and dialing the fax machine is eliminated;

Waiting at the fax machine is eliminated;

Distractions and side-trips are eliminated.

(ABConsultants and Nuntius Corporation)

Fax Reception Truths

Businesses need to tread cautiously in the area of electronic fax reception. Have strategic reasons to do it. Test it out with end users.

What issues do network administrators have to consider if they implement electronic fax reception?

Receiving faxes electronically means that bit-intensive fax images cross the LAN, potentially slowing overall response times in high-volume fax environments. It also means that received faxes are temporarily and/or permanently saved, which can eat up tremendous amounts of memory; some file-deletion/memory-management system must be implemented.

Finally, receiving electronically means maintaining an inbound routing table, including adding and deleting names as employees come and go.

The truth about the benefits of electronic fax reception (EFR)?

1. **Saving Trees.** Many buyers expect electronic fax reception to eliminate the printing of received faxes onto paper. That would save trees and toner, both morally sound objectives in an ecologically-conscious world. But the economics aren't all that compelling: consider that at 3-cents-per-page, eliminating the printout of 10,000 fax pages saves a mere $300. Meanwhile, the extra time it can take to read faxes on screen (and then to print them, e.g., under Windows) can wipe out those savings in a hurry. And here's the real-world truth: when end users display faxes on screen, more often than not they end up printing the faxes anyway, especially when multi-page faxes are involved. So, the reductions in paper and toner usage tend to be substantially less than what many buyers anticipate.

2. **Receive Alerts.** When faxes are received electronically into user fax mailboxes, they trigger visual and/or audio alerts at recipient PCs. Thus, awareness of the fax's arrival isn't delayed until the user goes looking for it at a fax machine or until the mailroom delivers it.

3. **Confidentiality.** Received direct to memory rather than sitting openly in a fax machine output tray, electronically received faxes may remain confidential. That is, if they are automatically routed (e.g., via DID or sub addressing) rather than manually routed. In any event, fax mailbox security usually isn't some kind of rocket science, but merely simple password protection.

4. **View & Delete Junk Fax.** Unimportant faxes can be viewed on screen, quickly identified as junk faxes, and deleted before wasting paper and toner. Well, keep in mind that "quickly" is a relative term. Imaging a received fax page (text-only) on screen can take from a couple seconds to more than a minute, depending on the amount of text per page, PC power, the operating system, and the power of one's graphics card. The soft labor cost to view a page is often greater (sometimes far greater) than any hard paper and toner costs which electronic viewing is expected to eliminate.

5. **Attach a Distribution List.** Users can attach a distribution list to a received fax and easily forward it to anyone else with a "need to know." This is good unless it is overdone to the point it starts wasting people's time because people start "cc:ing" all their faxes. .

6. **Annotate.** Full-featured computer fax software supports the keyboarding of annotations onto received faxes in the process of viewing them. Computer-resistant workers may never master this, however. And, unless skillfully done, computerized annotations can be harder to notice than handwritten annotations (though easier to read once noticed).

7. **Read Faxes on Screen.** Users can read faxes on PC screens so they never have to leave their desks (of course, when mailroom workers or receptionists deliver paper faxes, workers never have to leave their desks either). But reading faxes on-screen can turn into an EFR anti-benefit. Faxes are hard to read on screen, even on today's "high-resolution" (super VGA) PC screens (7,699 dots per square inch versus 19,894 dpsi for a "low-res" fax;. Moreover, faxes often must be reduced in size (the print too) to fit the width of a PC screen, a situation made worse because faxes often are a bit blurry and skewed from phone line noise and scanning.

All this can translate into **lowered productivity**. Davidson Consulting conducted a study wherein a dozen people read three-page faxes both on screen and on plain paper -- and it took them 83% longer to read the faxes on screen (and reader comprehension was better with paper too). Short faxes may be easily readable on screen, longer ones usually aren't. And some people can't read faxes on screen at all.

8. **Remote Access.** With remote access, users on the road can dial up the fax server to retrieve faxes received in their fax mailboxes. Just how effective this remote access is depends on specific user needs at any given moment and if one-call and/or two-call remote access is provided. For instance, if a traveler only has immediate access to a phone, the need is for two-call access so he or she can call, direct faxes to be forwarded to a hotel or branch office fax machine, then hang up and have the fax system make the call.

But, if in a hotel room late at night when the hotel business office is closed, a two-call system is useless (i.e., a second call to one's room gets blocked when the hotel operator answers and, and upon hearing fax tones and not knowing to which room they are directed, hangs up). In that situation, a one-call solution is necessary so the user can dial up from a PC in one's room, access the fax mailbox, and retrieve the faxes back to the PC all in a single phone call.

9. **OCR.** A received fax can be processed by optical character recognition (OCR) software, which converts text in the fax image to editable computer code, eliminating the need to re-key the data. OCR works reasonably well with crisp laser-printed 300-dpi copies and platen scanners -- 99% to 99.5% accuracy, or about 5 to 25 errors per double-spaced page. But with 200x100-dpi normal resolution faxes, sometimes slightly skewed and blurred from scanning and phone-line noise, OCR more likely registers 80% accuracy rates -- or a whopping 400 errors per double-spaced page!

(Davidson Consulting)

Fax Server Productivity Enhancement

It's a simple matter of tracing the steps we go through when we wish to send a fax to someone or fulfill a literature request. For most companies, a labor-intensive series of tasks are what's ahead to carry-out a simple communication. Consider a breakdown of the time and steps it takes to send a two-page fax to a customer without a fax server:

Print the Document. Most of the time, this will involve a shared printer located outside the sender's workspace. By the time the employee walks to the printer, waits for another job to print, and then gathers up the document, at least two minutes have gone by. (This does not include distractions along the way to the printer.)

Create a Cover Sheet. Unless there is an automated means for creating a fax cover sheet, the employee must then fill one out with names and other identifying marks. For most people, this takes about a minute.

Fax the Document. If a fax machine is available and ready to use, feeding the documents, dialing the number and waiting to ensure the job goes through can take over a minute per page -- three minutes, in this example.

If another job is in the machine, or a line of co-workers is waiting for their turns to fax, more minutes are added. (Even if one individual is assigned to collect faxes and send them on behalf of everyone else, this is highly unproductive use of that person's time.)

At best, then, it takes about five minutes to send a two-page fax. A twentieth of an hour may not sound like a lot of time, but multiply that by the number of faxes that are sent by everyone in the company each day, and you'll quickly see how much manpower can be saved by automating these tasks. (Much the same series of events happens with a standard literature fulfillment request; The only difference is that instead of faxing the literature, an employee will have to print out a mailing label or envelope, and then affix postage and so forth).

Now let's retrace these same steps in a fax server or fax-on-demand environment Assume an employee has just completed the same two-page letter as above. With a fax server, all that need be done is "print" the document to the fax server. A cover page is automatically appended to the communication based on pre-configured data (e.g., sender name and extension). Recipient information is entered once and then stored in a *phone book* for future use. Total time: perhaps two minutes, probably less.

(ABConsultants and Nuntius Corporation)

Fax Server Traffic Volume

Fax server capacity hinges upon many factors:

1. **How many phone lines** one fax server node can support?

2. **How reliably fax phone calls can be initiated** and maintained so they don't have to be retried?

3. **How quickly fax phone calls are actually completed?** This largely depends on whether fax boards or modem support (1) 14.4 or 9.6 Kbps transmission speed (28.8 Kbps is coming in the 1996/1997 time frame), (2) whether they support faster G/3 compression methods: MMR is the fastest, MR is in the middle, and MH is the slowest, (3) whether they support on-the-fly bit stuffing, and (4) how fast they handshake and retrain. The difference can be huge!

4. **How fast the fax server can convert computer files?** ...to G/3 fax format, which among other factors has to do with the fax server software, the fax board, the operating system and microprocessor speed and power.

5. **How efficiently the fax server handles busy/no-answer calls?**

6. **How many servers can be installed** on one LAN and if load balancing between them is supported?

Although vendors often specify fax server nodes as supporting an unlimited number of users ,the actual number effectively supported hinges on the amount of fax activity on a LAN/network, upon the number of phone lines supported by a server, and upon system assets such as available memory and true multitasking.

Why would I need more than 1 phone line? Multiple phone lines support several fax server applications: simultaneous send-and-receive, simultaneous (as opposed to sequential [one call at a time]) broadcasting, multi-line receive and, where DID inbound routing is implemented (see below), multiple lines are required to send faxes because DID lines are receive-only.

How many phone lines are needed? Essentially, this an Erlang table issue. How long do your fax phone calls last on average? How many fax phone calls are sent and received during your peak hours? Grab the telecom manager and use the tables he or she uses to figure out the number of PBX lines needed. There are also some rules of thumb (which can work out just fine or make your fax server act as if its "all thumbs"):

1. One-line fax machines are typically shared by 10 to 20 users.

2. Small businesses might want to estimate the time to fax a page at 1 minute. If you send 10 4-pages faxes per hour, that's 40 minutes, which leaves only 20 free minutes.

If you receive the same number (another 40 minutes), you've run out of time and calls will be blocked unless additional lines are installed.

(Davidson Consulting)

Fax-On-Demand Benefits

Fax-on-demand delivers information to anyone with a fax and a phone. It is available 24 hours a day, every day. This simplicity of getting information and continuous availability is a boon to many companies. These companies are supporting software and hardware computer products and distributing product literature.

Catalog-based businesses are providing additional information to prospective buyers. The pervasive presence of fax terminals provides businesses with new opportunities to give better customer support, stay close to their customers and expand sales. These are great benefits in any part of an economic cycle and are invaluable in sluggish economic times.

Low cost delivery. Low cost delivery is one benefit of providing information by fax-on-demand. Intel Corporation, a pioneer in providing customer service and product information, discovered that fax delivery costs one tenth of the cost of mail delivery. In addition, fax delivery is far superior in its ability to provide the information when the customer or prospect wants it, not days or weeks later.

This table compares the costs of mailing literature to prospects and customers with the costs of using a fax-on-demand. The following assumptions are used in the comparison:

For U.S. Mail:

Mail weighs less than 1 ounce...
Postage: $0.32
Document cost: $0.20 (a printed brochure frequently costs much more)
Envelope cost: $0.05
Labor: $1.25
(five minutes at $15/hour for addressing and stuffing the envelope)

Total cost per item: $1.95

For fax-on-demand:

Telephone charges per minute: $0.25
Average transmit time for one page: 40 seconds
Transmission time including a cover page: two minutes

Total cost per item: $0.50

Easy Updating of Information. Updating literature can be costly, as it is difficult to forecast the number of brochures of each type that will be mailed during the life of the product or before a brochure is revised. Rather than have shortages, the tendency is to order more than is needed. Outdated brochures and papers are therefore frequently discarded in significant quantities at the end of the product's life or when features are changed.

With fax-on-demand systems, however, keeping documents current for distribution is easy. A single original in the database is changed when the document is revised or deleted. It overcomes the problem of purging documents from the system that are obsolete.

The expense of up dating the electronic version of the brochure is small, as final copy produced with desktop publishing software easily converts into fax format. Sometimes the original is scanned for entry into the system. This is the least desirable way of entering information into the system. The digitizing process during scanning reduces the quality of the document, as compared to converting the document in electronic form to a fax format.

Improve Service. Fax delivery of information improves customer service and keeps customers close to the company. Customers appreciate the availability of the service as they get the information they want at any hour of the day. When properly organized on a menu of documents, a system or service provides the answers to a caller's questions without requiring the caller to wait for a person to interpret the questions. One company found 70 percent of the callers who needed product support information could get the information they needed from its fax-on-demand system. In addition, this company satisfied an increased demand for customer support services without adding personnel.

Consider an instance of a customer having a product problem while out in the field, when it is after hours and no technical support staff are available. If the customer can use a touch-tone phone to access a catalog of technical support tips, he or she will likely perceive your company as highly responsive. Similarly, the ability of customers to have product pricing and description information faxed to them if all the lines in your call center are busy will be perceived as highly responsive.

Fax-on-demand allows you to maintain this level of service and responsiveness for customers and prospects across the country and around the world0in any time zone without extended staffing hours for your service desk.

Limit Staff Increases. Phone support grows with the increase in advertising and the product population at customer sites. Companies often find their supports lines are busy during certain periods during the day, which leads to customer dissatisfaction. A solution is to add phone support personnel which reduces the waiting time during the one to three hours each day when there is peak demand on the system. Fax-on-demand can reduce the demand on the staff by providing an alternative to talking with a support person. This reduce the growth rate for support personnel and improves customer service.

Increases Sales. One offers product literature to every prospect and customer who call. . Of those callers who received product literature by fax, 30 percent purchased a product based on the information received. This company believes this is an effective way to increase sales, reduce the time between advertising and receiving orders and reduce the cost of literature distribution.

Support Space Advertising. Fax-on-demand is an effective way of evaluating an advertising campaign. A document number and the system phone number can be printed in an advertisement. Early responses to the advertisement, as measured by calls to retrieve a document, show the interest aroused by the advertisement and the publication.

Advertisers frequently use a different document number for each publication when placing advertisements so they can evaluate the effectiveness of each placement. A quick and easy measure of the effectiveness of an advertisement is by counting the responses generated by the advertisement. Most of the responses will be generated in a two-week period following mailing of the issue of the magazine. This quick response helps in efforts to improve the advertisement message.

Can Give a Competitive Advantage. Another driving factor behind the popularity of fax-on-demand is the fact that it offers companies a means to distinguish themselves competitively. The competitive edge of an enterprise is often measured by its ability to deliver accurate and timely information because customers and prospects alike are looking for clues about how your company is different from your competitors.. These include their perception of your responsiveness, flexibility and stability. By providing access to important product literature around the clock, you can increase your responsiveness.

Some creative managers can make their fax service an integral part of their operations and provide a level of service above their competitors. Customers prefer to buy from such business, as better service helps resellers easily plan for the promotion and sale of both new and established products.

(ABConsultants and Nuntius Corporation)

Fax-On-Demand Success Tips

1. **Promote, Promote, Promote!** Tell everybody the application is live. Send out press releases, New Birth Announcements. Promote regularly, not just at birth. Take clever ads. Pick a unique color. Use it again and again.

2. **Put the App's 800 number on all letterhead**, business cards, catalogs and direct-mail. Paint it on the side of your building in letters 50-feet tall. Create a company jingle that sings the 800 number and record it on your auto attendant.

3. **Be sure your sales, customer service and receptionist have the 800 number** tattooed on their arms, legs and other parts. Tell everyone in your company. Encourage employees to test the system -- before it goes live -- to familiarize themselves. it. Then they should tell all.

4. **If you've got an auto attendant**, offer the FOD App as an option.

5. **Make your App user-friendly**. Remember: Repeaters Repeat. If your App has too many levels, is confusing or too complicated, repeaters won't call again.

6. **Put someone in charge.** Like a "sysop." Fax on demand systems need constant checking. Like every day. They need constant updating. Their menus needs to be cleaned, made clearer. Unpopular stuff should be erased. This is not a trivial job. It needs attention. It needs to be someone's responsibility. (Editors note: Computer Telephony Magazine took our Fax on demand system temporarily off the air because the person in charge of it wasn't.)

7. **If you're selling information, give some away free**. Nothing generates call volume like free stuff. If you're selling a report, provide the Table of Contents and one chapter for free. You'll be amazed.

8. **Add coupons and special offers that change often**. This is guaranteed to turn first-time callers into repeaters.

9. **Promote, Promote, Promote... again!** Send out another round of mail, faxes, etc. Smart marketers know that the key to advertising/promotion is repetition. Keep your app's number out in front of your audience and they will call.

(Computer Telephony Magazine)

Fax On Demand Integration With Your WWW Site

Why do I need fax on demand if I have a WWW home page? This is a interesting question. Today the Internet is hyped as the ultimate solution for Information Distribution.

The simple fact is the best estimate of the number of people using the Internet is 10 Million people, while 95 percent of all working adults have access to a fax machine as some time during the day.

This is not a argument that people should not put up Web pages for their product information and literature. This is a statement that a modern organization should any and all means to reach its customers and prospects.

Several approaches for combining fax-on-demand and the Internet are currently being implemented by "trend setting" vendors, service providers and users:

One innovative fax-on-demand user has combined on one Novell LAN: a fax-on-demand system using a separate interactive voice server and a fax server; a bulletin board server; and an e-mail gateway. This system allows documents to be requested by: interactive voice, by e-mail via Internet or any of the popular on-line and e-mail services, and by employees from their desktop PCs on the LAN. The selected document(s) can be automatically delivered by fax to the requester or to a stockholder, customer, prospect, or other person. Also, documents can be retrieved by e-mail for e-mail delivery or from the bulletin board.

WWW Home Pages are being designed to allow searching for and selecting of documents from fax-on-demand databases. When the title or identifying number of the document or information file is located using normal Internet and Home Page techniques, the information can be downloaded via the Internet to the desktop PC or send as a document "off line" to a fax machine so that the requester does not have to wait at their PC for many minutes or hours it may take for the download.

With fax, a PostScript quality document can be delivered with no significant increase in transmission time or cost compared to delivery of ASCII text. How-ever, e-mail and Internet frequently delivery documents as ASCII text in order to keep transmission time down to an acceptable level and because it is easier.

A service provider delivers WWW Home Pages and documents linked to the Home Page by normal fax-on-demand techniques. Each level of the linked Home Page documents requires one fax-on-demand call and retrieval.

One leading fax-on-demand software publisher announced that this capability would be available as an option with a future release. Others are expected to follow. One service provider accepts fax messages with an associated broadcast list via the Internet (usually from International or long distance locations) to avoid long distance transmission charges. The service provider then broadcasts the fax using "free" or at least lower cost dial-up telephone services.

(ABConsultants and Nuntius Corporation)

Fax-On-Demand & The Web -- Planning

The merging of fax-on-demand and the World Wide Web is an exciting prospect, but there are some limitations you should consider when planning for your own “web-to-fax” service. The Web is big, with over 10 million users, but there are five times as many people who access information via fax. So why not use the same documents for Web and Fax? This is a great way to provide access to old document libraries and it’s a handy feature if you are on slow Internet connection – since FOD is not real time.

Figure 1.3 - Volume Of Web And Fax Users

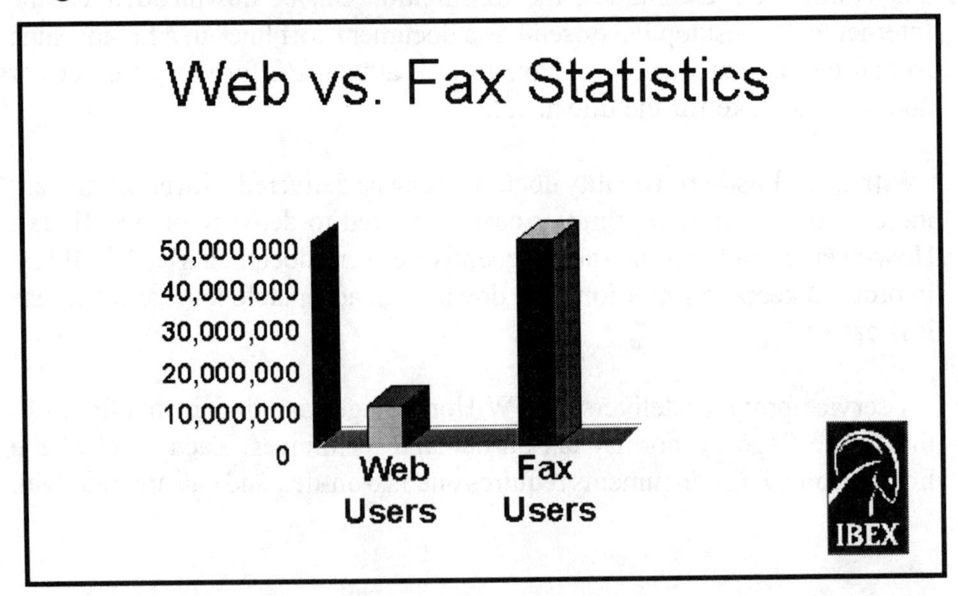

How it works. An ideal fax-on-demand system uses one integrated index of documents. Callers have no idea what the source of the document is or where it is stored. They simply know what information they want. A caller's touch tone input is sent from the FOD system to the Web server, which handles the request as from any Web browser. Using the NetScape API, the server passes a fax request to the FOD system. The document is downloaded thorough the API and the FOD system converts the image into a faxable format. The document is then queued and routed to the fax phone number input by the caller.

Figure 1.4 - Mixed Information Stores Feed FOD

Macromedia is one company that combines FOD and the Web. They design press releases and scheduling information that is authored on the Web. Callers who use their fax system have easy access to their Web documents.

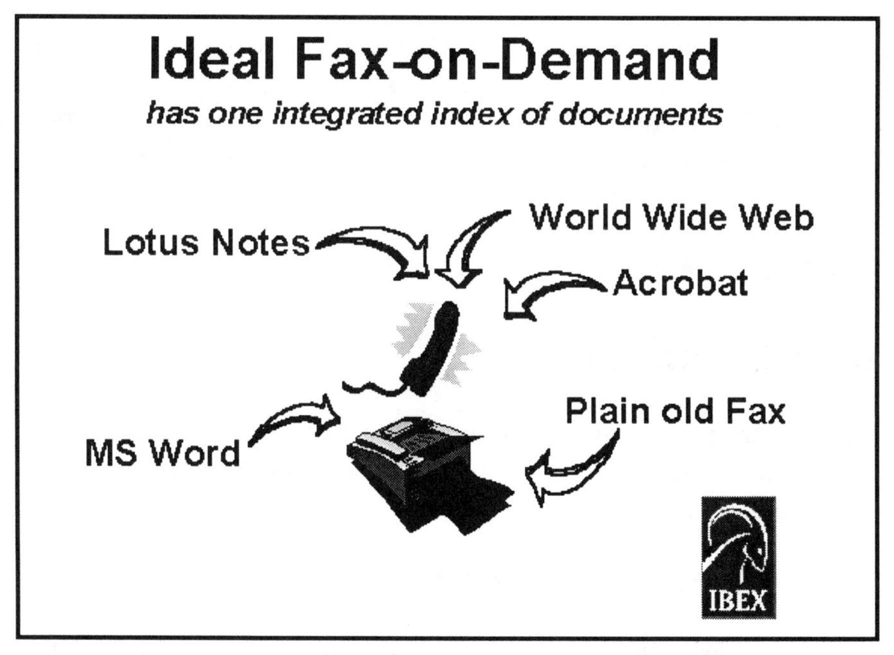

In addition, the company's technical notes are authored in Lotus Notes, which are also used to submit faxes to callers. All of their marketing literature is added directly using Lotus Notes, so there's one integrated index of documents.

Here are two top considerations before you bolt Fax onto your Web page.

Designing your Web. Users don't want to "surf" with FOD. Over-the-phone information systems need to deliver fast, efficient information or else you will lose the caller. Therefore, if you design your FOD application to use the Web as an information source, be sure to reference content pages only, not navigation pages.

Adobe Acrobat. Converting HTML to Fax images is not a perfect science. HTML is screen design language, not page layout language. It uses a fairly simplistic layout scheme that does not take into consideration the look of a printed page. Since there is no rich formatting, many web documents rely on illuminated (backlit and colorful PC) screens and imported graphics to look appealing. Unfortunately, many web documents are hype or just used for navigation, but have no valuable content.

Consider using Adobe Acrobat for complex Web documents. Acrobat is the coming standard for use on Web. It uses true page layout and retains the fonts and formatting of a printed page. You can use Acrobat for both WWW and fax-on-demand – so it's a great cross-platform product. With Acrobat, you can create documents from both PC & Mac.

(Ibex Technologies, Inc.)

Faxing Quality Tips

1. Never print anything promotional -- a catalog, a sales sheet, etc. -- without first faxing it to yourself. Amazing how many color combinations end up black on faxes.

2. Never put screens over forms that you expect faxed back to you.

3. Print everything large. Faxes shrink originals as much as 6% and 15%. And they don't do the shrinking proportionately.

4. Never underline in faxes. Horizontal lines print crooked.

5. To highlight a point, **bold** it for best effect.

6. Some photos fax. Some don't. Check.

7. Send non-urgent faxes after 11 PM, when rates are real cheap.

8. Always mail contracts, orders and other important documents you've faxed. Just for backup and because people like the originals.

9. If you take orders by fax, send confirmations immediately. Customers likes to know if they will receive what they just ordered.

10. Keep your fax machine clean. Fax yourself a copy to see what your clients see. If you see vertical lines or smudges -- a blow dryer works great! Clean monthly for best results and longer life.

11. Faxing internationally? Put the cover page in both languages. (Fax attendants don't always read English.)

12. If receiving faxes from overseas, get yourself a long paper tray.

13. Fill your paper trays every evening, and especially on Friday.

14. As your first-in-the hunt group (if you're not using a fax server), buy a decent, heavy-duty fax machine. Cost: at least $3250. Worth every penny.

(The Kaufmann Group)

Fax Machine - A "Keeper"

Fax servers are great tools, but don't throw out your fax machine just yet. A fax server per se can only fax out computer files, so paper documents can't be faxed unless one has a scanner (even then, it's easier faxing paper documents from fax machines than from general-purpose scanners). Also, fax machines are very reliable for receiving faxes, whereas some fax servers, especially those based on inexpensive fax modems, may have problems receiving faxes.

And, believe it or not, not everyone uses computers all the time. Finally, computers are known to be relatively unstable; when a fax server or network crashes, businesses appreciate having fax machines around.

(Davidson Consulting)

Fax Broadcasting Guidelines

1. **Fax broadcasting should not be shotgun marketing** (like when an office supplies retailer faxes to every company within a two mile radius -- that's trouble). Broadcast to targeted audiences that you have an established relationship with and who know your company -- for example, a cruise line broadcasts last minute specials to travel agents. That's great! But never fax directly to every machine in hope of finding an interested traveler. In general, never use fax to prospect.

2. **Like telemarketers, keep an internal Do Not Fax List**. Remove anyone who doesn't want your faxes. This is smart and will save you money.

3. **Add a message to the bottom of your fax**: "If you want to be removed from our fax distribution list, call us or fax." Your customers will appreciate this.

4. **Schedule your faxes at night**. Phone rates are lower and you won't interrupt busy machines. **Warning:** This is true, unless you've targeted the SOHO market. With SOHO, you must broadcast during the day, or expect nasty calls from those nice people you've woken up at 2 AM by their in-home fax squealing.

(The Kaufmann Group)

Fax Statements

Phone time is cheap and snail-mail is expensive (plus not totally reliable). If you are currently doing business with other businesses, you should start sending your invoices and monthly statements by fax.

Since most companies now have fax machines you can send these documents by fax and save $$. A one page fax sent long distance is less than a minute of connection time. Sending by mail costs $0.32 plus the cost of the paper, envelope, and the time to stuff it. So say stuff-it to snail mail and go to fax.

Another benefit of fax is that the receiving fax machine provides a confirmation of a successful receipt of the document. No more of "I did not send the check because I did not get the invoice. It must of have gotten lost in the mail". Now you can tell the deal-beat that he/she received the document on February 1, at 12:01 AM and the check better be in the mail.

(Optus Software)

Fax Service Bureau Considerations

There are plenty of reasons an end user should hire a service bureau to handle his enhanced fax application needs. Here are 12 of them. So you're ready to make your sales literature available 24 hours a day using fax-on-demand or launch a "Special of the Week" fax broadcast to your 5,000 customers. The application makes sense and the funding is available. So what's the next step?

Usually, it's deciding whether to buy versus build the system. Do you contact system manufacturers and VARs or call hardware, software and component makers to piece together what you need?

But there's a third, quite compelling, approach: outsource.

Since the first one was launched in 1989, the Enhanced Fax Service Bureau industry has grown steadily in the US. Top fax bureaus have made large capital investments and employ the programmers and engineers necessary to impress, well, even me.

No longer do six ports and two guys working from their garage equal a bureau. Today's bureaus are sophisticated operations, running multiple to a dozen T-1s.

They offer features like on-the-fly document creation, ANI, live feeds into news and wire services and credit card capture -- all necessities in many of today's FOD applications. Fax Broadcasting, too, has grown-up. It now entails embedded, multiple field personalizations, watermarks and intelligent retry functions. How long would it take you to build this stuff in-house? Here's why you'd want to hire a fax service bureau:

1. Seventy-five percent of all FOD applications are maintained by service bureaus.

2. The start-up time is minimal to negligible.

3. No capital outlay.

4. No hardware to install, maintain and upgrade.

5. No new software to learn and upgrade.

6. No need to hire additional expertise (programmers, engineers or administrative).

7. No down time.

8. No additional work for your network.

9. Emergency services available.

10. Competitively priced services.

Look who else outsources: AP, CNN, HP, IBM, S&P, etc. With a bureau, you try it before you buy it.

(The Kauffman Group)

Integration of Data, Voice and Fax

How Much Can LAN Integration Save? Integrating Voice/Fax With LAN-WAN services can save hundreds (even thousands) of dollars each month. Take two remote sites, each with 75 minutes/day of voice/fax toll charges to headquarters and 30 minutes/day between them. That's 3960 minutes/month or $515/month at13¢/minute.

Assuming you already have a WAN installed to carry data traffic, you can virtually eliminate the dial-up fax and voice call costs. By using NetRunner I-Routers (same cost and LAN-Only Bridge/Routers - $7500 for 3 Units), you can get a payback in 15 Months and then save $515/Month thereafter. The integration routing saves $6,180 each year ($23,400 over 5 years).

This is achieved by transmitting voice and fax signals over Public Frame Relay as shown in figure 1.5. This is also now becoming available with ATM devices.

(Micom, Inc.)

Figure 1.5 - Frame Relay Carries Voice, Data and Fax

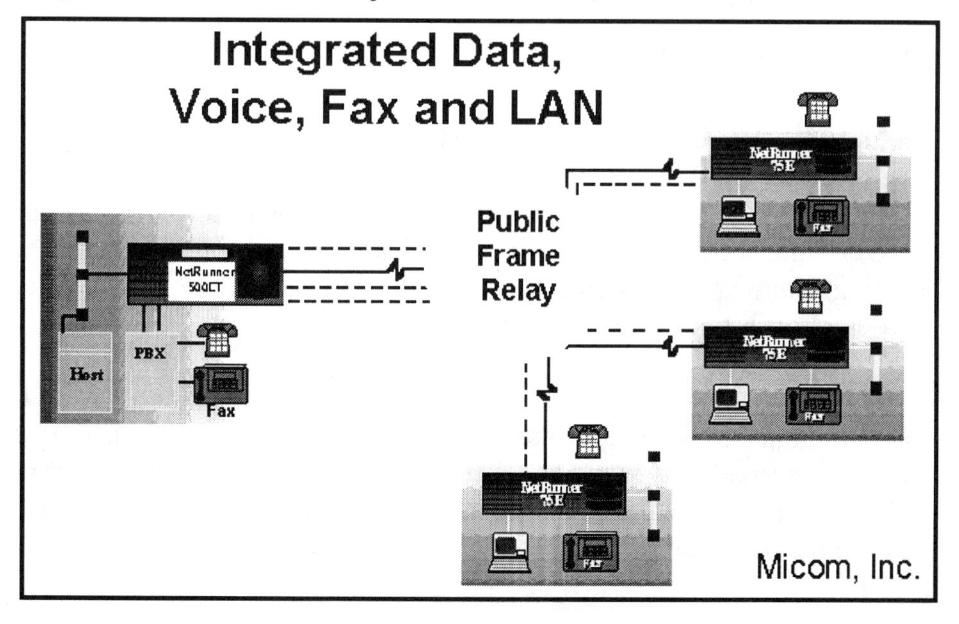

Debit Calling Service Planning Tips

Promotions. The beauty of debit systems is that because they use IVR you can run promotions every time a caller dials in to use his card. Don't forget to use this to your advantage in promoting the special of the week to customers. For example, cable companies can offer a listing of major pay-per-view programs for the day.

Collector cards. Creating specialty cards can make a promotion much more effective -- many people who keep the cards long after they've used up the value of the card (or they may not use them at all, meaning additional revenue). Popular themes are sports stars, cartoon characters, etc. Commemorative cards for special occasions like a company's anniversary are another way to add business or boost usage. But remember, if you're using trademarked art, you'll need to pay royalties.

Other applications. If you're already in the business of international callback, depending on your system, adding debit calling service can be easy. For providers of cellular services where subscribers may default on payment, the addition of debit-calling systems makes it possible to collect revenues up front.

(Harris Digital Telephone Systems)

International Call-Back Service Planning

Calling the U.S. from abroad is often three times the cost of calling from the U.S. Thus the new industry: International Callback. Call the U.S. Wait one ring. Hang up. A stateside machine calls you back with U.S. dial tone. You then touch-tone your way to savings.

Here's how it works. I'm at my branch office in London, and I want to call San Francisco. I dial a Dallas number, hear the number ring, and hang up. Within seconds, my phone rings, and I pick up. I hear a computer prompt, dial my PIN number and in response hear a dial tone: I can now dial my party in San Francisco.

International call-back exploits the low-cost long-distance and international toll charges available from US carriers. Deregulation and increased competition have dramatically reduced tolls, especially for large-volume customers. The service provider can pay for two calls: the call-back to the subscriber and to the called party, add a mark-up and still save the subscriber money compared with the rates that would be charged by the subscriber's national phone carrier (PTT).

A trick of the trade avoids a bill for the original subscriber's call to the service provider's switch. DID service will send the dialed digits to your Dallas switch even if the call is not answered. Each subscriber is assigned a different number to dial, all these numbers terminate on the same trunk, usually a digital T-1 or ISDN line. The computer captures the DID digits (typically the last four or seven digits dialed by the subscriber) and uses a database to find the subscriber's phone number, sort of a home-brewed "Caller ID".

Using the DID digits to identify the caller works fine when the subscriber calls from a fixed location. If the caller is on a road trip, he or she will need a way to keep the switch appraised of a call-back number. One option goes further with the idea of not answering a call. On a digital trunk, there will generally be a talk-path (audio connection) even before the called switch goes off-hook, allowing the switch to play ring tones or a busy tone if the desired DID extension is not available. This can be used in the other direction too: the subscriber will be able to dial DTMF tones to the call-back switch, informing the switch either "I'm at the same number as last time" by a short code such as a star key, or "This is my new call-back location" by dialing the number.

The call-back problem is even trickier when the subscriber's number cannot be direct dialed, for example in a hotel room or at a PBX extension without direct dial capability. In this case, the call-back switch can dial the receptionist and play a recording such as "This is an automated call for Harry Newton, please transfer now". Remember that several different languages may be required, even in a single such as Belgium or Switzerland where several languages are spoken. If there is an auto- attendant, the switch may be able to transfer by dialing the extension after the voice mail answers.

To support these "roaming" and "indirect dial" features, the switch will need a facility to allow the subscriber to register a call-back method and to record individual audio segments, such as a message asking a receptionist to transfer or a fragment like the name "Harry Newton" to be slotted into a pre-recorded phrase.

If you're planning to become a call-back service provider, don't underestimate the problems of collecting on your bills. Receivables can be a problem in any business, especially when the customers are overseas. Several providers have found themselves in financial difficulties because they failed to make accurate projections of their collection process. That's why debit card services, where the subscriber pays in advance for a number of units, is becoming an increasingly popular vehicle for delivering call-back services. Debit cards are printed using attractive designs with printed instructions on the back, including a number to call to activate the service by giving a credit card number.

On the subject of billing, you will need to consider carefully how you will bill your subscriber for the call. Not only are there the usual issues: first- and additional-minute rates, peak- and off-peak hours, one-minute, twenty-second or one-second increments, but when do you start charging? You are paying for an international call as soon as the subscriber answers the call-back, but the call will not be completed until several seconds later. In a competitive business, this may make all the difference.

Your computer telephony switch can easily be enhanced to provide value-added services which will increase your chances of success. As many as 40% to 50% of attempted calls will not be completed due to a busy signal or a ring-no-answer. This is your chance to bill extra for a voice messaging service: offer the caller the option of recording a message which your switch will make repeated attempts to deliver by calling back the dialed number. Remember that you may need voice recognition capability to verify that you have reached the right person ("This is a message for Bob, to accept please say Yes now.."), your party may not have touch-tone, especially if the number is outside the US.

Adding a switch in the country where the call originates adds new possibilities. Many countries, particularly where the telecom infrastructure is underdeveloped, charge a premium for international calling access. We are told that international access on a domestic phone in Russia can cost as much as $1,000 per month.

By providing a switch in Moscow, your subscriber can dial a local number to request an international call which can be completed through a call-back switch, thus requiring only a single out-bound data line capable of reaching the US.

Often it's difficult or impossible to get a domestic phone, in which case a simple voice mail system provides an attractive alternative: your subscriber can give out a number which terminates on your local switch, he or she calls in to collect messages from a payphone or other telephone when convenient.

Do your research carefully. The success of these types of services has already impacted the market, and rates to the US are coming down as a result, reducing the toll rate differential on which the business depends. The PTTs are aware of the lost revenue, and may not be inclined to approve any equipment you install in their country.

One solution: offer them a piece of the action in a joint venture deal. Phil Chakiris of EarthCall, a successful player in the call-back and international resale business, offered the following advice when we interviewed him: find a strong local partner who understands the telecom market in their home country and knows how to market the service successfully.

Phil also told us to be careful of using ITF (International Toll Free) numbers, which can easily be blocked by the local PTT. Critical to your success will be a strong marketing effort: find a niche with airlines, tour operator, long-distance providers or other businesses with a relevant customer base.

(Parity Software)

Digit Assignment

If you use 3-digit extension numbers in your company and are putting in an automated attendant, think how long before you outgrow it and need a four digit scheme. If you think you'll ever outgrow 3-digits (and you will), change now.

The cost to re-educate all your customers after they've learned to use your automated attendant is gigantic, much greater than just printing costs.

(Teleconnect Magazine)

Disaster Prevention

With so much importantly business being conducted via voice mail, even a very brief failure of the company's voice mail system can be costly and disruptive. Examples:

The company can lose customers whose recorded orders go unfulfilled.

Prospective customers turn to competitors when they can't get through an inoperative voice mail system.

Managers and other employees lose -- perhaps forever -- important messages that have been stored in voice mail.

Heart of the Problem. With proper setup and careful management nearly all potential voice mail failures an be prevented.

Key: A voice mail system is essentially a computer that's programmed to handle and tore messages. One of the basic elements of this computer is a hard disk that an store hundreds -- or even thousands -- of incoming messages.

For many companies, the disk is becoming increasingly important as employees, room maintenance personnel to CEOs, are using voice mail as a virtual file cabinet to store messages, often for months at a time.

Like a company's other computer disks, the disk in the voice mail system is subject to failure. In fact, at some point it will, almost certainly, do just that.

Unlike other hard disks, however, the disk that stores voice messages is all too rarely backed up. And even if it is backed up, in many businesses the backup system is inadequate. Follow these easy-to-understand tips:

1. **Don't rely on automatic back-up devices.** Several types of mechanical devices can now be installed to automatically copy the hard disk once every 24 hours -- usually around midnight. Problems: Backup systems are difficult to set up -- and use. And, sooner or later, the backup system itself -- especially tape drives -- will fail, and no one will know until after the damage is done.

Best backup strategy: Instead of relying on auto0matic devices, instruct an employee to physically handle the backup. If the company has its own computer or management information systems department, a qualified employee should be easy to find. In smaller companies, non-technical employees can learn how to do it.

Example: An employee links a desktop computer to the company's computer network, which typically has access to the hard disk of the voice mail system. At whichever time interval the company chooses, the employee transfers data from the hard disk to the other computer.

Before relying on this procedure, however, test it by first copying the data from voice mail to the backup computer and then copying it back. If the process works, go with it. Also, make sure there is someone on staff who knows how to re-install the voice mail software after a crash. Smaller companies may want to rely on an outside computer technician or a vendor representative.

2. **Change the hard disk in the voice mail system once a year** -- even if there is no apparent need. The cost will usually be less than $600 -- and it will decrease the odds of a crash in the next 12 months. Even that cost can be partly offset by putting the used hard disk on a less important computer -- after data have been transferred, of course.

Don't believe claims that one voice mail system is more reliable than others. There's not much difference, if any, among the dozens of PC-based voice mail systems.

In fact, even if a voice mail system touts easier backup, that alone is probably not enough reason to buy it. Reason: Other features are more important. Examples:

3. **Use software that lets the voice mail system run smoothly on a local network**. The ability to let users get messages from their computers as well as their telephones. This is useful in case you phone system goes down – you can still pick-up and respond to messages by e-mail (or at least hack your way through your messages).

(Harry Newton)

Redundancy Tips

A key part of many system architecture designs, especially for wireless infrastructure, is based on the availability of redundancy capability within the PC environment. That often means hot-standby capability for the voice processor board, the power supply board and the network interfaces.

Multi-Chassis MVIP provides solutions for applications with redundancy requirements. A two chassis solution allows for hot swap to a 2nd chassis. An N+1 scenario or N+m scenario is justified for large systems that require multiple chassis for basic traffic handling. Also, there are some companies creating special adapter cards and/or backplanes with ISA buffering to provide hotswap in a box.

A number of companies do build (sometimes proprietary) multi-chassis solutions right now. For example, Centigram Communications Corp. (San Jose, CA) has an architecture that uses multiple boxes in a load sharing fashion with the ability to take down any one box while keeping the rest of the system (especially the file system) intact. Boston Technology (Wakefield, MA) has an architecture that uses a front end switch to distribute traffic to multiple voice processing units.

Most of the industrial chassis vendors offer an option with redundant power supplies in a single chassis. Hot swap of CPUs is a more significant issue - both electrically on the ISA or PCI bus, but more importantly in a software sense. As an alternative, you can get two complete 10 slot passive backplane computers in one industrial PC box (one 20 slot backplane split 10 & 10). These boxes can also come with redundant power supplies.

Hot swap of T1or E1 interfaces is less important than having redundant T1 or E1 links. I.e. the thing that fails most often is the T1/E1 link, not the interface board. To maintain continuous service, the system design needs to be large enough to include at least two T1/E1 links, preferably terminating on two separate T1/E1 interfaces controlled by two separate CPUs. If these CPUs are located in a single split-backplane industrial PC chassis, then the chassis should include separate, or joint but redundant, power supplies.

Various redundant software architectures are possible depending upon the applications and their requirements upon the underlying files and databases. Generally, files and data fall into three classes:

1. **Read mostly** (like programs, prerecorded prompts and configuration information) - updated only at OA&M (Operations, Administration and Maintenance) time. These can be stored with a master copy plus read-only copies cached at each computer in the multi-computer system.

2. **Write-mostly** (like traffic and billing information). The on-line CPUs can broadcast this to one or more independent systems that record data for later billing or analysis.

3. **True Read/Write information** (like voice messages in a voice mail system). Here is where system architectures differ. The basic choice is whether to move the call to where the data is (using MC-MVIP or a front end switch) or to move the data from some (fault tolerant) file system to the CPU that is handling the call (using a high speed LAN for example). There are commercial examples of both approaches.

(Natural MicroSystems)

Failure Planning For Networks

Make sure you have some plan for network failure. Each of the major long distance companies -- AT&T, MCI and Sprint -- will suffer major network failures. On the day of the failure, it will be too late to plan. You must plan now. The more redundancy you put into your network, the better.

(Teleconnect Magazine)

International Service Planning Lessons

1. **Control of your own source code is important.** Depending upon the country, it can take a lot of money and time just to discover exactly what code will talk to what switch.

In developing countries, switch manufacturers sell on soft loans. This means that the country of origin for the switch underwrites a 20-year low-interest loan. This also means that a given developing country can have different brands of switches -- each with their own deviations from standard.

Once again, this means time and money to discover how to do it. If you then rely on a third-party board manufacturer, someone else's "E-1 driver", an applications generator, or by whatever other means, you must reveal the exact details of the protocol outside your own organization. I suspect you will soon see competition riding in on your investment. Call me paranoid.

2. **Don't believe it.** Most switch engineers have only a vague idea of how the bits change during a call. Establish hooks in your own protocol to capture and log bit transitions. Buy a good E-1 tester and use it. Be prepared to spend anywhere from three days to three weeks to get the project working.

3. **Don't believe it again.** Nasty surprise. Software upgrades on a CO switch can affect your protocol. When a switch goes down (this happens more than a little) and then reloads, the programming for your lines can get screwed up.

Be prepared to go in and do it all over. If you have gone through an approvals process with the PTT, do not expect that the protocol that was approved will be the one that actually answers the call. Expect to be told the local equivalent of "Mayo" when you wake up one night to a frantic call from your customer that all the lines stopped working.

4. **Integrate the switch protocol.** When you use an external driver, or an executable, you risk not having enough resources to perform the basic program your customer wants. You do not have to be a programmer to understand this.

To answer one call takes a combination of 27 bit transitions, tone acknowledgments, or what have you. Depending on the timer, you may be so busy just answering calls that the added overhead of an external driver or executable will not allow you to do the more complex and process or intensive applications. An example is using the silence bit to do grunt detect, which causes delays in the queue and thus can cause the application to hang.

Given that most customers in E-1 territory want everything including the kitchen sink for their system, processor use is a real issue.

Picture the longest "wish" you ever received from a "suspect." Add three items they left out. This is the typical configuration request from E-1territory.

WTS actually has installations that combine speaker-independent voice recognition with voice mail, faxback, IVR, Audiotex services, conferencing and outcall -- all running on the same E-1 node and all running while answering calls as previously described.

5. **Competition (and thus confusion) is coming to the world of computer telephony.** Telephone companies are responding worldwide by looking for added-value services. CO-based voice mail, information services, debit card, "expert" hot lines, virtual phone/fax and a host of others. Given the capital requirements of customer premises equipment solutions, most telephone companies are looking for "things" they can base at Central Offices or tandem switches to provide multiple applications simultaneously on the same set of nodes.

Switch manufacturers with (take your pick) a vast ignorance of how to do it, or a sly plot to maximize their own position and income, are developing a set of conflicting and probably self-serving alternate standards.

DSC is working on something called "Code Set Six." Siemens has their own protocol "under development." Excel and Summa each have their own. Some of he protocols are designed to work on extended versions of PRI. Others through an interface to a serial port. All just to answer a call and do something with it. Result: things are getting even trickier.

(WTS Bureau Systems, Inc.)

Intranet Planning Tips

By creating an interface of interlinked Web pages for your company, your business must now "make sense" -- gray-area employee policies made explicit, workflow defined, timeliness of updated information enforced. This requires inter-departmental cooperation -- good luck.

1. **Employees should be made aware of the Web's potential.** Give them lots of hands-on experience. Listen to their ideas. You should concentrate first on what you want to do with the Web, not the technical details of how to do it.

2. **If the technical stuff (programming and site hosting) gets too intense, outsource it.** Also, get someone knowledgeable to flesh out the security issues – then get a second opinion.

3. **Secure your company's DNS name** so no one else on the 'Net causes name / trademark infringement troubles.

4. **If your Web server will also be connecting to the Internet**, create a Web Home page that captures the spirit of your business and its vision -- make it look good.

5. **Develop a corporate standard guide for creating and storing documents**, extracting information from them and making them accessible in Web form when necessary.

6. **Look into increasing your network bandwidth.** With multimedia, data conferencing and videoconferencing coming, study-up on Isochronous Ethernet, ATM, servers with PCI buses, etc.

(Computer Telephony Magazine)

Middleware And Computer-Telephone Integration

In a typical call center CTI application, several applications must cooperate to provide the desired capabilities. An example would be an integrated messaging application. Notice the modular structure in figure 1.6; no single subsystem controls everything. The subsystems must constantly communicate with each other even if all subsystems are off-the-shelf, they must be correctly arranged ('integrated') to satisfy a particular customer's needs.

Figure 1.6 - Computer Telephony Middleware

Middleware: The General Idea
Middleware Application Fetches Screens As Needed
Application Systems
Middle-Ware
CTI API
PBX or ACD
CTI Link
CTI Server
Application Systems Need Not Be Changed
Data Network
New capabilities without disturbing existing systems
DIALOGIC
CTI Division

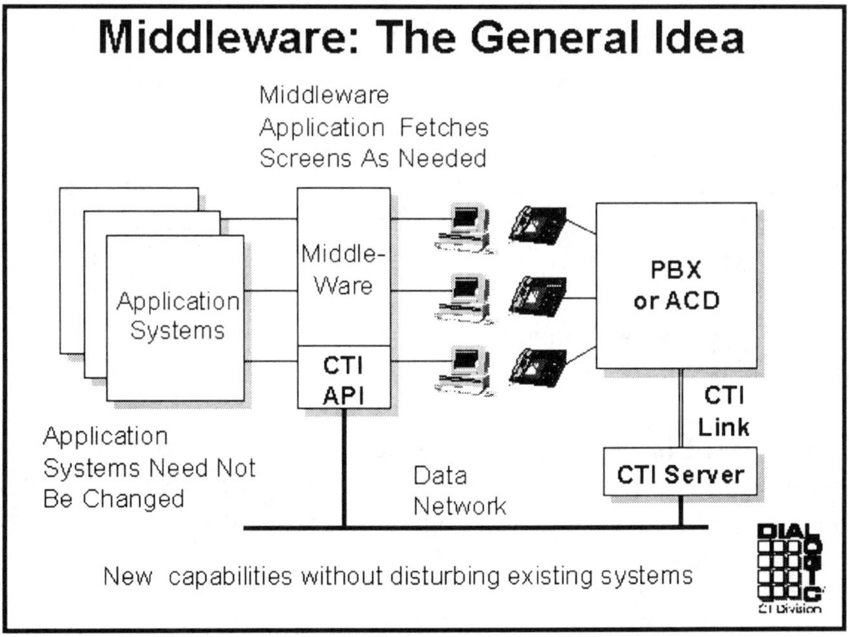

This is all part of the client-server world. According to the world's foremost authority on CTI integration, Carl Strathmeyer (CT Division Dialogic Corporation), client-server computing is "organizing a system as a cooperating group of independent, modular subsystems."

With client-server, I.S. functions can reside anywhere; all they need is a network interconnection. As a consequence, it is much easier to integrate desktop, voice and fax systems to provide CTI features.

Host-based vs. client-server systems. With host-based systems, virtually all of the functions were handled by one system (implemented as monolithic software modules). These systems were hard to change without creating side-effects and had to be tightly managed. On the other hand, client-server systems divide functions across many independent modules.

Modules are defined by what they do, not by what's inside them. This allows the 'plug-in' substitution of modules. You need to be careful to manage these modular resources carefully, because they can get out of control if not managed.

With client-server implementations of CTI, there are desktop and server-based ways of doing things. Each method has its advantages.

Desktop interfaces. These are easier to install, however, they do not support all call types. For example, TAPI-based systems are generally not as robust when it comes to third party conference calls and supervised transfers, where as TSPAI (server-based) solutions generally are.

Desktop applications require hardware at each desktop and are therefore more economical for smaller workgroups.

Server-based interfaces. These support advanced application functions, but are more complex to install. In addition, server-based CTI solutions often require expensive switch upgrades. They are more economical for larger workgroups.

Desktop interfaces vs. desktop applications. The physical interfaces themselves have been the subject of much discussion and hype, but the physical interface is not the important issue. How it works and how easy it is to use is much more important.

In some cases, desktop CTI application software is easier to build than host-based CTI software, and desktop applications can run with either type of physical interface. This is where the idea of "middleware" comes in.

What is middleware? Middleware is a client-server application which responds to telephone events. It fetches the required data on behalf of agent positions and can be part of an application or a separate system. Middleware can be implemented in desktop systems and it can do screen-pop and voice/data transfer without modifying existing applications. Some examples include Aurora Systems' FastCall, SoftTalk's Phonetastic and Answersoft's SixthSense.

With middleware, the general idea is to make voice processing and fax functions just another subsystem. The important skill in making this all work is application integration. There are lots of elements to be coordinated, and automation must be specific to the user's desired work flow. When planning for such a system your automation must adapt to the user's business needs. In other words, each business situation is unique, so be careful to identify work patterns and workflow on a case-by-case basis.

Many components, like the software mentioned above, are available off-the-shelf. You need to configure each one to fit the circumstances. Use this checklist before you implement any middleware package:

1. **Make sure a switch upgrade is not required.** If so, you should broaden your scope to include more than desktop-type solutions.

2. **For small workgroups, consider the use of analog lines** (maybe two or more per station) in lieu of a switch upgrade. Consider also the use of Centrex lines (either behind your PBX or standalone). You can do call transfers to other workgroup agents behind Centrex without effecting the PBX.

3. **Survey each of your agents or workgroup employees** and get them to separately write down the type of calls they make and take during the work day. Review this data before buying anything. Discuss the results with each vendor.

Questions to ask on the survey:

How many calls do you take each hour, on the busy hour?

How many calls do you make, etc.?

How many conference calls do you do?

What are the number of new callers?

What are the number of callers who are already in you customer database?

How many calls do you conference, and with whom?

How many calls do you transfer? Inside, outside, to whom?

How often do you put people on hold?

How often do you lose callers who were on hold?

(CT Division Dialogic Corporation)

On "Hold" Buying

A survey of consumers who order by phone found that, when placed on hold, 85% preferred to hear messages about the company's products **rather** than canned music...and between 15% and 20% of those who heard such messages made purchases.

(Teleconnect Magazine)

Peer To Peer Versus The Internet

As the demand for remote digital file transfer continues to grow, understanding the different solutions to facilitate the transfer maybe the most difficult task to overcome. This is especially important for Prepress, Printing and Publishing Companies. As different solutions are promoted, confusion seems to follow. Ranking at the top of the list is the subject of how to connect to the remote site. The question arises, should one use point to point file transfer or the Internet.

In fact, it is not actually a question of which does one use, one or the other - There are times when both methods are attractive to the user. What must be decided is which type of communication should be used. As ISDN rapidly gains momentum in the US market - finally catching up with the rest of the world, where it has been widely available for several years, it is becoming acknowledged as the most practical way to transfer data electronically. Significant advantages are obtained in utilizing the speed of ISDN - turnaround times on customer jobs can be slashed without FedEx in the middle of the workflow. Though being rolled out later in the US, it is anticipated that the growth rate will be significantly faster than that in other markets, due to the fierce competition that thrives in this industry.

Now, once it has been decided that ISDN is the technology to use, there is still the question of *how is it best used?* ISDN has become the backbone that now supports the infrastructure of the Internet - Rather than access the Internet with analog speeds, ISDN is quickly becoming the access mechanism. So, it is now more of a question of when to use a direct dial-up ISDN link as opposed to an ISDN Internet connection for file transfer? This is where the confusion often lies. The Internet, though ubiquitous, may not be all it's cracked up to be, but it has it's time and place in the use of an ISDN system.

A company must consider what their needs are and which transfer method fits those needs. The easiest way to help end users position each of the technologies and help evaluate how and when each should be implemented is to use a general example.

Internet Access. At present, the main option are high speed Modems or an ISDN device. To date, modem users still have problems with full color graphical pages, with some home pages taking many minutes, sometimes hours to download to provide information. ISDN is becoming the chosen connection mode for most serious users of the Internet.

Communicating the message. The Internet has become a worldwide shopping window allowing companies to display their wares to millions of potential customers at minimal costs. This new medium can be exploited by the design world as they are able to create interesting places to visit with imagination and creativity. To display pages graphic designers need a WWW of their own. The two most popular options are renting space on an established server or purchasing and running a Web server. The latter requires the graphic designer to establish communications links with the outside world. Again they could choose between ISDN or modems but the more people you want connected viewing pages at the same time the more likely you would choose ISDN.

Production and delivery of goods. For publishers and graphics companies, this whole process can know be done electronically from taking digital pictures to placing pages on the WWW. All the process in between can also be electronic and even if the finished article is hard copy paper the majority of the process is performed on computer.

It would seem as though the Internet would fit snugly into this process as a delivery mechanism between the process of image capture, image manipulation, page layout and final output. Unfortunately this isn't the case.

There are three fundamental reasons why the Internet is not a viable tool for large file transfer. The first reason is the issue of speed. Graphic designers use BIG files - often in the region of 50-100MB. With an Internet connection, even if you have an ISDN line to your provider you would be very lucky to get over 14.4 rate across the net at any given time. One reason behind this is that once into the Internet, the dedicated throughput of the ISDN line is no longer relevant as your transfer pipe is shared with all others surfing the net.

This leads to an unacceptable situation where transfer times actually vary with how busy the 'Net' is. Think of it like estimating how long it will take to drive across town, depending on what time of day it is!

Big Files over the Internet? At a recent trade show, the question was asked, "Has anyone tried sending a large file over the Internet, say 40MB?" One person, out of about a hundred, put their hand up to say they had done this.

> "Did it work"?
> "Sort of."
> "How many times have you done it"?
> "Once"

Using a dial up application for point to point file transfer, over the same ISDN line, you will be guaranteed 10 times the speed of a regular modem, on the most basic system. Also, as a scalable technology, ISDN can also be aggregated to reach delivery speeds of up to 10MB a minute.

Internet Considerations. It seems that those who are regular users of both Internet and point to point file transfer systems, have recognized many of the issues that face the new user. In addition to the obvious issue of speed, there are often two other concerns that the Internet presents, when considering the transmission of large amounts of data - often on behalf of a client, and with a deadline.

1. Security is a significant issue. Connecting the network (in other words creating a Wide Area Network) to the Internet always brings up the concern of opening up your network, servers and confidential data to the outside world. This involves erecting "firewalls" around your sensitive information, which is a costly endeavor, and requires high maintenance. To date, this method is still in its infancy and has not been proven to be as reliable as it should be; Internet hackers and crackers prove this every day.

2. Ownership. The Internet offers little or no ownership security and with designers relying on the uniqueness of their work this is a key aspect of using a dial up service.

With a direct connection, the line is 'owned' for the duration of the transfer - error checking occurs during transmission, something that packages (like 4-Sight's iSDN Manager) can implement on the direct dial, but are unable to follow through if the Internet is used.

Overall, the Internet today is unmatched for browsing, for information, opportunities, and general promotion. However, with regard to file transfer of graphic files, there is no substitution for point to point file transfer.

(4-Sight, L.C.)

Reseller Or Distributor CT Tips

Conduct an internal audit. You need to set the correct expectations for your business. For example, what is the acceptable ROI? What level of risk and resource will you put into the sale of computer telephony to your customers?

Often, your business model will dictate the acceptability of risk in this venture. Are you set-up to meet, greet, and sell solutions to your base of customers? Will they be open-minded to your approach -- or is this a whole new venture that will cause you to redefine who you are selling to. Also, consider how much easier it is to sell a new product or service to a customer you already have a relationship with.

Product & Service mix. Does your current mix of offerings lend itself to computer telephony? For example, if you already sell telephone switches or install computer networks, 0r sell long distance service, computer telephony may well be the perfect fit. If, on the other hand, you have no complimentary products, it may be tough to get going.

Capabilities & Expertise study. Take a close look at the people who make-up your company. How many of them can approach customers, or speak with them over the phone about how their business runs and how they deal with transactions? How many of them will be able to help customers either face-to-face or over the phone with technical support.

How many of them understand or have the capacity to learn the basics of telephones and computers? If you're coming up short, you may need to train heavily or hire some more people. While computer telephony has become more "shrink wrapped," it still takes a lot of basic knowledge of how calls are made and received over the network to sell effectively.

Partnering abilities. Asses how close you are to other companies who are related to computer telephony in some way. If you are in the computer business, consider joining forces with a telephone system (interconnect) company. Consider working closely with local and regional consultants. Speak to your suppliers about what kind of strategic partnerships they have forged and how successful they are. It's not typical that anyone very successful in computer telephony does it all alone. At the very least, you can establish a lead referral relationship with companies with complimentary services and products.

Solution choice market considerations. There are plenty of applications that you can sell. Consider the most popular ones and think about your customers and the problems they face each day. Some solutions are easy to sell and install, but then – the easier they are to sell and install, the more competition there will be. Here's a snapshot of some of the popular applications and some considerations for each one:

Fax-on-Demand

Type of Reseller: All Types

Benefits: Short sales cycle, Easy to implement, Customers understand benefits

Interactive Voice Response

Type of Reseller: VARs who sell database driven applications; Imaging VARs

Benefits: Many applications in market, Customers understand benefits

Integrated Messaging

Type of Reseller: LAN integrators; Groupware VARs

Benefits: Strong turnkey applications available; High demand

In-bound/Out-bound call routing

Type of Reseller: Strong understanding of both data and telephony; Strong consulting services; High risk/high reward mentality

Benefits: High demand - few Resellers able to deliver. Products like Sun XTL, TAPI, TSAPI making this easier.

Vendor Selection Strategy. After you have defined the solution you think your staff would do the best selling and supporting, begin to compare the overall capabilities of your soon-to-be suppliers of the technology. Your potential partners will be vendors, consultants, systems integrators and VARs. You should conduct a straightforward vendor analysis and take. It's a good idea to get involved with a supplier that has some kind of VAR Program. A good VAR program will provide your company with sales training, technical training, sales leads, joint sales calls, co-op advertising, trade show support and assistance in getting editorial placement. Make sure you set the correct performance expectations with the supplier. Don't over-commit to large sales volumes. Ask for a "ramp-up" period to reach the expect sales levels.

Create a sales methodology. Make sure you are providing the proper incentives to your sales staff. You should construct a separate sales compensation plan in keeping with the skill level required and complexity of the product or service that you're selling. Remember that a direct sales force's performance (and compensation) will be effected by: 1. How well they are trained; 2. The quality of the sales leads you give them; 3. How well you and your supplier supports them on conference calls and joint sales calls and 4. Their territory limits.

Sales methodology ground rules for success. First, productize your offerings: Hardware, software & services. Make sure you "fold in" new computer telephony products into what you already sell. Make sure the new products get part numbers, price sheets, brochures, white papers, and all other corporate "branding" that your successful lines get.

Second, sell your capabilities to your partners. Work closely with your strategic partners and let them know what you are doing. Give them a demo. Visit them and show them how exciting this stuff is. See if you can help them at their business with computer telephony. Third, you should establish reference accounts early. Put triple the resources into your first dozen or so accounts so you can establish a good rapport and get all of the kinks out of your sales, installation and support processes. Let these accounts know right up-front that you want to establish a "reference account" rapport and relationship with them. It will be much easier to sell if you've got some happy (first) customers under your belt. Consider waiving certain support charges, installation charges, etc. to "sweeten the pot" for some of these customers.

(Access Graphics, Inc.)

Traffic And Port Sizes

All IVR app developers and users must figure out how many ports they need. Here's a real simple way. Figure your total minutes of calls each day by multiplying the number of calls by the average length of each call. As for fax ports, calculate the total number of fax pages sent per day by multiplying average length of fax by average number of fax transmissions per day.

This is NOT a precise way of figuring ports, since it makes no assumptions about when it all happens (in one hour or spread evenly through the day) and what the callers do when they get a busy (hang up, call back?). A more scientific way is to consult a book on traffic engineering, e.g. "The Complete Traffic Engineering Handbook" by Jerry Harder $59.50 -- 1-800-542-7279 or 212-691-8215.

Voice Port Sizing
Total Minutes -- # of Ports

300 -- 2
600 -- 4
1,200 -- 6
2,100 -- 8
3,600 -- 12
5,100 -- 16
6,600 -- 20
8,400 -- 24

Fax Port Sizing
Fax Pages per day - Est. # of Fax Ports

650 -- 1
1,300 -- 2
2,600 -- 4
5,200 -- 8
7,800 -- 12
10,400 -- 16

(Computer Telephony Magazine)

Traffic Thumbnail Sketch

This topic is presented here because the determination of how many phone lines will be needed by the system is an integral part of the application planning process. If enough lines are added to a system so a caller never experiences a busy signal, the system is probably over-sized (and thus too expensive). In this case, many of the lines may be idle a good percentage of the time.

To properly determine how many lines are enough, you must decide on a blocking factor and know the average length of each call. This blocking factor is the probability that a caller will experience a busy signal.

A typical blocking factor to use is 1% (that is, one out of 100 calls will experience a busy signal), and will usually adequately handle peak hour traffic.

Briefly, here is how to estimate how many lines your system should have for a 1% blocking factor. First, determine the average length of each application call and then estimate the number of calls expected during the peak hour (sometimes you have to take an educated guess). The following equation will give an approximate idea of how many lines are needed so that very few callers will hear busy signals. Figure 1.7 gives an example of a "thumbnail traffic analysis." More sophisticated traffic modeling for high-density systems can be done with the aid of traffic tables and books on the subject. The *resources* section of this book provides several publications on the subject that provide great detail.

Figure 1.7 - Thumbnail Traffic Analysis

Lines = Calls/hr * Avg call * 0.023 = 8

where:

- ***Lines = total telephone lines required*** 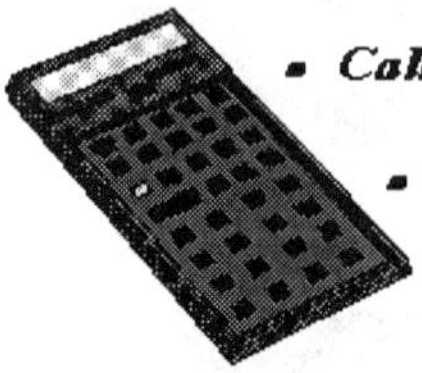
- ***Calls/hr = number of calls per hour during a peak hour***
- ***Avg call = length of an average call in minutes***
- ***0.0238< = constant value***

For example, if 250 calls were expected during the peak hour and the average call was 3 minutes long, the number of lines needed would be:

250 * 3 * 0.0238 = 17 (or about 18 phone lines).

Once you have calculated how many lines you will need for adequate call handling, you can purchase your speech cards and related hardware.

(Telephone Response Technologies, Inc.)

Voice And Fax Impact On The LAN

If you are having nightmares about adding voice and fax messaging to your LAN, you can put them to rest. While voice and fax produce message files that are larger than basic text messages, a voice message only requires 4 Kbytes per second for storage. That translates to 240 Kbytes for a one-minute message. Assuming an average message length of 90 seconds, with individuals in an enterprise receiving approximately 5 voice messages per day, that's less than 2 Mbytes of storage.

Figure 1.8 - Impact Of Voice On The LAN

Storage of Voice on the LAN

- **Size of a one minute voice message**
  - **24 Kbits/sec (encoding rate)**
  - **or 3 Kbytes/sec (divide by 8)**
  - **or 180 Kbytes/ min. (multiply by 60)**
- **5 (daily average) 1 min. messages = < 1Mbyte**
- **Impact is minimal as most voice messages are transient**

OcTel

Understanding that, like e-mail, most voice messages are transient—people don't usually save all of them. So if you're a system manager, your storage and networking requirements will not exponentially increase simply by adding voice messaging capabilities.

(Octel Communications Corporation)

Chapter 2

Ordering Telco and Carrier Services

With the dizzying array of services now available from local and long distance companies, it's hard for today's telecom manager to stay ahead of the game. There's fast packet network services, leased lines, DID service, ANI and good 'ole POTS (Plain Old Telephone Service). You'll see that the best advice from the experts is: "*don't assume anything*" and "*ask a lot of questions.*"

These ideas certainly hold true with ordering service. Most telco service center employees are trained to take orders over the phone, but they assume you know exactly what you're asking for. Many service center agents don't have the knowledge to stop you when you're about to make an expensive mistake. So study-up first. There's a couple ways of doing this. First, you can get good books (1-800-LIBRARY) on the subject of T-1, frame relay and saving money on long distance carriers. Second, make sure that when you order service – you have all of the information about your equipment handy.

When calling for repair, make sure you tell them exactly what type of line your reporting trouble on (they might send the wrong repair person to fix the problem). Above all, ask questions. Some of them are listed here for you. If your contact at the phone company does not have the answers to your questions – rest assured that someone there does. Ask for a business office supervisor or network services manager if all else fails.

Caller ID And Centrex.

A small or medium-sized business that can't afford to set up a PBX for Caller ID applications may opt for central-office switch services, namely *Centrex*.

Every RBOC has its own name for Centrex, including BellSouth's Essex and Southwestern Bell's Plexar. Centrex services now include Caller ID.

Octus (San Diego, CA -- 619-452-9400) makes OctuLink and OctuBox -- sold through the RBOCs -- which connect one or two Centrex lines to a PC communications port. These products work with Caller ID and the Windows-based OCTuS Personal Telecommunications Assistant (PTA) software. PTA displays contact information from an OCTuS address book. (Starting at $279 for the package plus a $59 voice-mail line). OCTuS also sells their products separately for the home and SOHO market.

(Computer Telephony Magazine)

Dedicated 800 Lines vs. Virtual Lines

When 800 numbers were first introduced, the only way you could get one was to buy a leased line -- a special phone line dedicated **only** to 800 calls. Now you can get "virtual" 800 numbers. What happens with virtual lines is that someone calls your 800 number, your friendly long distance carrier looks up that number in a gigantic "lookup" table and finds your corresponding "normal" phone number, which in TELECONNECT's case, is 212-691-8215 and direct dials that number.

If you don't receive a huge number of calls, then it may be cheaper to switch your 800 calls to virtual lines. You may also be able to handle more calls simultaneously.

(Teleconnect Magazine)

Detecting Loop Current Reversal

Loop current or battery reversal is a special service (Reverse Battery Supervision) which signals a caller when the remote end has gone off hook. The battery may remain reversed or there may be two transitions depending on the service provider. This is typically used for determining when to start a billing cycle or when to go to a live operator. Make sure you ask your telephone company how they will provide "off hook" detection. If you can get this service, let your programmers know so they can build it into the application.

(Dialogic Corporation)

DID Or DNIS Digits

If you are setting up a phone circuit using DID or DNIS and find it ringing busy, check the number of digits the circuit is delivering. If you are expecting a different number of digits than the line is configured for, it will often ring busy.

(Computer Telephony Magazine)

DNIS And DID Digit Passing

In a computer telephony integration, will your PBX pass three or four DID or DNIS digits to your telephony server? Can the computer be properly set up to handle digits received? Look into these special services from your local and long distance carriers. Make sure you speak with your switch and IVR vendors before ordering them. It's not all automatic.

DID means "direct inward dial" and DNIS means "dialed number identification service." DID applies to your local telephone lines and DNIS applies to 800 numbers. In both cases, when a number is dialed, the telephone switch sending the call to your system repeats the dialed digits. This allows your own telephone system to direct the call. These digits can also be sent to your computer so it will bring the appropriate record up on the computer screen at the same time the incoming call arrives at the desktop.

(DIgby 4 Group, Inc.)

Document Your Telephone Service Changes

1. **Keep the first few pages of each telephone bill** you receive for at least five years (and have your bill audited at least that often).

2. **Ask for a detailed breakdown** of your fixed monthly charges at least every six months.

3. **Every time you place an order to add, change or delete telephone service, confirm the request in writing**. Include the specific request, dates, order number, circuit number(s) and quoted pricing. Keep these letters in a file for at least five years. If you find overcharges on a bill audit, you have evidence on how far back it goes. You get a bigger refund. Don't assume the phone company has historical records. Most of the time, they're lost them.

(Teleconnect Magazine)

Fax On Demand Services

Long Distance Carriers. AT&T, MCI and Sprint all offer some enhanced fax services. Their primary focus is on the broadcast service, as it's most in demand. Their services are targeted at large volume customers. As a result, these companies do not offer as many services to support broadcast and fax-on-demand. For example, these companies do not offer desktop publishing services to improve the quality of the fax documents delivered.

Their strength is in the size of their operation and their reliable service. These companies ca handle as large a job as any customer may bring to them.

Regional Bell Operating Companies. The Regional Bell Operating Companies (RBOCs) have been slow to get into enhanced fax services. Many are now planning fax mailbox and fax overflow services. They are expected to add fax broadcast and fax-on-demand service in 1996. US West Enhanced Services has offered fax broadcast service for several years. Pacific Telesis is adding fax services via the Information System Group, which offers voice mailbox and other enhanced voice services.

Independent Service Bureaus. Most of the companies that offer enhanced fax services are small independent companies. Many of these were started in the last three years. There are a few that have added fax services to established businesses. PR Newswire is one of these, and Swift Global Communications is another. The independent companies frequently offer more services in order to make it easy for customers to use their offerings. These companies made investments in equipment to add services such as call forwarding and to accept redirected calls from service centers.

(ABConsultants and Nuntius Corporation)

Installation Bills On New Lines

Carefully check the installation charges on your telephone bill when you have new outside lines installed. They are often **wrong**. When you order the lines, ask what the installation charges will be and confirm it in writing. Then check the bill and compare it to the original price quote. One mistake we frequently see is a charge for an "RJ21X" which is basically a jack (i.e. an amphenol connector) that handles up to 25 outside lines. You may have room on your existing RJ21X, but when you order new lines, your telco representative may not know this and you may be charged for another one (even though it may never be installed!) Cost is about $160 per RJ21X in New York City.

(DIgby 4 Group, Inc.)

Installation On Time -- Getting Your Phone Lines

1. **Plan on the phone company being late with their lines**. If you can, give them an installation date **at least a month** before you actually need the lines.

2. **Don't get personally upset** or physically violent if they continue to mess you up with stupidity, slowness and *"that's not my responsibility"* type arguments. Play nice guy. Remember, it will all be solved within a month. (Remember the month? That was Point #1.)

3. **Get to know how your local phone company works** and work within its incredibly bureaucratic framework. Typically they move your installation order laboriously from one department to another sequentially. Phone companies assume everything will go perfectly. They don't look ahead to see if there are any bottlenecks. Your job is to help them plan your phone job by checking over each aspect of your job **at the very beginning**, not at the end when it's too late. The Squeaky Wheel still gets the most attention with your local phone company.

(Teleconnect Magazine)

ISDN Is Real

ISDN is available today. There is extensive ISDN service availability of both PRI and BRI. In the U.S., over 90 million RBOC circuits are ISDN-ready. Most major metropolitan areas have near 100% circuit availability. ISDN is also readily available from long distance carriers (e.g. AT&T, MCI, Sprint, Wiltel, etc.). In Europe -- France, Germany, and the U.K. have had 100% availability for several years - but with national variants. Euro-ISDN availability for both PRI and BRI will approach 90% by mid 1996. See figure 2.1. Uses for BRI include telecommuting, videoconferencing and Internet access. PRI is used for concentration, multimedia & videoconferencing applications, leased line backup & overflow and Internet service providers.

(Xircom Systems Division)

Figure 2.1 - Global Spread Of ISDN

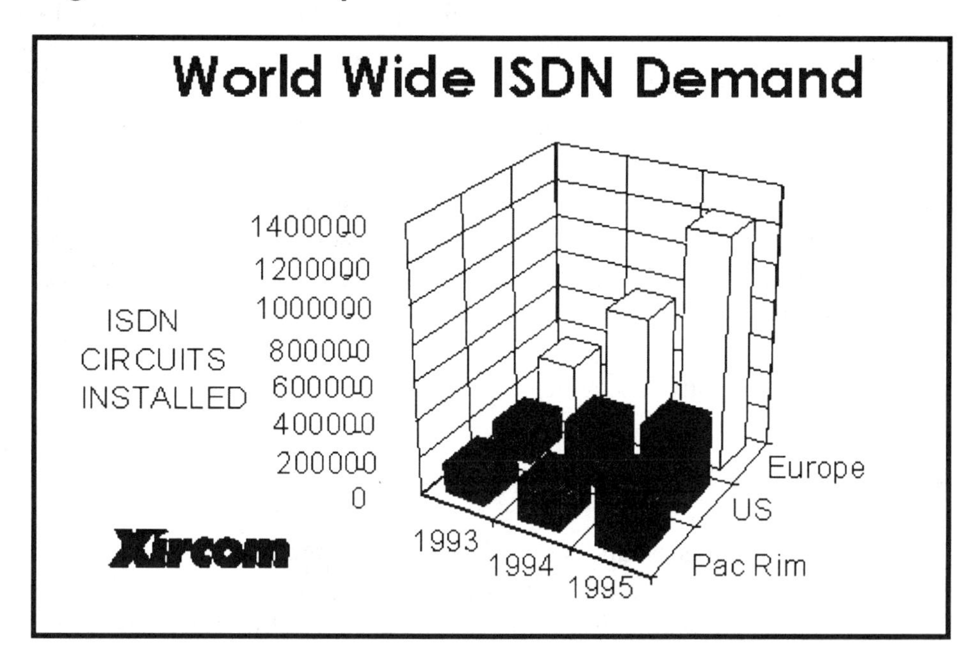

ISDN And Why To Use It

Besides speeding up inbound transmissions, ISDN quickens the outbound side as well. An outbound call is placed by sending a single packet message to the central office containing the number being called. Although a T-1 trunk is capable of carrying 1.544 Mbps of digital information, making an outbound call still involves digitizing DTMF tones and transmitting them inband, which is only marginally faster than an outbound manual dial on a plain old telephone. With ISDN, not only is dialing eliminated, but the telephone network can interpret the call request quicker and route the call faster.

ISDN support on PC-based CT systems also lets service bureaus take advantage of features like AT&T's Vari-A-Bill. Vari-A-Bill allows the charge for a 900 number call to be varied during the call.

Callers can select features they desire using touch-tones and will be billed as appropriate for the service selected. Vari-A-Bill also allows for fixed charges, providing an alternate way of billing for things such as a magazine subscription.

To save on ISDN access charges, developers can also take advantage of ISDN Non-Facility Associated Signaling (NFAS). It's NFAS which allows all of the signaling information, for up to eight trunks of 24 channels, to be passed through a single signaling channel (D channel), thereby saving on access charges (because D channels carry a separate *expensive* charge when ISDN is provisioned).

(Computer Telephony Magazine)

ISDN Standards

The ISDN network is a standard worldwide. Irrespective of which territory you are in, ISDN is delivered to your organization in the form of multiple B channels. The standard flavor of ISDN is the Basic Rate Interface (BRI). This is entry level ISDN, which can give you two B channels for data service.

PRI (Primary Rate), which is a higher bandwidth service, gives 23 B channels for North America and 30 B channels for the rest of the world. In any of these scenarios, the B channels are all common. Today, ISDN is currently available and implemented in 37 countries.

(4-Sight, L.C.)

ISDN Is Cheaper Than Switched 56

ISDN BRI is cheaper than Switched 56 with many local phone companies. ISDN BRI is **compatible** with Switched 56, i.e. you can call back and forth. Better yet, you get **two** 56 Kbps lines with ISDN, not one.

(Teleconnect Magazine)

ISDN -- Frequently Asked Questions

What ISDN service do I need to order? First, don't order ISDN until you know precisely what you want to do and which precise ISDN equipment you are getting. Then ask the ISDN equipment maker for full specs on what ISDN line you should order. Give him the name of your local phone company. Then go back to your phone company and find out if they have that configuration.

Once you know your local telco has what you need and your ISDN maker can ship what you need, then -- and only then -- should you order anything.

What about changing later to new equipment? We've actually unplugged a one ISDN piece of equipment and plugged in another into the same line without any telephone company changes. National ISDN-1 in its "Intel Blue" configuration is pretty "standard." We've also had good success with AT&T Custom.

How long will it take to get my ISDN line and equipment up? From one day to an eternity. Do not be surprised by rampant stupidity. It's getting better. We've actually installed ISDN equipment in under three minutes and had lines installed in less than two weeks.

How many SPIDs do I need? Order two. That way you'll get maximum flexibility. Your first app may work on one, but your second ISDN app may need two. Your first app will also work on two. So get two. If you ask, your local phone company will change the specs on your ISDN line most ways from Sunday. But they'll charge. Boy, will they charge. Nynex wanted $150 to add a second SPID to one of our ISDN lines.

What equipment do I need to connect to my ISDN line? All ISDN BRI lines need an NT1. About $200. Sometimes the NT1 (pronounced N-Tea-One) comes separate. Sometimes, it's built into the equipment. It's cheaper if it's built in. But it will be less flexible in the long run, as devices are increasingly connected behind ISDN switches. Those devices won't work if they have the NT-1 built in.

How does Bonding work? Bonding lets you join both B channels to transmit a single 128,000 bit per second data stream. You dial a call normally. Devices like Intel's ProShare and Vivo's Vivo 320 put two B channels together, dial on both lines and get 128,000 bits for video, etc.

With more sophisticated equipment, you can join several ISDN lines together and get as high as 512,000 bits per second. Some phone companies let you bond your two B channels into one 128 Kbps channel. Some do. Some don't. Most are now allowing it. Some may charge more. Ask.

If I bond two B channels into one 128 KBPS line, what's my long distance phone bill going to look like? You'll pay for two phone calls. Don't believe otherwise. There's a story that AT&T hasn't figured out how to bill for long distance ISDN calls and doesn't. Don't believe stories. You want to pay our AT&T ISDN long distance bills?

Can I "talk" to my Switched 56 DSU or modem with ISDN? It depends on the equipment you have. Some ISDN "modems" will only talk to other ISDN "modems." They won't talk to analog modems. Some will. Needs checking.

The Adtran ISU products (and some others), with internal V.32 bis modem option, allows the customer to communicate with ISDN terminal adapters, SW56 DSUs, and analog modems.

What specific ISDN services are there? Circuit-switched data, circuit-switched voice, and packet-switched data are all supported by Basic Rate ISDN. Tariffs vary.

What applications are well-suited for ISDN? Popular apps are high-speed Internet access, collaborative computing (also called document sharing), videoconferencing, digital audio, fast file transfer, remote PC/server access (such as work-at-home), credit card authorization (because ISDN dial up and connect is usually less than a second) and medical imaging, which need big bandwidth.

What's H.320? H.320 is the most common videoconferencing standard. If your ISDN equipment supports H.320, you can talk to and see anybody anywhere on an H.320 compatible device -- whether it's PictureTel, or Vivo, or Intel ProShare, or British Telecom (Tandberg), etc. H.320 is poor man's video, but it's a lot better than straight voice. Video does add meaning to a conversation.

What happens if my ISDN software gives trouble? Get to know the designer. This ISDN business is so small, he or she will write a patch for you or tell you a way around.

ISDN is supposed to provide 64 KBPS service per B channel. Why are some areas limited to 56 Kbps? The 64 KBPS rate is referred to as "clear-channel" because the signaling is done outside the 64 KBPS. 56 KBPS circuits steal 8 KBPS for signaling. An intelligent out-of-band signaling system is needed to guarantee that call are routed over the appropriate 64k trunks. Such a signaling system, SS7, is currently being completed by the local and long distance companies. At this time, most metro areas have full 64 KBPS.

(Adtran, Inc. and Computer Telephony Magazine)

ISDN (BRI Versus PRI) For Call Routing

Two success keys: capacity and speed. Did you know that an ISDN BRI line can have up to 16 or 63 calls ringing into it at any one time? Some Central Office switches can't handle more than 16 appearances per number, but the ISDN standard quotes up to 63 at once. And an ISDN BRI circuit can also transfer a call to another number without actually answering the phone. Capacity.

You could take a fast PC and pack it with a few BRI cards, write a little database code and before you know it you have a product locator system. Every time a call rings into the system an ISDN data packet comes in on the ISDN D channel containing the calling number and who they called. ISA bus ISDN BRI cards are less than $1,000 each, not including an NT1 (about $165) per BRI line.

You can build a system that receives the call data, finds out where the caller is located, where their closest dealer is and then tells the CO to forward that call to the dealer's phone number. Once the caller has been transferred, the line is available to accept the next caller. You can handle a call in around three seconds which translates to 20 calls per minute per card or 4,800 calls per hour with four ISDN BRI circuits. Capacity and speed. Assuming a business rate ISDN BRI line costs $125 each, the monthly operating costs (not including long distance) would be about $500 per month, which is less than a single T-1 line.

The forwarding system would incur any charges associated with transferring a call to each destination. This may seem high, but at least it does not require a huge number of lines to support the inbound and outbound circuits for each forwarded caller. If four ISDN BRI circuits won't handle the call volume, Dialogic, NMS, PRI, Rhetorex and several other vendors offer ISDN PRI compatible cards that will handle 23 inbound channels per card.

The advantage to using a PRI connection is that you have a single card supporting lots more lines plus most of them offer voice record and playback features as well. This would allow our product locator to act as a voice mail system for all the participating stores after closing time.

Instead of forwarding the call to the store only to have it ring with no answer, have the system pick up the calls and greet the customer, offing to take a message to be delivered when the store opens again. Billing would be done by getting the phone bill delivered on a disk. This would let another small application apply all the charges to the appropriate store that received the call for collection purposes. If the inbound call was made on a toll free number, the forwarded call duration would be used to bill the 800 time used.

Each store record would include all the details about that store. The opening and closing times for each day, all the observed holidays and any "midnight madness" sales would be added to their database entry. If an area was covered by two stores close to each other and their closing time was different, the system could route calls to the nearest open store until there aren't any stores within a preset distance from the caller. Once the distance is over the specified limit, it would drop the caller into the nearest store's voice mailbox.

The system would use the same database to know what stores have just opened and would take a line or two to start delivering voice mail to each store manager. One of the limitations of this system is that non-answered calls would be lost completely. There is no way to tell when the call is not answered, after it is forwarded to the destination. Another limitation is the fact that once a call is forwarded, the call is being charged to you, and you have no control over the call duration. It would be nice to have a message sent back to the system once the call was complete but that isn't possible at this time.

(Computer Telephony Magazine)

ISDN -- BRI Lines And How To Order Them

Step 1. Call your local telephone company. Figure 2.2 shows a list of ISDN Help Lines. If your telephone company is not listed in this chart, contact the company which serves your location or call 4-Sight and ask for assistance. The telephone company may require the following information to install your ISDN or Switched 56 line.

Figure 2.2 - ISDN Help Lines

Telephone Company	Service Name	Number
Alascom	ISDN BRI	800-252-7266
Ameritech	ISDN Direct	800-832-6328
Bell Atlantic	IntelliLinQ	800-570-4736
Bell South	ISDN BRI	800-428-4736
Cincinnati Bell	FlexLine	513-566-3282
GCI	ISDN BRI	800-800-4800
GTE	ISDN BRI	800-448-3795
Nevada Bell	Centrex ISDN	702-688-7124
NYNEX	ISDN BES	800-438-4736
Pacific Bell	Centrex ISDN	800-472-4736
Rochester Telephone	The Explorer	716-777-4501
Southwestern Bell	DigiLine	314-235-9553
Southern New England	Digital Enhancer	800-430-4736
Sprint LTD	ISDN BRI	708-768-6043
Stentor	MicroLink	800-578-4736
US West	Single Line Service	800-289-9091

Have the following information available before you call your local telephone company:

Location Information. Contact name, company name, address, telephone number and fax number.

Billing Information. Billing name, billing address, billing contact name and contract length.

Customer Listing. Type of business, service to be listed under and additional listing.

Step 2. Is ISDN available in my area? Once you have contacted the telephone company, request a loop test on your location. A loop test is used to determine if ISDN is available at your location. A loop test may take some time and the phone company may need to call you back.

Step 3. Ordering your ISDN line. Most of the telephone companies understand the codes developed by the North American ISDN Users' Forum (NIUF). When you order your ISDN line, request Capability M with 2 directory numbers (one per channel) and call forward on busy from the first line to the second line.

Capability M - (2B) includes alternate voice and circuit-switched data on two B channels. Data and voice capabilities include Calling Number Identification.

Directory Numbers - Telephone numbers used to identify different telephone lines.

Contact your service provider to choose a long-distance carrier.

Step 4. What if ISDN is not available? Four alternatives may be possible if ISDN is unavailable in your area. When contacting your local telephone company, inquire if one of the following alternatives to ISDN is available.

Virtual ISDN - Requires installing a new board in your local phone switch to make it capable of re-transmitting your signal to an ISDN compatible switch.

Foreign Exchange (FX) - Requires installing a new board in your local phone switch to make it capable of handling an ISDN line run from another ISDN compatible switch.

Repeaters - If your switch is non-compatible, repeaters can be installed every 3.4 miles from your location to the nearest compatible phone switch. Repeaters retransmit your ISDN signal.

If one of the above alternatives is available, refer to Step 3 to place an order. If one of the above alternatives is not available, inquire if Switched 56 is available.

Switched 56 - A single channel alternative to ISDN that operates at 56 Kbps.

Step 5. Ordering Switched 56. When ordering your Switched 56 line, specify that you require a 4-wire circuit. No other switch settings are required when ordering Switched 56.

Datapath by Northern Telecom is the exception which uses a 2-wire variant of Switched 56 that requires different hardware. Contact your service provider to choose a long-distance carrier.

Step 6. Questions to ask your phone company. If your telephone company does not understand Switched 56, refer to Switch Setups below.

When will my ISDN line be installed?

Will my ISDN line be installed directly into the room where my computer is located (recommended)?

What are my Service Profile Identification (SPID) numbers?

Not every installation will use SPIDs. SPID numbers are like telephone numbers and should be written down in the order given to you. SPIDs will be used during software installation.

If your ISDN line does not function. In the unlikely event your ISDN does not function correctly, try the following solutions.

Ask the telephone company to perform a loopback test on your ISDN line. Observe your NT1 during the loopback test (On 4-Sight gear, the U Loop-Back light displays during a loopback test).

Ask the telephone company to check the configuration of your ISDN line and refer to Switch Setups below.

Call 4-Sight Technical Support for assistance between 8:00 A.M and 5:00 P.M. U.S. CENTRAL TIME. (1-515-221-3000).

Switch Setups. In most cases AT&T 5ESS, Northern Telecom DMS 100 or a Siemens central office will be what you're hooking-up to. Here are some pointer on how to "set up" these switches for use with your gear:

Capability M. Request Capability M with 2 directory numbers (one per channel) with call forward on busy from the first to the second.

Capability M - (2B) includes alternate voice/circuit-switched data on two B channels. Data and voice capabilities include Calling Number Identification. No other features are required from your telephone company.

The following points are used by your local telephone company during installation to configure various switches. If your telephone company did not understand Capability M or the description, inquire what switch type will provide your service.

Provide your telephone company with the switch setting corresponding to your specific switch type (AT&T 5ESS, DMS 100 or Siemens) and installation (Custom or NI1).

AT&T 5ESS Custom PRE-NI1. When ordering AT&T custom Pre-NI1 implementation, ask for:

Point-to-point connection.
Terminal type = A.
NO EKTS.
No calling restrictions on the line.
The line must be configured for Circuit Data Only.
Call Forward on Busy from one channel to the second.
Call Line Identification, if available.

Note: You will not be issued SPID or TID numbers. Both of your ISDN channels have the same number.

AT&T 5ESS NI1. When ordering AT&T NI1 implementation, ask for:

Multipoint connection.
NO EKTS.
No calling restrictions on the line.
Terminal type = A.
Call Forward on Busy from one channel to the second.
Call Line Identification, if available.
You will receive 1 or 2 SPID numbers, depending on your switch type.

Northern Telecom DMS 100 PVC1. When ordering Northern Telecom PVC1 custom Pre-NI1 implementation, ask for:

NO EKTS.
No calling restrictions on the line.
Circuit Switched Data = Yes.
Call Forward on Busy from one channel to the second.
Call Line Identification, if available.
Two SPID numbers are necessary.

Northern Telecom DMS 100 NI1. When ordering Northern Telecom NI1 implementation, ask for:

Multipoint connection.
Circuit Switched Data = Yes.
NO EKTS.
No calling restrictions on the line.
Call Forward on Busy from one channel to the second.
Call Line Identification, if available.
Two SPID numbers are necessary.

Siemens NI1. When ordering Siemens NI1, ask for:

Multipoint connection.
Circuit Switched Data = Yes.
NO EKTS.
No calling restrictions on the line.
Call Forward on Busy from one channel to the second.
Call Line Identification, if available.
Call Forward on Busy from one channel to the second.

(4-Sight L.C.)

ISDN Versus T-1

A big question you need to address is whether you should use T-1 or ISDN PRI circuits? If you will ever need more than 16 simultaneous outside conversations, a digital circuit will pay off. The price difference between single T-1 with 24 combination lines and 24 analog lines in a rollover is enormous. In one comparison we did, you would pay for the T-1 installation and all additional hardware required in eight months of use. The other reason to go digital is reliability. If you ever had to track down the source of a noisy line, you know. On a digital circuit, all the lines work or don't work. The T-1 provider can troubleshoot all of their hardware from their CO to your premise without sending a technician out. Unless it's a problem with your equipment.

You also have a choice between T-1 and ISDN PRI circuits. Both are digital circuits carrying multiple voice channels. On PRI, one of the 24 channels acts as a signaling channel. The good news is that if multiple PRI circuits are used, one PRI circuit can do the signaling for many (figure up to 8 PRI T-1s) thanks to Non-Facility Associated Signaling (NFAS). This means you have all the advantages of ISDN circuits (out of band signaling, speed, ANI, DNIS, etc.) without the penalties.

With ISDN, the D-channel carries call control and information, and the B-channels carry payload (voice, data, video). With ISDN you can get Vari-a-Bill service from AAT&T so you can change the billing rate during the call. You can also get Non Facilities Associated Signaling (NFAS) – it lets a single D-channel control B-channels on multiple spans.

But even if you do lose a voice channel on the first PRI circuit, you get a much faster method of dialing. Instead of using MF or DTMF to dial the phone to set up a call, an ISDN connection sends a data packet telling the system to setup a call from point x to point y. This makes a lot of sense if your system places many outside calls. Instead of taking 15-20 seconds to determine if a number is busy, on an ISDN call, you typically know in fewer than three seconds.

The power of PRI is its concentrated bandwidth and fast call setup (300 milliseconds vs. 3 to 5 seconds for T1). With H0 (384K) and Multirate can get you greater bandwidth without Bonding. In general, ISDN PRI gives you more information on the call than traditional T-1 service.

Of course, some areas are not yet ISDN equipped and would charge a mileage based premium to get ISDN from the nearest ISDN equipped central office. In such a case your decision is virtually made for you. T-1 is also often less expensive to lease than ISDN. One example of a monthly quote from BellSouth was $917 for T-1 and $1,367 for ISDN.

(Computer Telephony Magazine)

ISDN PRI Versus T-1 Speed

PRI:

64 KBPS on each of 24 channels (channel 24 reserved for signaling).

T-1:

56 KBPS per channel (8 KBPS per channel "robbed" for signaling).

ISDN PRI Versus T-1 - Reliability

PRI:

Any channel can be used for inward calls, outward calls, WATS calls or 800 calls on a call-by-call basis.

Channel is not allocated if the destination number is busy (look-ahead routing)

T-1:

Channels are pre-assigned for type of calls, so some channels can sit idle while other channels ring busy.

Channel is allocated for the call while it checks to see if destination line is busy

ISDN PRI Versus T-1 - Cost

PRI:

Allocate channels by time of day, e.g. more voice calls during day and high-speed data transfer at night. 25% more efficient use of channels (average).

T-1:

Channel is allocated for the call while it checks to see if destination line is busy. Voice and data channel assignments are fixed and can't borrow from each other

(Xircom Systems Division)

Loop Versus Ground Start

PBXs often have switches on their trunk cards reflecting whether they're connected to ground start or loop start trunks. If your switch is wrong, the trunk won't work. Vibration (and old age) can cause the switches to change themselves. Check there first if you're having trouble with certain trunks.

(Teleconnect Magazine)

T-1 Carrier Lines And How To Order Them

When ordering T-1 service from a carrier, you can arrange for the carrier to coordinate access, make the arrangements yourself, delegate parts of the process to a subcontractor, or you can hire a consultant. The more information you can supply to your carrier about your installation and the more specific you can be about requesting the type of service you want, the better chance you have of avoiding costly delays that can plague T-1 service turn up. That knowledge can help make T-1 service installation proceed smoothly.

T-1 Carrier Provided Service. Many carriers provide some form of end-to-end coverage. They will assume responsibility for coordinating service between their lines and the local exchange carriers. This kind of support can be especially useful if you are installing equipment as part of a private network where several companies may own the different lines you are using. This ownership patchwork can be a headache when you are trying to troubleshoot problems at service turn up. The advantage of end-to-end coverage is that one party is responsible for all the details of providing service.

Arranging for Service Yourself. When arranging for service yourself, your carrier will ask you a number of questions based on the type of service you are ordering such as the location of the demarcation point in your building, the name of your local exchange carrier, the type of service you require, and the type of equipment at your premises. Procedures and requirements will vary from carrier to carrier. What follows is a checklist of points you should keep in mind when talking to a service representative.

Winking Protocol. Be aware that some aspects of your program design will depend on carrier network specifications. For example, each carrier can use a different winking protocol, and you may have to customize the values.

Digit Length. If your application makes use of ANI or DNIS digits, you need to make sure that tone length and interdigit values you are using for DTMF and MF digits are compatible with the carrier's specifications.

Ask for Details. Find out as much as you can about details of the service turn-up process. For example, some carriers commonly activate a single timeslot on a T-1 line while the line is being tested. The customer must call back to have the carrier activate the rest of the line after verifying that the installation operates correctly with a single timeslot.

Service Options. Be as specific as you can when describing the kinds of service options you want. Your carrier may offer certain options that the carrier representative did not mention. For example, a carrier may allocate calls to multiple-line sites using some kind of least-busy-line-first scheme. You may want call allocation to be assigned sequentially to certain lines. This allocation scheme is available as an option, but you have to ask for it.

Points of Responsibility. Understand what aspects of service your carrier is responsible for and what aspects you are responsible for. In many cases, different companies control access to different parts of a T-1 line. You may encounter difficulty during the troubleshooting phase of the service turn-up process when the source of a problem appears to be at an interface between two carriers. Unless you have previously determined who has responsibility in these situations, you may experience delays in service turn up or, worse, you may be unable to get anyone to provide a solution.

Be sure you find out who has responsibility at each point for the service you are ordering. For the customer premise equipment you are installing or that is already at a site, have the manufacturer's name, equipment numbers, and equipment registration numbers ready to provide to the carrier.

T-1 Carrier Third-Party Options. Consider hiring a telecommunications or telephone consultant to coordinate service with a carrier. Although you incur additional costs, enlisting someone knowledgeable about the service order process and about the service options available can streamline the ordering process. You have one party responsible for coordinating service between the various local and long distance providers.

Another approach is to delegate parts of the service acquisition process to others. In some cases the company that sells or rents the Channel Service Unit (CSU) will install it as well as verify that you have a functional T-1 connection with the carrier. Some facilities management companies and service bureaus will sell or lease access to a T-1 line as part of their service package. By dealing with such a company, you can bypass most of the T-1 line hookup and service ordering process.

Supply T-1 Equipment Information To Your Carrier. When ordering T-1 service from a carrier, you must provide them enough information about equipment at your installation for them to give the proper service.

Manufacturer of Device
Model number
FCC registration number
Ringer equivalence (0.0A, etc.)
Framing type(D4; superframe, ESF, etc.).
Interface (DSX-1,etc.)
T-1 signaling (Robbed bit, etc.)
Supervisory signal (2- or 4-wire E&M, modified ground start, etc.)
Other supervisory signaling (DPO, DPT, DX, ETO, RPT, SDPO, etc.)
Channel order (1-24, etc.)
Wink (wink, double wink, etc.)

T-1 Equipment. Other equipment that you will need to familiarize yourself with are CSUs (required in most cases when connecting directly to the public T-1 network), DSUs (depends on capabilities of CSU; usually not required), a CSU-to-T-1 card cable (in most cases must be built by customer). If you are using built-in multiplexor gear, a channel bank is not required .

CSU. The Channel Service Unit (CSU) is usually the first piece of equipment on the customer premises that connects to the T-1 line. A CSU is required anytime metallic T-1 service is ordered from a local exchange (all the Regional Bell Operating Companies) or interexchange carrier (including long distance carriers like MCI, Sprint, and AT&T).

Metallic T-1 uses wire to carry the signal instead of optical fiber or microwave radio. The CSU connects a DS-1-level line to DSX1-compatible customer equipment. It may be a stand-alone unit, or it could be integrated into a PBX. The CSU provides functions that are critical to the network. It provides loopback capabilities for running diagnostic tests to locate where problem conditions are occurring on the T-1 chain. Some CSUs have dialing capability for performing remote diagnostics.

The CSU is the last point where the signal is regenerated as it leaves the network. It also provides line equalization, provides voltage protection for telephone company and customer premise equipment, monitors the T-1 line for violations, and can generate "keep alive" signals and error condition messages. A "keep alive" signal is a bit stream of all "ones" used to power repeaters in the circuit when there is no activity on the line.

If you will connect a CSU to a Dialogic DTI/1xx, it must support superframe (D4) format. The DTI/1xx does not directly support Extended Superframe (ESF) format.

Some CSUs have the ability to convert superframe format to ESF format and the DTI/1xx will function properly with them. However, CSUs with ESF ability can be up to three times as expensive as regular superframe CSUs.

CSU Companies

Larse Corporation
Santa Clara, CA
408-988-6600

Kentrox Industries, Inc.
Portland, OR
1-800-824-4510

Larus Corporation
San Jose, CA
408-275-9505

Verilink Corporation
San Jose, CA
408-945-1199

Digital Link
Sunnyvale, CA
1-800-441-1142

Telco Systems, Inc.
Fremont, CA
1-800-776-8832

(Dialogic Corporation)

T-1 Line Installation

It's always best to install T-1 lines directly on T-1 trunk cards fitting directly into PBXs. It's cheaper. It's easier. It seems more reliable. It seems to work a lot better.

(Teleconnect Magazine)

T1 Localized -- A Long Distance Bargain

If you have lots of long distance traffic -- at least $5,000 a month -- consider renting a T1 line from your office to your chosen long distance carrier's office, also called their POP (for Point Of Presence). Your long distance company will be able to drop its per minute charges to you based on the fact that it will no longer be paying heavy per minute access charges to your local phone company. You can carry up to 24 phone conversations on one T1 phone line. And you can carry more if you install special multiplexing equipment.

(Teleconnect Magazine)

Touch-Tone Charges -- Don't Pay

Don't pay for touch-tone charges on outside lines that are used only for incoming calls such as 800 numbers (their associated local lines) or DID (direct inward dial) lines.

(Teleconnect Magazine)

Traffic Tips

In your computer telephony application, it's critical that you have enough phone lines so that none of your callers get a busy signal. Here's the simple formula to figure out how many you'll need: Estimate: (a) how many calls you expect to service during the busiest hour of the day; and (b) how long in minutes the average call will be during this period. Then multiple a x b x .0238 (calls per busy hour x average call duration in minutes x .0238). This defines how many simultaneous phone lines your system will need to make sure that 99 out of 100 callers will get through.

For example, say you expected your busiest hour to handle 45 calls with an average length of 3.5 minutes. 45 x 3.5 x .0238 = 3.74 lines. A four-line system could handle that.

This basic formula assumes a "smooth" distribution of calls. One other thing you must consider is what kind of calling "pattern" you expect. Some systems are more subject to "random" hits, like those that are advertised with a mass medium like TV or radio or that deal with a time-sensitive subject, like automated "sports-scores" or "movie-line" information hotlines. If you're pattern is going to be random, you need more phone lines (up to 25% more), especially for a service where a busy signal equals a lost customer.

For non-critical applications, like a company voice-mail system, this formula may yield more lines than you really need. In other words, if you don't mind that some callers will get busy signals, you can drop the number of lines this formula spells out.

Line considerations are important for overall quality and cost of a CT system. The latter can't be dismissed. Ensuring extremely high levels of peak-hour capacity also means more cards, more phone lines and (probably) higher software costs. After you're up an running, ask your local phone company to provide you with reports that show how many incoming calls get busies. They will do this. They would love to sell you more lines.

Pick analog or digital service. If you need less than 24 simultaneous phone lines, use normal POTS ("plain old telephone service") analog phone lines, the same ones you use in your residence or business.

All of the two- and four-port voice cards come with onboard line interfaces to connect to these lines. You can stack them in a single PC to handle the amount of lines you want, though there are voice store-and-forward-only cards that can connect to separate line-interface devices (for network-interface flexibility -- DID connections, loop-start connections, ground-start connections, E&M connections, etc.).

Some newer "high-density" analog voice cards with built-in line interfaces can break you out of the two- and four-port doldrums in a single slot (Dialogic's D/160SC-LS, for example, is a single-slot voice card with an analog loop-start interface for 16-ports).

If you need 24 lines or more, you have the option of using T-1 service. The advantage of using T-1 is that you can pay significantly less per line for your monthly phone line charges as well as your voice-card hardware. In fact, several speech card vendors now offer a single voice-store-and-forward card that services a single T-1 span (some with fax too).

(Telephone Response Technologies, Inc.)

Virtual Nets Cheap

The biggest long distance bargains are the virtual private networking services from the long distance carriers. Rates have dropped substantially in the last several years as intense competition among the big three -- AT&T, MCI and Sprint -- has intensified. Today the VPN customer typically pays 50% of what a "retail" long distance customer pays. A virtual private network typically consists of a dedicated T1 line from all your offices to the local offices of your long distance carrier. V-nets will complete on-net and off-net calls you make. The on-net calls will be a little cheaper.

(Teleconnect Magazine)

Chapter 3

Application Development And Operating System Tips

You may be one of the most brilliant programmers of this modern age - but there's plenty of "gotchas" and snares in the wild and woolly world of telecommunications. Add to that its merger with computers and you've got *computer telephony* – an ever-changing discipline that takes patience and a lot of good, sound programming talent. But don't be afraid of computer telephony, as there is plenty of good help out there. For example, Dialogic Corporation (Parsippany, NJ -- 201-993-3000) even has a "*CT University*" complete with classrooms, workbooks and hands-on workshops.

But you'll find plenty right here. This chapter represents one of the largest and most eclectic lists available on great CT programming tips. Most of them are not source code tips, but common sense approaches to the design and development of platforms and solutions. Share these ideas with your product management and programming employees before your next big project -- you'll all benefit by saving precious time and money.

ADSI Information Sources for Developers

There are quite a few developers building ADSI capability into their applications and tool kits. Most of them have the standard document suite from Bellcore, but that leaves a bit to be desired.

Nortel and Philips, the two major suppliers of ADSI phones have a number of services for developers. The Nortel Manager of ADSI Applications is at Northern Telecom (Ottawa, Ontario – 613-763-7083) has a 2 day interactive course with course notes and example code, which takes the place of the Bellcore specs and then some. They sell the course notes for $500 and can be contracted for consulting. Philips ADSI Marketing at Philips Home Services (Burlington, MA – 617-238-3400) has an Authorized Developer program.

(Dialogic Corporation)

Answering Consistency

If both a telephone instrument and a computer will be on the desktop, be clear as to which functions will be controlled by the telephone versus which will be controlled by the computer keyboard or mouse. It should be logical.

For example, to dial an outgoing call do you use the a touch tone dial pad on the telephone or on the computer keyboard? To answer an incoming call do you press a button on the telephone or a function key on the PC keyboard? The point is to be consistent so that the person using the system is not jumping back and forth between the telephone and the computer to complete a transaction.

Our recommendation is to have the computer keyboard control all of these functions since most of the screen entry and scrolling also uses the computer keyboard.

(DIgby 4 Group, Inc.)

Application Specification And Development

Now is the time to translate the list of required features into a more complete road map of what the application will look like and how it will appear to callers. Without creating this specification, not enough information is available to adequately estimate how long it will take to develop the application. There are three essential parts to this specification: the Application Functional Details, Application Block Diagram, and Application Design Details. Let's explore each in more detail.

Application Functional Details. The purpose of this portion of the application specification is to describe, as accurately as possible, how the application will appear to the caller. This section is not intended to nail down technical details about how it will be implemented, just a clear description of how the caller will be allowed to interact with the application. One approach to preparing this section is to write it as if it were a manual that every caller could read to familiarize themselves with the way things should work when they call in.

The Block Diagram. Constructing the application block diagram is the next step after completing the functional details. This diagram should show, in reasonable detail, all steps a caller can take in traversing the various menus in the application. The time to develop this diagram is before a single screen or line of application code is written. It will, in effect, be your primary guide as you begin to create the various sections of the application. When completed, it will also give you a good visual feeling of how easy or complicated it might be for callers to navigate the system - that is, how difficult it might be for a caller who wants to do a specific task to actually get there and start doing it.

Drawing the Diagram. There are several ways of actually creating this diagram. One is to get a large piece of heavy paper, a sharp pencil, and an electric eraser. The electric eraser will be needed for the many changes that will be made as you begin to build your application concept. The preferred way, however, is to use a graphical diagramming program on your PC, such as allCLEAR by Clear Software, Inc. (Newton, MA -- 617-965-6755).

This program is moderately priced and very simple in concept but at the same time very powerful.

Choosing Box Names. Choose descriptive names for each block diagram box. This makes it possible for the names to also be used as Labels when creating the application - using either an application generator or a script language.

Labels are the identifiers that help to quickly locate where a particular application statement or screen is. For example, to represent the main menu in an application on the screen, "Main_Menu" would be a good choice.

Application Design Details. After completing both the Functional Details and a Block Diagram, it is a good idea to review them with the customer or the ultimate users and get their written approval before continuing.

The next major step is to develop the Design Details section. This section's purpose is to describe how all the application's features will be implemented - that is, at a very low level describing the individual specifications and design methods to be used. Keep in mind it is vital for the development schedule to be as accurate as possible - for the customer (so he knows when to expect the delivered system) and for the developer (so you know how much to charge for the project).

So what kinds of details should be developed in this section of the specification? To put it simply, anything that is not fairly obvious (to you) how it will be done. Taking one item again from our requirements list, let's see how it could be expanded: "Maintain a database of orders that will be transferred to the host computer once per day."

This example requirement is leaving out many important details - depending on the answers the task of implementing it could take anywhere from hours to weeks even when high-level development tools are being used. Figure 3.1 lists several questions that you need to answer before moving ahead on the CT application.

Figure 3.1 - Database Questions For Application Development

Question	Database Fundamentals
1	What is the format of the database? (including the database record details, number of fields, etc.).
2	How are orders written to the database? (one record per order, one record per item ordered, etc.).
3	Is the database on a LAN or only local to the voice system PC?
4	If it is on a LAN, will other LAN-based processes be reading or writing the database at the same time as the voice system? (if so, this will require record locking to prevent data corruption from occurring.)
5	Is there only one order database file? If yes, what happens if you run out of records or disk space? If not, how many are there and how are they used by the application? (e.g., a different one used each day, etc.)
6	How will the file be transferred to the host? (e.g. RS-232 serial port, copy operation on a LAN, etc.)
7	What times should the file be transferred? (e.g., once per day at midnight, etc.)

There are probably several more questions that should be answered before expanding on them in this specification section. After generating the expanded list of questions, consult with the customer and resolve each question. Then take all the information you have gathered and write the application design details section.

(Telephone Response Technologies, Inc.)

Application Generators – Good For Simple IVR

Let's say you want to develop a CT system that speaks the contents of a company database to callers. It will contain stock numbers, unit pricing, available stock quantities and the date of last order. How do you get started?

You'll need a PC with at least voice-store-and-forward cards in it, the software to make it bark, and the phone lines to run it on. If it's a simple application, the biggest time-saving hint is:

Use a CT application generator. Merely employing your own code-level programming skills with the primitive development tools bundled with CT board-level erector-set building blocks, it would be time-consuming and, depending on your CT knowledge, perhaps downright impossible to build even this relatively "simple" IVR app.

Probably the biggest challenge would be to set up the app to speak the information "conversationally": Stock number four-five-one-three-two is available with one-thousand two-hundred fifty units in stock. Your last order for this part was on October thirtieth, nineteen ninety-five."

Your computer telephony application development environment should allow you to easily piece together different speech prompts to construct the full sentence of information the caller is requesting. In the above example, you would need to record a series of numeric and date prompts (depending on how your CT software speaks these quantities). In addition, you would record the following separate prompts:

"Stock number..."

"...is available with..."

"...is not available."

"...in stock."

"Your last order for this part was on..."

With the help of an application generator, you would prompt the caller to enter the stock number, which you could then save into a variable named STOCK_NUMBER. Then you would do a database search for the appropriate database record.

Because CT application generators specialize in accessing a wide range of different database formats, you should have no problem dealing with your particular format. You would then read the necessary database fields into variables you specify with easy-to-remember names like QUANTITY and LAST_ORDER.

Finally you should be able to specify a single step in the app that speaks each component of the response in rapid succession, thereby creating a full sentence of information the caller can understand. It won't be too difficult to create an application like this. Like anything else it will take you longer the first time you do it. But after that, it's a matter of applying the same principles and powers of your application generator to a different database.

(Telephone Response Technologies, Inc.)

Application Menu Design

A caller menu is a list of choices, any of which can be selected by pressing a single digit. Menus are used frequently in IVR applications. They provide a point where the caller can hear all his options before deciding how to proceed.

There are two common uses for menus in an application. The first use is to present a large amount of information to the caller. The second is to determine how next to route the call.

Menus that are used to present a large amount of information to the caller can use looping to allow the caller to hear several different pieces of information. For example, suppose you wanted to provide callers with three pieces of information: pricing, quantity and lead time. Some callers may want all of that information; others may want just one or the other. Rather than playing one prompt with everything, construct a menu that says:

"For pricing, press 1 For quantity in stock, press 2 For lead time, press 3"

The system can loop through the prompt, giving the caller the chance to select a different option each time, or to write down information before proceeding. An exit option can also be added (e.g. "To enter another part number, press 4" or "To continue, press #").

Menus that are used to determine how to route the call must present choices that are clear and mutually exclusive. It is very frustrating for a caller to hear a menu in which several choices apply to him.

"If your'e a dealer, press 1. To place an order, press 2."

Well, what if you are a dealer and want to place and order as well?

Always present menu choices in 1-2-3 order, with the most common selection presented first. Don't try to get fancy with menu options, like "For pricing, press P." Simple 1-2-3 menus are the easiest for everyone to understand, to anticipate and to use.

The way the data is ordered in the database should not influence the order in which it is presented to the caller.

Menu prompts should say the choice first and the keystroke second (for example, "For pricing, press 1"). Callers tend to listen more intently after they hear what they want to do. That is the time to tell them how to do it.

Each menu should present no more that 4 new options to the caller. These options can be in addition to options that are presented on every menu. For example, you might want to use the 9 key to replay the choices. Thus, every menu would end with "To replay these options, press 9." If more than four new choices are needed, say "For more options, press 4" - then present another menu.

(Voysys Corporation)

Area Code Look-Ups

If you are developing an application that needs to match Caller ID or ANI with routing information, or if your application involves referral services, you should consider the purchase of an area code database. *Area Codes Plus* by Info Tech Marketing (Littleton, CO -- 303-979-9318) is one of those incredibly simple products that you find yourself scrambling for.

In short, Area Codes Plus consists of three ASCII files on a diskette containing area code information for US, Canada and Puerto Rico. You can merge these files with your own applications, pull them into a database or spreadsheet or just print them out as a reference.

The fields include Area Code, Area Code Description, State Postal Code, Country, State FIPS Code, Census Division Code, Census Division Description, Regional Bell Operating Company for that area, Time Zone Descriptor, and 32 fields containing "time elements" (time zone offsets, opening, noon and closing times for all of the zones in relation to each other) in HHMMM format. The time elements are based on Standard Time and are generally good during Daylight Savings Time (three states -- Arizona, Hawaii and Indiana -- don't observe Daylight Savings Time).

For Canadian area codes, FIPS code, census division information and RBOC's are omitted. Also Puerto Rico data has similar omissions, except that FIPS code data is included.

Each file has the same order of fields but in differing formats. ACFLAT.TXT for example, is a traditional "flat file," while ACTIMER.CSV is a comma delimited file with time expressed as military time. The last file, ACTIMEP.CSV, is a comma delimited file with time expressed as a percentage of seconds in the day (8 AM = .33333; 8 PM = .83333).

(Computer Telephony Magazine)

ASR - Application Design Tips

Successful use of ASR (Automatic Speech Recognition) requires more than just raw speech recognition technology. You need to establish a dialogue with the caller that helps you to quickly and accurately identify the caller and then his or her requested information.

Of course, any good application will require substantial interaction with a common database or databases. The general strategy in developing an ASR-based application is to ask a series of questions to identify the item from your database. For example, you can further qualify a customer record by asking first for the Customer-ID and then a name or company. This ensures an accurate match and helps to establish confidence with the caller:

System: "Please speak your 5-digit badge number"
Caller: "09459"
System: "Now say your first and last name"
Caller: "Mike Phillips"
System: "Thanks. Please say your company name"
Caller: "Applied Language Technologies"
System: "You are Mike Phillips from Applied Language Technologies, is this correct?"
Caller: "yes"

(Applied Language Technologies, Inc.)

ASR And Vocabulary Break-Down

A large vocabulary task can usually be broken up into several medium/small vocabulary tasks with clever application design. This allows for a larger vocabulary and better recognition accuracy.

(PureSpeech)

Auto Attendant Critical Questions

When programming an auto attendant, ask some questions:

1. Can your callers reach a real human being by dialing "0" at **any** time during the conversation?

2. Can your caller punch in the first three or four characters of the last name and be transferred to your extension?

3. Are all your messages under 15 seconds?

(Teleconnect Magazine)

Backup With XCOPY

XCOPY is a wonderful DOS utility that allows you to do the simplest imaginable backup. Here's what we do to back up our file server "F" that has no tape drive to our file server "H" that has a nightly automatic tape backup. We go in as Supervisor on both servers, and:

XCOPY F:\ H:\ /S /E /M

This command means copy everything on the F drive to the H drive, with directory tree structure starting from the root (XCOPY F:\ H:\) including everything in all subdirectories (/S), and any empty subdirectories (/E), but only if the archive bit on the source files is set -- and then turn off the archive bit for the files that have been copied (/M).

The "archive bit" on DOS files is set "on" when you change or initially create a DOS file. Depending on the utility you use to look at it, it reads "A" or "w" or some such. With this command, XCOPY will only copy files that are new or changed since the last time an XCOPY was made, following the source tree structure from the top. And it turns "off" the archive bit as it does so.

If you carry a laptop, you can keep all your data in a single subdirectory, and make periodic XCOPY backups to a 1.44 meg floppy. As one fills up, start another. You'll never lose anything. Run it from a batch file and never worry again!

(Computer Telephony Magazine)

Caller ID Problems

Calls being placed from cellular phones are reported using the number of the line where the call enters the phone network or no number at all. If your customers often use cellular phones, Caller ID won't be of much use to you. In the future, cellular phones will be able to transmit their number with the call, but so far they don't.

When you place a call through a PBX, it normally selects which outside line to use. The outside line number is the one that is reported on Caller ID. This means your Caller ID database has no way of exactly matching a particular person with a company's generic phone number. The best that anyone can do is show you a site name and perhaps a list of known people at that address. In general, Caller ID works best when the callers are calling from their home to a business. Most homes have one or two phone lines and this makes it easier to keep up with. One point: When designing databases to "screen pop," keep in mind that you may need multiple phone number fields to pop up the same record.

(Computer Telephony Magazine)

Caller Input Quality -- How to Solicit It

The uniqueness of IVR is its ability to accept different input from each caller and act on that input. Input from the caller is generally either used to identify the caller (an account number, social security number, telephone number, etc.) or to specify a particular item (a part number, flight number, stock symbol, P.O. number). In either case, the entry is used to look up a record in a database and give the caller some information from that record.

Numeric data, such as a telephone number, is the simplest type of input. IVR systems can accept alphanumeric input, but it is more complex to enter. The caller is instructed to press the key corresponding to the letter. But there are three letters per key, so there is often a conflict. For example, part numbers A17, B17 and C17 would all be entered as 217 on the keypad. In these cases, the caller must enter two keystrokes per letter (e.g. A = 21, B = 22, C =23). This gets very hard, very quickly, especially when more than one letter is involved. Try to avoid conflicting alphabetic entry whenever possible. If not, then provide callers with a written user guide that shows the keys that should press for each letter.

In cases where letters do not conflict, the caller can enter one key per letter. For example, a company has part numbers that begin with C, G or T. The C is the 2 key, the G is 4 and T is 8. Thus, the system knows it should translate 2233 into C233.

Ask the caller to enter a # to complete their entry. This positive acknow-ledgment ensures that both the caller and the system recognize that the input is complete. The * key can be used to cancel input and start over. This is especially useful when callers are entering a long string, such as a credit card number. When the input is canceled, the system should re-prompt the caller to enter the data, so they will know they are starting over (not just erasing the last digit).

The next challenge is verifying that the caller entered the correct data. In some cases, you may want to repeat the entry back to the caller, then give him the option to confirm or change it.

This takes time, and should be done only when security is involved, or if there is a high risk for error. Next, check the length of the entry against known bounds (e.g., an account number between 5 and 8 digits). If it is outside the bounds, flag an error and go to error handling. This will save a step in checking the database with a known invalid value.

Caller Input Standards. Here are some things that are becoming de facto standards in IVR applications. Try to follow as many of them as possible:

1. **Use the pound (#) key to terminate** all multi-digit entries from the caller. If all entries in an application are of a fixed length, the keystroke can be eliminated. You should use only one method throughout the application.

2. **Use the pound (#) key to skip** a prompt or continue to the next step.

3. **Use star (*) key to cancel** an entry or to back up to a previous menu.

4. **Use the zero (0) key for help** prompts or to reach live assistance.

5. **Present new menu options in 1-2-3 order**. Follow them with standard options (e.g., press 9 to end a call).

6. **Play the opening greeting only once** at the beginning of the call.

7. **Play the closing only once** at the end of the call.

Caller-Recorded Messages. Using recorded voice messages in an IVR application presents special challenges. On one hand, it allows callers to deliver information that they cannot with the telephone keypad. On the other hand, it means that the caller's transaction is not complete when he or she hangs up. Someone must check the message and act on it. This takes time and effort. And that could create a problem because the completion of the transaction and the labor savings are two of the main selling points of IVR.

Messages should be used sparingly, and only when there are enough resources to retrieve and act on them. Examples include: collecting additional comments for a survey, letting callers record a problem description as part of a help desk application, or allowing a vendor to make changes to a standing order by leaving a message. Be sure to only ask the caller for a few pieces of information. For example, say: "At the tone, please record any additional comments. When you have finished, press #." It is difficult for the caller to remember lots of items, such as "Record your dealer ID, P.O. number, invoice number and amount, shipping address and bill of lading requirements." Close your eyes and try to remember all of that!

Callers often want to replay or re-record their messages, so you may want to add those capabilities after each re-recording. If you think that callers are going to be hesitant to record messages, then give them the option not to record one, and be sure to let them know up front that they will be able to review the message.

There will need to be a separate application for retrieving the messages. The application will need to notify someone that there are messages to be picked-up. Use outbound calling to reach the person. Make it simple. Don't try to emulate their voice mail user interface. When a message is played, preface it with the caller's identification number. If possible, create a report that shows when messages were recorded and when they were retrieved so that management can monitor the activity.

(Voysys Corporation)

Channel Buffer Size Under SR4.1 for OS/2

Some applications need to play or record large VOX files thus making the application I/O intensive. When this happens it may become necessary to increase the size of your applications channel buffers to avoid underrun or overruns. To accomplish this edit your config.sys and find the line for your OS/2 driver, this line will look something like :

```
DEVICE=DXGENDRV.SYS CFG=C:\DIALOGIC\CONFIG\DIALOGIC.CFG
```

To increase the size of the channel buffers add X=xx (where xx is the size in K bytes) to your driver entry. For example to increase the channel buffers to 32K your entry in your config.sys would look like :

```
DEVICE=DXGENDRV.SYS X=32 CFG=C:\DIALOGIC\CONFIG\DIALOGIC.CFG
```

By default the Dialogic OS/2 driver(DXGENDRV.SYS) will allocate a 16K buffer for each voice channel in the system. These buffers are used by the driver for the transfer of data during a play or record. For example, to play a file the driver must fill the channel buffer in PC memory with voice data for transfer to the voice board so that play over the network can begin.

The driver makes a request for more data so it can fill the channel buffers in PC memory with more voice data. The data in these buffers are then transferred to the voice board to be played out. As the channel buffers in PC memory are depleted the driver requests more data with which to fill the buffers. Each of these requests for data results in a disk access. In an application that utilizes large VOX files the access time can slow down the system and result in poor response time.

(Dialogic Corporation)

Conventional Memory & Event Queue Optimization

The number of channels that can be in a system is limited by the amount of conventional memory available. D40DRV has a memory saving option -m, an option in particular is the -m4 option. One of the methods it employs to obtain the maximum amount of conventional memory is to enable event queue optimization. The result of this option is a greater channel density.

In order to save space certain initialization code is overlaid with the event queue since the initialization code is used only when the driver is loaded. In general with D40DRV, code and data must fit within 64K. So without the -m4 option a memory map might look like :

```
        CODE
  ------------------
     EVENT QUEUE
  ------------------
  PER CHANNEL DATA
```

The code , data and event queue are all within the same 64K. When the driver is loaded and the initialization code is no longer needed, the event queue is fitted into the hole left by the initialization code. This allows for 168 entries in the event queue. When -m4 is used the event queue is moved outside of the 64K block and the per channel data is moved up to start where the event queue used to start, so now the code and the per channel data are all within the 64K block.

The net result is more space for the per channel data structures which allows more channels to fit in the 64K segment. The memory map of this would look like:

CODE

PER CHANNEL DATA

EVENT QUEUE

With the code and data in the same 64K block and the event queue outside of this block, the event queue grows to hold 260 entries. While this does achieve higher channel density the tradeoff is the fact that the size of the TSR will grow.

(Dialogic Corporation)

Credit Card Software For IVR

There are many legitimate applications that could be vastly more valuable if they had an easy method of integration for a reliable system for handling credit card transactions.

IC Systems' (Oakland, CA -- 510-339-0391) IC Verify ($349 - $499) has several features that are interesting for developers. They offer a Terminate but Stay Resident (TSR) mode that waits on specific files to appear in a directory. Once a file is created, the TSR reads the file in and uses the information to start a credit card verification. Once the results of the transaction is received, it writes another file for the foreground application.

It can also collect information off the screen that can be used to do a credit card transaction. It also logs all credit card transactions to disk and printer for legal recording purposes. Optional signature capture and credit card readers are also supported. The downside for IC Verify is the memory requirements. Unless you have expanded memory available on the machine, it probably will use most of your memory for the credit card verification.

Their background mode takes about 30K of memory for background operation over a LAN. This would let multiple users on the LAN do credit card transactions without requiring a modem on each machine.

IC Verify makes sense in many cases instead of a traditional credit card verification station. It can cut down on fat finger transcription errors by taking the information straight off the screen. By supporting Debit cards, ATM cards and traditional credit cards through a single system, it opens the door to many other methods of payment. It still requires you to have a merchant agreement with each credit card company accepted by your system.

The other route to take is to call Verifone (San Francisco, CA -- 415-598-5540) and get a Verifone PNC-330 ($327) terminal. The credit card authorization terminal has a card reader, a keypad and a fluorescent display. The internal processor uses a downloaded program to connect to different credit card processors using its internal modem.

Instead of using the card reader (since a telephony application won't have access to the caller's card), you can send the information down the serial port and have the box submit it you. Once the transaction is approved or denied, it would send back a response packet through the serial port.

Since the processor in Verifone's box takes care of all the credit card functions, all your application has to do is send a request and wait on a reply. It can't affect any of the other voice lines in the application.

All of the protocol problems are handled by the clearinghouse. They (or Verifone) write a little script that is uploaded to the box to take care of any twists in the protocol. Any updates in the protocols can be downloaded using the same modem that sends the verification requests.

Any other process that interferes with the timely handling of the callers on other lines won't work. The Verifone PNC-330 box is the best solution. Only the serial communications to and from the unit are required by most applications.

(Computer Telephony Magazine)

Dialing International Numbers

When dialing an international number from the US, dial a pound (#) at the end of the number. The call will be completed much more quickly. This is because the US switch handling your call doesn't know how many digits to expect, so it doesn't even start making the call until there is a long pause without a digit, unless you dial pound to mark the end of the number. This works for manual dialing as well as computers.

(Parity Software)

Document Imaging Enriches Fax Communication

Fax is an effective delivery technology, but the way most people use it -- fax only delivers documents. You are also limited as to a fax's automated routing and indexing. Another problem is that document flaws are retained.

Consider the use of imaging technology to make faxing intelligent. With it, you can clean up and route faxes, index and store faxes, extract data from faxes and even trigger associated workflows.

The key to enhancing fax communication is recognition by using machine printed text (OCR), hand printed text (ICR), forms (forms id & removal) and glyphs (smart paper). These recognition technologies add fax management capabilities, data extraction from faxes and integration of faxing with other activities. Consider some of the following challenges associated with faxes and contact the vendors who can help you to correct the problem. Many of these solutions work alongside fax servers.

Fax clean-up options. You can correct "upside-downness" of faxes, enhance degraded resolution and "deskew" skewed documents automatically. An example is Sequoia Data's Scanfix.

Automated fax routing. You can use recognition-based routing to identify recipient by reading the name and checking it against a database. OLE is then used to embed the fax in an e-mail message to the recipient. An example is Nestor's N'route.

Fax indexing and retrieval. You can use structured data such as OCR fields to create relational records. Even with unstructured data, you can use OCR to index the entire document and then retrieve a fax based on its content. Much of this process can be automated. Data extraction from faxes uses software that analyzes repetitive forms (uses ICR, OCR and glyph recognition)

Forms processing of faxes. Forms can be used so data input on the forms can be removed and used for automated data entry. First, the content is recognized by the forms software. Further actions include database lookup or transfer to an operator, for example. You can use forms to recognize for structured data. This assumes that you know where the data is, so you can then verify the data against legal values. In order for this technology to work best for you, you can take steps to reduce the variability. Since forms recognition requires controlled parameters, use software that's compatible with Glyph reading (Xerox smart paper).

Faxes and workflow. Automated fax servers can decode faxes and trigger workflow. For example, product inquiry from a faxed-in form could cause a confirming fax or e-mail to be sent to the customer. Service technicians could be dispatched or make a call from a help desk. This enhances customer service. Since imaging and telephony development are converging, you should work closely with software and tools vendors who can help you automate imaging at your call center.

Many vendors use high level development tools such as Visual Basic, Delphi and PowerBuilder. These development tools use component software called VBX's and OCX's. The benefit of using these kinds of software components is their basis in a standard architecture that works on common development platforms. In most cases, it is straightforward to integrate vertical applications with your line of business data using these tools. Some influential pioneers in this area include Stylus Innovation, Cardiff, Nestor and Diamond Head Software.

(Diamond Head Software, Inc.)

Error Handling

Error Handling For Databases. Sometimes the caller enters a valid record number, but the record is locked of the entire database is unavailable. Check for availability of the database at the beginning of the application, before you ask the caller for input. If the database is unavailable, ask information of the caller and end the call. If it becomes unavailable during the application, do the same. Say: "We're sorry. The arts database is currently unavailable. Please try again later. Goodbye" and end the call. If a record is locked, say "We're sorry. Information for part number 1234 is not available at this time." Then continue with the same prompts as if the information had been delivered (e.g., enter another part, end the call).

Error Handling For Caller Input. When the system asks the caller for input, the caller can make several kinds of mistakes. They can:

Not enter anything (time-out).

Select an option that was not presented in the menu (invalid menu option).

Enter a value that is too short or too long (invalid multi-digit entry).

Exceed maximum recording time for a message.

Each type of error is handled differently, so that the caller is clear what they did wrong and how to correct it.

For time-outs, simply repeat the request prompt ("Please enter your account number"). Make sure you are giving the caller enough time to start. Five seconds is average.

For invalid menu options, say "Sorry, that is not a valid option. Please try again." Then repeat the menu of choices.

If a caller enters and invalid multi-digit entry, say, "We're sorry, that entry is not valid." Depending on the situation, you can then provide more customized help, such as "Account numbers must be seven digits long."

Then repeat the original request prompt. Again, make sure you are giving the caller enough time to complete their entry. The interdigit time-out value should be about 3 seconds.

If the caller enters an invalid record, say "We're sorry. Part number 12345 is not in our records. Please try again." Then repeat the original request prompt. By repeating the number to the caller, you allow him to determine if he entered the number he meant to enter.

In an application where security is an important consideration (such as Bank-By-Phone) you may not want to give the caller much help. If the caller fails at some point, simply say, "We're sorry, the information you entered is not valid." Don't give any hints as to which part was wrong. If the caller is recording a message and exceeds the maximum recording length, interrupt him with the prompt "You have reached the maximum recording time for a message" then proceed as if the message had ended normally.

Error Handling -Three Strikes. In all cases, give the caller three total chances to enter the correct data. you may want to customize the second error prompt to provide the caller with more instructions. After three failures, end the call. Play a prompt such as "We're sorry, we are unable to process your request at this time. Please try again later. Goodbye." Further explanation is not useful; however, you may want to provide a number the person can call to get live assistance (or route them to an operator).

(Voysys Corporation)

Expanded Memory Use

Expanded memory is a set of specifications to make that memory accessible to the CPU which would normally lie outside its address space. This is done by mapping the memory into the CPU's real mode address space. Expanded memory blocks are made available in 16K sections called *pages*, each *frame* of 64K containing 4 pages.

EMM386.EXE is an expanded memory driver provided with DOS 5.0. It is installed by including a line such as the following in the CONFIG.SYS file:

device=emm386.exe ram

For a complete list of the 17 switches and parameters supported by EMM386.EXE see your DOS manual. This line must occur after the line that enables HIMEM.SYS, and before any line that installs device drivers that use expanded memory. The main function of EMM386.EXE is to provide expanded memory services to programs requesting them. D40DRV is one such program, equipped to utilize expanded memory for its channel buffers. *This utilization does not occur unless explicitly enabled by its command-line option.* The command below will result in 256K of expanded memory being utilized for the channel buffers; if the **-e** option were not used conventional memory would be used instead.

c:\d40drv -e256

(Dialogic Corporation)

Extended Memory In DOS

Normally, DOS runs in conventional memory. However, most of DOS can be run in HMA. To load DOS in HMA, the extended memory manager HIMEM.SYS must be installed. HIMEM.SYS is included with DOS 5.0. Simply add the following two lines to your CONFIG.SYS file:

device=himem.sys
dos=high

The first line installs HIMEM.SYS, the second loads DOS into extended memory (the HMA). This alone frees up about 45KB of conventional memory, at a cost of 1.2KB that is used by HIMEM.SYS itself. The line for HIMEM.SYS must occur before any lines for device drivers that use extended memory.

Note: When DOS is loaded into HMA, it tries to load all of its disk buffers into HMA. However, if we request 49 or more disk buffers(using BUFFERS= in our CONFIG.SYS), all the disk buffers will reside in conventional memory. As an example, using BUFFERS=49, DOS took up almost 43Kb of conventional memory on a test machine. Using BUFFERS=48, however, DOS took up only about 18Kb. Therefore, it is best to use BUFFERS=(48).

(Dialogic Corporation)

Fax Blaster Software Features

Here's a checklist of some of the features you'd like on your next LAN-based fax blaster:

1. Should be able to **load more than 1,000 names** into a database and have the machine blast to everyone on the list. In short, no arbitrary size limits on lists.

2. Should have **mail merge capability**, even though it means multiple rasterization.

3. **Transparent compatibility** with leading word processors and other document producing software.

4. **Compatibilty with multiple database** formats for address lists. You shouldn't have to keep changing the format of your database. dBASE and comma delimited ASCII is no longer sufficient. There are other formats. These should be menu-selectable, without having to convert.

5. **Coherent (intelligible by normal human beings) real-time problem reports** on fax delivery.

6. **Simple architecture**. One server. As many clients as you like. No need to dedicate a client for long broadcasts. Blasting can be done directly from the fax server itself.

7. **A smart licensing scheme** that lets you add clients conveniently and cost effectively and that doesn't have breakpoints at certain points, like adding five more users you need a whole new software module.

8. **A sending process that is not destructive** either of the input database or of the documents and files involved.

9. **The system should remember what the sent document looked like**. You should be able to easily send again to Joe because he lost his first copy. Or send again to everybody from whom you got no answer the first time.

10. **Simple client end.** You need to be to easily create cover and attached pages in standard Windows graphics formats.

(Computer Telephony Magazine)

Fax-on-Demand Document Changes

Changing documents. When your client changes a fax document on a LAN-based drive that your Fax-on-Demand system can send, depending on how they do it, they could interrupt the transmission of a document. The safest way to do this is to have them rename the existing "live" fax document first, then generate the new one to the live document name. For example, say you had a one-page fax document called PRODUCT1.PCX and you wanted to replace it with a new version. First, rename the document to PRODUCT1.OLD, then generate the new one as PRODUCT1.PCX. This will allow any pending fax transmissions to continue sending the old document without interruption, while allowing any new requests to get the latest version. You can safely delete the old document when you are sure no lines are sending it any longer.

LAN Performance. Some corporate LANs slow to a crawl at peak usage times. If this is the case in yours or your customer's facility, there could be some negative effects on your fax-on-demand system if it "looks" for documents on a LAN drive. When a fax document is being sent, it is a "real-time" process, meaning that once the transmission starts, data has to flow to the fax card in "blocks" and without significant delays.

Worse case, a very slow LAN can cause fax transmissions to fail. The good news is this: most LANs are more than fast enough to handle this kind of application. Test your client's LAN and see how it performs.

If you try to install the voice prompts for the system on the LAN, too, you may have more of a performance problem. To be safe, I recommend keeping these files on your local PC drive.

(Telephone Response Technologies, Inc.)

Fax Server Integration Operations

Depending on the fax server, existing computer applications and resources can be integrated into your CT apps. APIs, embedded codes, DDE, OLE, NLMs, etc. Simple print-to-fax. Most vendors have a proprietary Application Programming Interface however there are several defacto industry standards that many vendors support including CAS the Communicating Application Specification, originally developed by Intel for the SatisFAXtion boards and MAPI the mail API for Microsoft Platforms. Check with the fax server manufacturer about APIs that are available before re-inventing the wheel in your application.

(Davidson Consulting)

File Handles - Opening More Than 20

In DOS, only 20 files associated to an application can be opened at a time. Among those 20 files handle, 5 are automatically opened by DOS for the standard input, output, error, aux and printer devices. This is a limitation for Dialogic applications using 30 channels, like E1 applications for example. Even if you set FILES=40 in the CONFIG.SYS file, you will still be unable to open more than 20 files at a time from your application. There is a solution to allow your application to open more than 20 files. The DOS interrupt 21H function 67H allows you to change the file handle count. Note that the upper limit will now be set by FILES=xx in the system file, CONFIG.SYS.

Solution: DOS Function 67H

The DOS function 67H works as follows -

BX = number of file handles
call interrupt 21H, function 67H.
The carry flag is reset if successful,
The carry flag is set in case of error.

The DOS error code is then available in AX

Sample code

```
/*********************************************************
* NAME: int set_handle_count(int n_files)
* DESCRIPTION: set the file handle count
* INPUT: n_files = max number of files open at a time
* OUTPUT: handle count is set
* RETURNS: dos error code. (0 if success).
* CAUTIONS: The hndl count is limited by FILES= in config.sys
*********************************************************
/
int set_handle_count(int n_files)
{
  union REGS regs;
  int ret;

  regs.h.ah = 0x67;     /* Function 67H: Set Handle COunt */
  regs.x.bx = n_files;    /* desired handles */

  ret = intdos(&regs, &regs);

  /* if carry flag is set, there was an error */
  return(regs.x.cflag ? ret : 0);
}
```

(Dialogic Corporation)

Forms-Based Application Development Tools

This section discusses what it is like using forms-based application development tools. These tools are advertised as the ones most likely to save you development time, some claiming to give you the power to create finished, polished applications in days. Do they, and who are they targeted for?

Forms-based tools encompass two fundamental types of voice application development environments:

- **Character-based application generators**

- **Graphical User Interface (GUI) application generators**

Each has its own strengths and weaknesses. What they both have in common is that to develop applications you do not write lines of C code or script - rather, you enter commands and parameters by filling in "forms."

With character-based environments you are usually using pull-down menus to access the various features and functions that allow you to create applications in a step-by-step format: each "step" that is created typically expands to a screen of fields that are filled in by the developer. Each of these fields controls some aspect of the application: what speech prompts to play, how long to record a message from the caller, how many touch-tone digits to gather from the caller, and so on. It is not uncommon for these screens to have integrated hypertext help and choice lists available to aid the developer in his or her field choices. With GUI tools, you usually use a mouse to click on a portion of the displayed application block diagram, revealing more details and ultimately bringing you to the same "fill-in-the-blanks" forms that are present in character-based environments.

Skill Level Required. Unquestionably, forms-based voice application tools are the easiest to use and require the least programming and technical background to be productive. Much more than script-only packages, forms-based tools can come with a significant amount of integrated tutorial help aimed at the beginner to computer telephony.

The level of skill required can be as little as having familiarity with PC spreadsheet and word processing programs. For the most experienced programmers, significant time can still be saved by developing the first few voice applications using forms-based tools. At some point, however, the forms most of these packages present may feel too "confining" to programmers. After developing a number of applications using "forms," some gravitate toward a script language package while fewer may opt to start building applications directly in C.

While the technical skills required using forms-based tools are the least of all methods, there still is a requirement for the user of the tools to think in a structured way when planning the application. Even when using forms-based tools, it is quite easy to get "lost" in the step-by-step details of constructing a large or complex application. In other words, forms-based tools do not dismiss the need to carefully follow the planning steps outlined earlier in this pamphlet.

Integrated Environments Save Time. A significant advantage of many forms-based software tools is that they are delivered as part of integrated work environments that contain virtually every utility and function needed to build complete voice applications. Tools are provided that allow the creation of applications, generation of documentation listings, editing of speech prompt scripts, and recording and editing of the speech prompts themselves. If the tools are designed right, they will be easily accessible when you need them.

This saves the application developer valuable time, which really is the whole purpose of using higher-level tools rather than coding applications in C. If you are interested in going with a forms-based development package, be sure you check to see if the work environment model suits your needs.

Using the Tools. So what are some of the typical features and commands you can expect to use when using forms-based application generators (AGs)? This section presents a list of typical operations you can expect to perform, day in and day out, while developing voice applications.

Editing the application block diagram (sometimes referred to as a flow diagram) is most useful when done as part of the application planning effort before the application steps and their associated details are created. While not all application generators have the ability to generate a block diagram, the application generators that do generate this diagram after the application is completed miss the point of this all-important step: namely, to allow you to think about what needs to be developed before the application building starts.

Editing the application steps is the process of actually creating the step-by-step instructions that will guide the caller from the very start of the call to when they finally hang up. When you are in any AG, this is where you will be spending most of your development time. Working with a detailed block diagram as your guide, you will enter the function details that each application section requires.

Some AG products have context-sensitive hypertext help that shows you very specific help text for the field your screen cursor is resting on. If you have a choice, select an AG that has all the required information on-line.

The task of recording and editing speech prompts (and their corresponding scripts) is usually interspersed with the editing of application steps. Whenever you create an application step that "speaks" a speech prompt to the caller, you may want to immediately edit the script for that message, then later record and edit the speech message itself at a later time. AG environments that allow this can save significant development time.

You will typically generate application listings throughout the development process. This is done primarily for the purpose of seeing certain types of application information in front of you rather than having to view it on the screen every time you need it.

Build Modular Applications. Many design and development rules are "universally" valid, and developing forms-based applications in a modular way is no exception. The same rule is quite true for script and C language development. What does this really mean?

Simply treat each functional application area as if it were a single stand-alone application. This is a helpful approach, because you can reuse application segments in other applications with little or no editing. In addition, you can fully test each application segment without reliance on other areas of the application.

To develop each segment in an independent manner, don't use application variables that are common to multiple segments - if you do, you may need to keep all segments together that use these common variables. Also, don't jump from one segment to locations inside other application segments; rather, if there are common tasks that many segments need to execute, create a common subroutine that all segments can use.

Strengths and Weaknesses. Since forms-based tools are especially strong at constructing "application frameworks" (menus, submenus, and caller error handling processes), if the application you are developing has many such menu "blocks" you can expect these application generators to save you a significant amount of time as compared to script or C development.

If, on the other hand, your application has to interface with one or more time-critical hardware devices requiring interrupt service routines (e.g., analog-to-digital converter cards, pressure transducers, etc.) and these devices are not supported by an off-the-shelf software module from your software vendor, you are probably better off doing one of two things:

1. **Write a special add-on software module in C** that deals with such devices. Several application generator manufacturers offer add-on libraries that allow new low-level functions to be added to the basic architecture.

2. **Write the entire application in C.**

Maintainability of applications is another key point of comparison. After the application is "done," months later you may need to add or modify several features quickly. Not all application development methods are the same when it comes to this need.

Applications built with forms-based tools are often the easiest to maintain. And because things are more "visual" with this method, maintenance is easier for individuals who were not the original architects of the application.

(Telephone Response Technologies, Inc.)

Forms - Save Them In Text Mode

If you're developing a Visual Basic application, always save your forms in Text mode (default is binary). This allows you to edit the .FRM file using a conventional text editor when something unusual happens which Visual Basic can't handle, like upgrading a VBX to an OCX with a slightly different set of properties.

(Parity Software)

ISDN DOS Driver Configuration File

In the DOS ISDN Package, running **isdndrv** without a configuration file name or with just the prefix of the configuration file name will not work. Specify the complete configuration file name. For example the default configuration file is c:\dialogic\config\dialgoic.is, to use this file type "isdndrv dialogic.is". The command "isdndrv" should use [dlcfgpath]\dialogic.is as the ISDN configuration file. Similarly the command "isdndrv prefix" should use [dlcfgpath\prefix].is as the ISDN configuration file.

(Dialogic Corporation)

IVR Programming Tips

Less is More. To improve VRU utilization and customer satisfaction, remove information that is accessed the least. Cluttered, complex IVR scripts frustrate even the most patient customers, forcing them to opt out of the automated system.

Simpler IVR scripts have fewer automated tasks. However, if designed properly, the ease of use will encourage high utilization for those tasks with an overall net effect of better utilization and customer satisfaction than obtained by including more tasks.

Balance Breadth with Depth. Frequently used tasks should be accessible within no more than three menu levels. At the same time, any particular menu should have no more than three choices. This three x three guideline will result in up to 27 automated tasks, which should be plenty! At the same time, relaxing this guideline is often appropriate. That is, adding four choices to a menu may eliminate the need for an entire level of sub-menus. Obtaining the best balance of breadth and depth often takes several attempts. The goal is to get the greatest number of customers to the desired information with the fewest numbers of keystrokes.

Place Specific Choices First. Since customers do not always listen to the entire menu before making a selection, put specific IVR choices first, followed by more general choices. An IVR menu might have the following choices:

- **Obtain information about an existing order.**

- **Place a new order.**

- **Obtain account balance.**

The "obtain information" choice is the most general. If placed as the first choice, customers might choose it to obtain an account balance. That is, it is reasonable to assume that account balance is one of the pieces of information that can be obtained. By placing "account balance" and "new order" as the first two choices, the general category of "obtain information" will be left to handle all other types of inquiries.

Access Popular Tasks From Various Menu Levels. Allow access to popular tasks from all logical IVR menu levels. For example, making a payment by credit card could be a prompt at the opening (main) menu.

It could also be a popular choice following information about account balance. Put it in both places!

Prompt Once for Account Numbers. Only prompt for account numbers to be entered once. Design the IVR to store the data so customers do not have to enter it repeatedly even if they navigate to other parts of the IVR before returning to account-specific requests.

Thoroughly Test The Application. Thoroughly test IVR application changes before putting the system in service. As obvious as this seems, a surprisingly high percentage of IVRs have "dead ends," "hang ups" and other "bugs." The reason? Customers do not use the system the way programmers want them to. A thorough test plan should test every possible combination of key strokes a customer could try at each menu, not just the ones included with the menu prompt.

Conduct A Usability Test. Conduct one-on-one or focus group tests for any significant IVR script changes. Customers are not insiders and do not always speak the same language. If you cannot afford a formal usability test, invite your friends, family and/or neighbors to try the system. Provide coffee or pizza and then, using a speaker phone, ask them to use the IVR test system to do a particular task (e.g., obtain account balance). Do not give them any instruction! If properly designed, they should be able to quickly and easily perform any task. Observe how they do. When they complete the task, or give up, discuss the experience. Was the task easy? Would they be inclined to use the automated system for this task if they were a customer? If not, change it.

(Enterprise Integration Group)

Japanese Disconnect Supervision

Here is a great tip for anyone thinking about shipping applications to Japan. Specifically, disconnect supervision (which is a pretty basic requirement of any IVR, messaging, or even fax app system) does not work as it does in North America -- you can’t rely on loop current drop.

Fortunately, Dialogic provides the GTD facility to allow users to configure a process for handling disconnects. The disconnect signal for the Japan network is 400 Hz tone of 500 msec on 500 msec off. You need to set up a tone template using GTD to detect this signal.

(Dialogic Corporation)

LAN Survival

Think survival on a LAN. Many computer telephony apps keep database files open whether calls are active or not. As a result, they can't be backed up -- manually or automatically. Each app should have a configuration option to allow the user to configure the system to support LAN operation.

One solution: Tell your program to close all its files and copy them to a specified directory and then re-open them. The LAN administrator can then tell the backup software NOT to backup the live databases but instead, backup the other directory. This way, the backup gets a closed, complete copy of the databases and not an image of a currently open file. It also prevents file locking problems with the backup software.

(Computer Telephony Magazine)

List Management

You never know where the next list will come from. It could be a purchased list, a list from former customers or almost any other type of list. The list manager should be able to import almost any type of database format into the system. Once the data is in the system, it should allow updating of that information as numbers change.

One example is in the Atlanta metro area: Atlanta has a perimeter highway around what used to be the outer part of the city. The highway opened up the city so that most of the growth of Atlanta in the last 25 years has been around the perimeter. Southern Bell has decided to split the 404 area code once more, this time using the perimeter as the line.

Anyone inside the perimeter will stay in the 404 area code, outside the perimeter will become area code 770. Let's assume we have all of the current 404 phone numbers entered in the system and this change takes place. The system operator would have to change all the former area code 404 numbers to 770 so the dialer would work properly. The easiest method of making this change is to keep a database of substitutions. Each substitution record would store the first six digits of a phone number (nxx-nxx) and include the replacement six digits and the date it goes into effect.

If you built the call center using this database, all new lists that are imported into the system could be updated as they are imported. The existing data could be updated with another program that scans all the data against the substitution list.

Some other systems offer a run-time replacement option. If a 404 number is called that results in a SIT tone, it could hang up and call again using the 770 area code. This would make the hit rate drop drastically, until most of the modified numbers had been hit and updated. The other problem with this method is that you get SIT tones on all dialing errors and, by switching the area code, you could be changing it when it didn't need to be changed.

The database must also be able to accept ad hoc operator queries to manage the number database. The operator may choose to call a specific zip code or geographic area. You can't build in all the ways the operators will want to use the data. The simplest method I know is to have a set of common functions and another list of user definable functions. In the user definable set, each one could be a different database language query. This would let someone who knows how to do the queries write them and any other operator use them.

Duplicate user checking is another problem. Since the calling lists come from almost infinite sources, you never know who is the most accurate. There is no 100% accurate method of deleting duplicates, but comparing names and addresses with a fuzzy search can find close matches. If the street address is the same and part of the name is the same, it could be a duplicate record. The last thing you want your system to do is call a person twice in a row due to slight differences in the database information.

Duplicate searches take quite a bit of network resources and should probably be done at the slowest part of the day. You could have the dupe checking function work only when system response time was faster than a specified point. This would let it use all the otherwise unused bandwidth to optimize the databases.

The best method of maintaining accuracy would be to maintain a known bad list and at regular intervals use an outside source like Metromail (Lombard, IL -- 708-620-3014) to verify and update these addresses and numbers. Metromail offers dial-up access to their databases.

Database segmentation is another minor problem. How do you keep the data from one list from mixing with another list from a different client? Some systems treat the databases as individual database files and never mix them, while others mix the data into a common database and keep a list of who belongs on which list.

Which method you use will depend on the database you build the system around. If the system handles lots of smaller databases easily, you may choose to create separate project databases instead of a master database. Once a project is done, you could delete the database and not worry with database packing.

If your database supports the reuse of deleted records easily and is a relational database, you may want to merge all the lists into a single database. A single database is the best method to handle inbound traffic, otherwise your system would have to search multiple databases with each inbound call.

If you do merge your lists into a single database, you will also want to be able to remove everyone in a particular list once a campaign expires. If the list is property of a customer, they may not want you keeping a copy of their list once the campaign is over. You may want to keep track of hit rates for each list, since most list providers offer rebates for errors. If you buy the list for a specific criteria, any errors should not cost you. Some of the better call center products keep lists of any errors that are found.

Skip lists are also important, since some states require a call center to not call a person's phone number again, once they have asked to be removed from their list. Skip list rejection can be done at list import time or when the number comes up for an outbound call. Either method meets the legal requirement and would prevent a call once it has been restricted.

(Computer Telephony Magazine)

Logic Flow Considerations

Design-Do research. Get experience with voice response units. Use them yourself whenever you can. Call your competitors and use their systems. Adopt the best approaches from other systems and obviously exclude those things that you find annoying.

Do it for your callers. A system designed without the customer perspective will only irritate the callers. A voice response unit should be an option for the customer, to provide better service to them - at their convenience. Systems that are designed to make life easier for the company inspire callers to find ways to circumvent them. Work with your customers to find out what THEY would like to be able to do. Customers can supply great ideas. Involving the customers can also give them a feeling of ownership n the system. Equally important, find out what your callers wouldn't use. Don't waste time designing something no one will use.

Let them escape. Always provide your callers with a means to reach a living, breathing, human being. And make it available at every level of the menu. While you are encouraging your callers to use the voice response system, you do not want to alienate them by forcing them to use it. Some customers will never feel comfortable using a voice response system. However, most customers will work through the scripts as long as they know help can be obtained if they need it.

Plan for errors. Your callers are human - expect them to make mistakes. They will. Provide error messages that help guide the caller to the correct response. Make the error messages polite and descriptive.

For example, if the caller needs to press the * key at the end of an entry, your error message should remind them to enter it. "Please re-enter your account number, and be sure to press the * key when you are finished."

Don't disconnect your caller! If you wouldn't hang up on one of your customers, why would you disconnect them from the voice response unit? It is hard to imagine any good reason to hang up on your customer. If a caller makes too many mistakes, transfer the call to a representative who can help.

Document, document, document. Produce a call flow diagram and a script for your voice response system. Constructing the call flow can show you where problems can occur, like dead ends that callers can't escape from. The script should be written out and perfected before it is recorded. Once the system is installed, keep the call flow diagram and the script updated with any changes that are made. Often, new menu items are added and no old items are removed, which result in a large, unwieldy system. An updated call flow diagram and script can prompt removal of unneeded or underutilized sections.

Watch out for dead ends. Some call flows create dead ends, which are areas that the caller can't get out of without hanging up. Be sure your caller can escape from any area they get into. Offer the main menu, the previous menu, and any likely option. After receiving the account balance, the caller may wish to return to the main menu, make a credit card payment, or speak to a representative. Offer logical choices and don't trap the caller anywhere.

Be consistent. Use the same commands throughout the system. The command to return to the previous menu should always be the same, where ever it appears in the menu. The command to transfer to a representative should always be the same. Also, be consistent with the descriptions used. Consistent commands and wording can help prevent confusion for the callers.

Verify entries. Because callers will make mistakes (knowingly or unknowingly), give them the opportunity to verify their entries and correct those mistakes:

"You requested a payment of: ____. If this is correct, press 1. If this is incorrect, press 2." The verification step reassures the customer that the transaction will be completed correctly.

Tell them how to disconnect. Callers may be concerned that if they just hang up, their transaction will not be completed or recorded. By offering a way to disconnect, they are reassured that their efforts will not be wasted. A simple method is to say: "To end this call, press 9 or hang up." The customer knows that either action is appropriate and feels more secure.

Make it speedy. Put the most frequently used menu options first. This will reduce the time most of the callers have to spend using the voice response system.

Monitor the usage. Build in counters to track the call volumes and usage. The usage record can facilitate decisions to remove underutilized options and reorganize the menus for ease of use.

(Enterprise Integration Group)

Mail Merge With Powerfax

MultiFax by Commetrex (Norcross, GA -- 404-564-5522) takes advantage of the open DSP architecture of Natural Microsystems' VBX and AG-class high-density "voice boards" by adding Group 3 (G3) fax capability to them via software. The PowerFax enhancement gives you the ability to add a GUI-based fax send/receive function to your voice application.

The PowerFax Developer's Kit lets you interface your app with PowerFax to serve as a connection management, notification or reporting agent. The PowerFax Enhanced Printer Driver enables any app to create a fax-ready file by "printing" to an OS/2 system printer. It's used when your app needs to dynamically create faxes from a database search or as part of a mail merge. The PowerFax Printer Driver looks for a file called FAX.CGS in the fax program directory when starting a print job. If it finds this file, it assumes that the file contains printer driver commands and it applies these commands to the print job.

For example, the line ">TO=,,555-5555" tells the driver to spool the fax for transmission to 555-5555. After faxing, the FAX.CGS file is deleted. This works whether the printing is from an OS/2, DOS, Windows app, or even a text file copied to the FxPrint LPT device. If any application can write the FAX.CGS file with a macro, it can now send a fax. A Microsoft Word macro file to execute a mail merge with PowerFax looks like this:

```
Sub MAIN

REM Get the first merge record on the screen and look at it

MailMergeFirstRecord

MailMergeViewData

prevrecord = 0

thisrecord = MailMergeGoToRecord()

REM While there are more records to send

While thisrecord  prevrecord

  REM Get the name, company, and fax number from the merge record

  name$ = GetMergeField$("FNAME") + GetMergeField$("LNAME")

  company$ = GetMergeField$("COMPANY")

  faxnumber$ = GetMergeField$("FAX")

  REM Write out the FAX.GGS file

  Open "D:\POWERFAX\FAX.CGS" For output as #1

  Print #1, ">TO=" + name$ + "," + company$ + "," + faxnumber$
```

```
  Print #1, ">INFO=," + "," + notes$ + ",,2"

  Close #1

  REM Print the file to the current printer

  FilePrintDefault

  REM Fetch the next merge record

  prevrecord = this record

  MailMergeNextRecord

  thisrecord = MailMergeGoToRecord()

Wend

End Sub
```

Here is a similar mail merge macro, this time for the DeScribe word processor and PowerFax:

```
VAR faxNumber  ! the merge data

MACRO FaxSend

  SET faxNumber TO MergeField !Get fax number from data record

  NewDocument  !Create a temporary document

  PUT ">TO=,," + faxNumber !Insert the >TO=,, faxnumber line

  NewLine !Put in carriage return and line feed

  Set FileName TO "C:\POWERFAX\FAX.CGS"
```

```
REPEAT !Wait until old file is used/gone

EXIT WHEN NOT FileNameExists

PAUSE 3

END REPEAT

SaveAsAsciiFile "C:\POWERFAX\FAX.CGS" !Write it to file

SystemTileClose !Close doc and macro

PAUSE 3

END MACRO
```

To do a similar "mail merge fax" with Delrina WinFax Pro 4.0, take a look at the macro descriptions in Chapter 11 of the WinFax Pro manual.

(Computer Telephony Magazine)

Memory Terms

The one megabyte limit of Intel CPUs running in real mode has produced a confusing zoo of memory technology and terminology. A quick glossary:

Conventional, aka Lower or DOS Memory. Memory with addresses from 0 to 640 Kb.

Upper Memory. Memory with addresses from 640 Kb to 1 MB.

Upper Memory Block. Extended memory which has been "mapped" by a memory manager to appear to a real-mode program in upper memory. Upper memory blocks are used to fill areas which are not used by expansion cards such as display adapters and voice boards.

Having upper memory blocks means that real-mode programs have more memory available for storing data. It may also be possible to load TSRs and device drivers in upper memory blocks if they are large enough and if the developers of the TSR or driver software have not made assumptions about where in memory they will be loaded.

Real Memory. Lower + upper memory, ie. memory with addresses from 0 to 1 MB. This is the range of memory addresses which can be accessed by programs running in 8086 mode, such as MS-DOS and most MS-DOS applications.

Extended Memory. Memory with addresses greater than 1 MB. This memory cannot be accessed directly from a real-mode application, such as a typical MS-DOS program.

Expanded Memory (EMS). Additional RAM up to 8 MB accessed through a window, the page frame, which is provided as a fixed 64 Kb range in upper memory. An EMM device driver is used to move the window to make different areas of EMS accessible through the page frame. Often, extended memory is converted to expanded memory by using a device driver such as MS-DOS EMM386 or as part of the functionality of memory managers such as QEMM and 386MAX. For software developers, the "Pro" of EMS is it makes more than 1 MB available to a real mode application. The "Con" is it's slow and complicated to program.

High Memory Area (HMA). A block of just less than 64 Kb above upper memory in segment hex FFFF. It's the first 64 Kb of extended memory. Only available if the CPU is an 80286 or later. The "Pro" is it makes another 64 Kb available to real-mode programs for storing data. The "Con" is it can't be used for any data used in MS-DOS services.

Memory Manager. Software which uses the memory mapping capabilities of the 80386 and later processors to re-organize DOS memory. Memory is "mapped" when it appears to the CPU to be at a different address than the physical mounting of the memory chips.

On a typical motherboard, memory chips are installed at addresses from 0 - 640 Kb and from 1 MB and above, leaving the upper memory range from 640 Kb to 1 MB free for memory which may be mounted on expansion cards such as VGA RAM. If areas are unused in the 640 Kb to 1 MB range (from hex addresses D000 to FFFF), a smart memory manager can "map" memory from above 1 MB so that it appears to a DOS program to be present within the upper memory range.

Memory managers can also use mapping to provide expanded memory services to DOS applications where 64 Kb pieces of memory above 1 MB are mapped to a fixed position in upper memory (the "page frame address").

(Computer Telephony Magazine)

Memory Saving Tips

If you're creating an application for MS-DOS, you can quickly run out of conventional (lower 1MB) memory. Here are five tips for saving memory:

1. Remove any device drivers and TSRs you don't really need. Some examples: SETVER.EXE, DOSKEY.COM, ANSI.SYS, FASTOPEN.EXE, SHARE.EXE, SMARTDRV (though be careful, some applications need SHARE to run correctly).

2. Set DOS=HIGH,UMB in CONFIG.SYS.

3. Use a third-party memory manager like QEMM instead of EMM386.

4. If you are using Dialogic, use the following command-line options to D40DRV: -e, -c0, -M4.

5. Re-map VGA memory. Chances are that your DOS app only needs text mode, so you can use VGA memory as an upper memory block. Your memory manager documentation will explain how.

(Parity Software)

Modem Hang-Up

When writing code for PC-based voice modems, be sure to supplement the normal hang-up termination with a time-out sequence. Many PC modems lack the more expensive circuitry needed to do good hang-up detection. If there is no hang-up detection, the system won't know to process the next call.

(PureSpeech)

Operating System Selection Tips

Windows 95? DOS? UNIX? Something else? Which operating system should you choose for your call-processing computer telephony PC platform? Here are some pros and cons of the different options.

In many cases, Windows 3.1 is not a good choice for a call-processing computer telephony platform. If you see an hourglass icon, your call-processing application is dead! Within a second or two at most, messages will stop playing or recording, touch-tones dialed by the caller will not be processed, incoming calls will be ignored.

For example, starting Word for Windows 3.1 usually produces an hourglass icon of 10 seconds or more. If you are running a call-processing app when you start Word, the app will simply freeze. Visual Basic exacerbates the weaknesses of Windows for call processing. So if you're thinking that a VB custom control is the coolest way to create your next application, you may want to read on.

In developing computer telephony apps, there are at least three platforms to consider: the server, the client and the development tool.

The Server. Voice cards are controlled directly by the call-processing server. Requests are received by the server from the client software, which are analyzed and converted to API commands to the voice card device driver(s).

The software running the server will be handling many simultaneous conversations. Often it will be running unattended in a phone room or basement with no human operator. It needs to be high-performance, reliable and robust.

In many cases, it will have no user interface or only a few simple commands (show line status, shut down). In some systems, the client and server software will be combined into a single program running on one computer. A typical auto-attendant, which will be a PC running voice-mail software with voice cards installed, is a well-known example.

The Client. The client is the rest of the application software other than the server. In a call center, the client will be an agent station which receives a notification from the server when an incoming call is received -- perhaps with the caller's number retrieved via ANI.

In an Audiotex service bureau, the client software may be used to install and configure the services which run on the server. In an integrated messaging system, the client will be electronic/fax/voice-mail management software. The client is generally the software that your end user sees. It therefore probably needs a user-friendly interface, at least menu-driven and perhaps graphical.

The Development Tool. If you are creating a call-processing system, you have a choice of development tools ranging from low-level coding using the C language, an applications-oriented language like VOS, forms-based/menu-driven "application generators" and "join-the-icons" GUI-type application generators.

It's important to realize that the style of development tool used to create the software has nothing to do with the style of the software which is created. A highly graphical, user-friendly interface can be created with an unfriendly procedural language typed into a character-mode editor (say, C++ for Windows NT). Remember that even with Visual Basic, you spend most of your time writing old-fashioned procedural code.

In most cases, you want to provide user-friendly client software for your agents and system administrators. This does not necessarily mean that your call-processing software development tool needs to be user-friendly or Windows-based.

Overall, developers need powerful and flexible development tools, servers need to be efficient and robust and end users need friendly, menu-driven and graphical interfaces. In a real computer telephony system, it might break down like this:

You build a call center. Each agent station has a DOS PC running Clipper apps for data entry and account queries. Incoming phone lines and agent headsets are connected to an ACD (Automatic Call Distributor, a special type of PBX for call centers). A Voice Response Unit (VRU) -- a PC containing voice store-and-forward cards -- is also connected to ACD extensions. All the computers are connected via a LAN.

Here the agent stations are the clients and the server is the VRU. Remember, when I say "server," I mean a call-processing server. This may or may not be part of the same computer which contains the disk server, database server, etc.

One of our customers runs a 180-line Audiotex/fax service bureau in Prague, Czechoslovakia. Six digital E-1 trunks each carrying 30 conversations terminate on PCs running VOS for DOS. These are connected via a fiber-optic LAN to a NetWare server and to a number of other PCs which are used as agent and administrative work-stations, most of them running Visual Basic applications under Windows 3.1.

A CT-based call center could have a NetWare file server running TSAPI to control a PBX. The agent stations might be running a Windows for WorkGroups application for database management. The database could be stored on a dedicated Oracle server. The applications software might have been developed using Visual C++ on Windows NT.

So what about the different operating systems? Which is right for what?

MS-DOS. The pros are that it's inexpensive, well-known, has low overhead – and is smaller and faster than any other O/S. But, there's no native GUI, no multi-tasking provided by the operating system, new APIs may not be accessible and server software is exceptionally hard to write without a good tool.

In many cases, DOS is still the best platform for call processing. It retains a market share in computer telephony that is much higher than in the desktop software world.

DOS takes very little memory and never takes a CPU cycle unless the application software specifically makes a request. Writing call-processing server software for DOS is hard to do with low-level tools like the C language because multiple conversations must be managed by a single program, requiring difficult techniques called "state machine programming."

The low overhead of DOS means you can run more lines with less memory and a slower CPU than any other operating system. The lack of a native GUI may be an issue if the client and server are in the same box. The fact that DOS cannot run more than one program at one time may be an issue for remote maintenance. If the DOS PC is on a LAN, remote control programs such as Co-Session can be found which can access other workstations. If the PC is standalone, this may be a problem.

And then there's new APIs (Application Programming Interfaces) such as MAPI (Microsoft's Messaging API) or ODBC for database access. They're often provided as Windows DLLs and are therefore not accessible directly from DOS. For smaller systems (say, 2 to 16 lines), a DOS server and Windows client can be combined into one box using OS/2, which can provide robust multi-tasking between a DOS session and a Windows session.

Windows 3.1 And Windows For WorkGroups (Win 3.11). It's the world's most popular GUI, your end-users are probably asking for it, APIs are available and a DOS box can run voice drivers. But multi-tasking has serious problems for call-processing servers, major voice-board vendors (including Dialogic) do not provide Win 3.x device drivers, software is hard to write (even harder than DOS) without a good tool.

Win 3.x is well-suited to support client software, as long as it is in a separate PC from the call server or in a separate session in a robust multi-tasking environment such as OS/2. If the server software which controls multiple conversations directly runs under Win 3.x, then the cooperative multi-tasking scheme used by Windows becomes a major problem.

Win 3.x applications must voluntarily relinquish control for another application to run. Well-behaved software will do this at regular intervals. But even if all your software is well-behaved, there are still times when Windows hogs the processor with a single process for a significant length of time, such as when loading a new application from disk.

When you see an hourglass icon, this means that Windows or your application is grabbing the CPU for its own exclusive use and does not relinquish control until the hourglass changes to another icon. Even if you never see an hourglass, Windows can still decide to perform time-consuming operations (e.g., consolidate unused memory areas) during which your call-processing application is stopped.

In these situations, the call-processing server software will be stopped and your application will die. Ironically, old-style DOS applications, unlike Windows apps, can be pre-emptively multi-tasked under Win 3.1. For example, VOS for DOS may run better in a DOS box under Win 3.x than the native VOS for Windows engine. However, even pre-emptively multi-tasked DOS boxes grind to a halt when you are looking at an hourglass, so this is not a true solution.

In summary: by all means, provide a Win 3.x user interface for your end-users. But don't run the voice boards with a Win 3.x control program.

Visual Basic. VB is a widely known and used programming environment (more than one million sold) and it's great for building client software. On the bad side, unless you've got the OCX version, it runs under Win 3.x and has same "hourglass of death" problem as any Win 3.x product. It's not designed for multi-tasking. The visual control/event-driven metaphor is not a natural fit for call processing and it requires a program instance for each conversation (in other words, 24 phone lines means 24 .EXEs running).

Visual Basic allows the developer to draw a user interface on the screen and to attach Basic code to events caused by interacting with user interface controls such as buttons and scroll bars. For example, the programmer can write Basic code to be executed whenever a given button is pressed.

VB is great for client applications, which are usually run by live users sitting at a PC. But server software interacts with the users via a telephone line and here VB is much less attractive.

There are no user interface elements to press or click. The call proceeds in a linear fashion from beginning (incoming ring or dial out) to end (hang-up), just like programs written in pre-Windows languages. The only way for software tools vendors to extend Visual Basic for call processing is to create a "custom control" (VBX).

The custom-control mechanism is for user interface items like buttons. It stretches this interface to its limits to make a custom control for call processing. More natural would be to define new functions and subroutines for the VB language, but there is currently no way to do this.

The call-processing control becomes an invisible element in the finished application. Actions such as playing messages and waiting for touch-tones have to be done by assigning values to control properties, a mechanism usually used for changing sizes, fonts, colors and so on in a visible control.

Touch-tones received and other notifications from the call in progress could be represented as Visual Basic events, but this would result in a complex state-machine programming model and is therefore rarely, if ever, done.

The one true asynchronous event is a caller hangup, which is awkward to deal with in Visual Basic. If a hangup triggers a special hangup event procedure which the user writes in Basic, it does not, as would be desirable, interrupt the Basic code that's currently executing. The only way to make your Visual Basic code jump somewhere when an event happens in the middle of a routine is to report an error condition to the Visual Basic run-time and to use the "on error goto" mechanism.

This can only happen when your Basic code references the call-processing control. So the hangup may go undetected for some time. Also any true Basic run-time error will be treated as a hangup, which makes debugging and testing the application more difficult.

Visual Basic is not designed as a multi-tasking engine. When Basic code is executing, it will not give up control to another program unless a special function (DoEvents) is called or a reference is made to the call processing control, which may then take the opportunity to a Windows API Yield() call, allowing another program to run.

These factors mean that Visual Basic code for a call-processing server must be carefully written so that any loops or other time-consuming operations are kept to a minimum or contain embedded calls to DoEvents() as appropriate. Otherwise other lines will not be serviced while this code is executing.

Calls to APIs such as ODBC may also "hang" other lines unless requests terminate quickly, which they may not. Sending a database request that requires a complex SQL statement may block the application for several seconds. If this proves to be a problem, there is no solution without re-writing the DLL providing the service (not for the faint of heart). Running multiple lines requires that multiple instances of the .EXE be loaded, a situation which Win 3.x does not handle very efficiently.

Windows NT. On the pro side, NT is a robust 32-bit multi-tasking engine, has a Windows GUI and is being adopted by many corporations as the preferred choice for server platforms. But it's big and slow, voice board drivers may not be available or are not as mature, it doesn't handle real-time and its DOS box cannot run voice-board drivers.

Unlike Win 3.x, Win NT uses pre-emptive multi-tasking, which means that an executing program will receive regular attention from the processor even if an hourglass is being displayed by another program or another time-consuming operation is in progress. This avoids the most important weakness of Win 3.x for call-processing servers.

Win NT is big (expect to buy at least 24 MB of RAM even for a small system) and slow (the pre-emptive multi-tasking and interprocess security features eat up a lot of CPU time). Win NT is finally getting attention from voice-card vendors and they are now providing drivers. This is likely to increase over the next couple of years as Win NT gains market share.

Despite the pre-emptive multi-tasking, NT is not a real-time system, meaning there is no guaranteed maximum time that will elapse from an event until it receives service by an application -- for example, responding to a touch-tone. This same comment applies to all the platforms discussed here except DOS, where it is possible (though difficult) to write real-time software, and QNX, which is a true real-time platform.

Windows 95. The good new is Win 95 will probably offer an environment for 32-bit programs with reasonably robust multi-tasking. It has a Windows GUI, it's smaller and faster than NT and it's likely to win wide acceptance on desktop. Not all 32-bit voice-board drivers are available (some future products announced) and its reliability in CT applications is not yet known. Win 95 is similar to Win NT in its ability to run 32-bit programs in a pre-emptive multi-tasking environment. Most of the same NT comments apply, except that Win 95 is smaller and faster since Microsoft has sacrificed some protection and other features for speed and size. Win 95 will also run 16-bit Win 3.x apps, but these will not be pre-emptively multi-tasked and will suffer the same weaknesses as running under Win 3.1.

Win 95 is likely to replace Win 3.1 as the platform of choice for client applications. It will yield way to NT in being a server platform. The industry has been quick to say they support the 32-bit multi-tasking features, but many are adopting a "wait and see" attitude.

OS/2. It's a robust, 32-bit, pre-emptive multi-tasking platform with a GUI and a DOS box will run voice-board drivers. Unfortunately, its long-term future in doubt. From a developer's point of view, creating native OS/2 apps is very comparable with Win 95 or Win NT 32-bit apps, with generally slightly better performance than NT.

OS/2 provides a robust environment for multi-tasking DOS and Win 3.x applications in one box. The question is, will OS/2 survive long-term? Microsoft clearly aims to position Win 95 as an OS/2-killer, and they have proved to be formidable marketers.

UNIX. On the pro side, it's robust and has 32-bit, pre-emptive multi-tasking. UNIX provides good connectivity for some specialized areas (Internet TCP/IP, databases such as Informix) and is a corporate standard for many phone companies. But it has poor integration with GUI, and the many different UNIX flavors are not totally compatible. There's a high license cost compared with competitors such as NT. Some popular UNIX flavors have slow process switching and inter-process communication.

There are many flavors of UNIX currently being used for call processing. SCO currently has the largest market share by most estimates, but SCO is losing ground to leaner and meaner competitors. QNX, for example, is a real-time UNIX variant -- probably the only real-time operating system which is a serious option for PC-based call processing. (See accompanying article.)

SCO does not yet offer multi-threading (where a single program can have multiple routines executing in parallel), only multi-tasking between programs. This means that process switching and inter-process communication is relatively slow.

Note: one benchmark at Parity Software showed that a widely-used vendor's UNIX driver was five times slower to process events under SCO UNIX than under DOS on the same machine. In addition to high license fees, UNIX also has hidden costs as expert UNIX programmers and system administrators often demand high salaries.

For us, DOS is still the platform of choice. You need a good reason to involve another operating system: connectivity via an API only available on Windows; access to an Informix UNIX database; or a requirement to integrate a graphical client and voice server into one box.

(Parity Software)

Operating System Tips For Fax Servers

The three primary choices are DOS, OS/2, NT and UNIX. Windows 3.x also can be used, but is widely considered a bit on the unstable side. GPFs in the middle of nighttime fax broadcasts or receptions can sour users on Windows 3.X-based servers in a hurry.

For the first decade of computer fax, DOS has been the most widely used fax server OS, with DOS-based servers capable of supporting up to 8 intelligent fax boards (assuming available slots -- and those can be multi-port boards. But DOS fax servers can bog down when lots of computer files are submitted for conversion to G/3 fax. DOS servers are not multitasking, and the file conversion process (usually the most severe bottleneck) can therefore impact all other system processes too.

OS/2-based fax servers have long had speed, power, multitasking, and ease-of-use advantages over DOS servers. E.g., you can have one OS/2-based PC run transmissions in the background and a second run file conversions in the background (both can be shared servers). Thus, the cost of a dedicated fax server is eliminated, file-conversion overhead doesn't impact the ability to handle transmissions, and end users can still get work done.

NT servers are now emerging to provide what is similar level of performance to OS/2 servers, with a key differentiator likely to be that NT will be used much more widely than OS/2. OS/2, however, is a much more proven OS than NT.

NLMs also bring the NetWare operating system into the picture by enabling fax server software to run directly on NetWare file servers, rather than on separate PCs. Again, this means no extra dedicated fax server PC -- and automatic fax support for all those NetWare goodies like security.

Finally, UNIX offers the most robust, most scalable, and most proven platform -- the whole AT&T phone network runs on it! -- but it tends to be more expensive than OS/2, NT and NLM-based solutions.

For workgroup fax servers, DOS is fine. In departmental servers, the choice is between DOS or the other three 32-bit multitasking operating systems (OS/2, NT, NetWare NLM). On an enterprise-wide level, it's a choice between the 32-bit-ters and UNIX.

The Mac also can act as a fax server, though one drawback is that it doesn't support intelligent fax boards (they all fit AT-compatible expansion slots). This limits Mac fax servers to use of fax modems (no microprocessor on board)

(Davidson Consulting)

Outbound Calling Considerations

Outbound calls are initiated by the IVR system based on some event. such as time of day or a value of a field in a database record. About two-thirds of outgoing calls will not be answered by the party they are intended to reach. You need to decide how to handle calls that are not answered vs. those answered by a machine vs. those answered by another person. Will you retry or just leave the information with whomever answers the telephone? How many times will you try? What happens if you don't reach them after the maximum number of retries?

The way you interact with the called party is different from an outbound application. Do you want the party to acknowledge the call before delivering the information? Do they need a security code? Is the party expecting the call?

Also be sensitive to the time of day you are calling. From 8:00 AM to 5:0 PM is considered normal business hours and 5:00 PM to 8:30 PM is evening hours (for consumer applications). Never cal after 9:00 PM or before 7:00 PM unless the called party has specifically requested it.

(Voysys Corporation)

Pager Tone Detection

Use a Global Tone Detection (GTD) template to detect the pager tone. The most commonly used (*) tone is a 3 beep single tone with frequency of 1400 Hz. The time on/off is around 80 milliseconds, but fluctuates, which is why the recommended "window" is 50-110 mSec. The function used to build the tone is the following:

xx_bldstcad(Tone_ID, 1400, 60, 8, 3, 8, 3, 2);

The prefix "xx_" is "dl_" for DOS, and "dx_" for UNIX & OS/2.

Setting the time-on/off to 80+/-30 mSec is "safe" enough to avoid false detection (since 2 repeat counts are required), yet wide enough to still detect the tone in spite of fluctuation/noise. There may be other pager tones which do not fit the GTD template described above. Additional GTD templates may be needed to detect these tones.

There are two main types of pager numbers:

1. A pager that has a voice prompt for a PIN (of the paged person). This tip doesn't apply to this type of pager service.

2. A pager number that has one "subscriber" (paged person) associated with that number. This is the type of pager that is discussed in this tip. When this service answers a call, only one RingBack tone cycle is heard, followed by 3 short beeps of 1400 Hz, followed by silence.

A Dialogic voice board that makes a call with Basic Call Progress Analysis (BCPA), and is answered by this type of pager, will terminate the call with a "Connect" event after it times out waiting for an additional RingBack tone cycle. But using ECPA, the RingBack tone cadence is only "established" with the second ring cycle, which is absent here. Since no recording is played, PVD will not be detected either, so eventually the call will terminate with a "No RingBack" event. Setting the above recommended tone template will ensure that this event is detected.

1. The OS/2 & DOS APIs allow termination of the dl_dialtcb & xcallp respectively with a GTD event as set in the TCB & XRWB respectively. However, under UNIX, the function dx_dialtpt() which will terminate on events specified in the TPT, is not yet available (scheduled to be available with System Release 4.2). Therefore, the only way to detect the pager tone under UNIX (before SR4.2) is with independent (asynchronous) GTD event. Upon receiving the GTD event, the application will then have to call dx_stopch() to stop the dx_dial() function.

2. Under OS/2, if this GTD tone is only needed as a termination for the dl_dialtcb(), be sure to call dx_distone() (both TONE_ON & TONE_OFF) after calling dx_addtone() which adds the tone to the channel's GTD "repertoire". The reason is that if it's not disabled as an asynchronous event, each time it's detected, it will be registered in the Event Queue (on top of terminating the dl_dialtcb function), and unless dl_getevt() is called to clear it, the Event Queue will eventually overflow.

(Dialogic Corporation)

PBX Integration With D/42-SX or -SL Boards

When connecting either a Mitel Superset 4 or a Northern Telecom SL-1 Digit Display to their respective PBX switches, a delay exists before the switch actually establishes the communication link with the station set or phone. When developing applications using the D/42-SX, D/42D-SX, D/42-SL or D/42D-SL Dialogic boards, the application must wait 60 seconds after issuing the startsys() command *before* issuing any other driver calls or receiving any incoming calls.

Programmers must ensure that the D/42 application waits 60 seconds after issuing the startsys() command *before* issuing any other driver calls or receiving any incoming calls. This delay will allow the Mitel or Northern Telecom PBX sufficient time to establish communication with the corresponding D/42 channels. Failure to observe this wait cycle can produce unpredictable results. In particular, when using the CPC utility to customize call progress parameters, the user should wait 60 seconds before initiating the first dial command; i.e., before entering the phone number to dial.

Note that the user should not use the CPC -n command line option, which automatically sets the number to dial from the command line.

(Dialogic Corporation)

Prompt For Zero Messages

Record a distinctively different prompt for an empty mailbox, such as "Your mailbox is empty". If you use "You have zero messages", the start of the message is exactly the same as "You have twenty messages", and the caller has to wait longer to find out if he/she needs to select an option to hear messages.

(Parity Software)

Silence Is Golden

Every time a caller has to make an entry, their input should be followed by silence to allow callers using hand-held phones to get the phone back to their ear before the system starts talking again. Most of us test on speaker phones and don't realize the frustration this can cause. Of course, if you expect all the callers to use speaker or desktop phones, this doesn't apply. However, if the users are people calling from home, it is likely they will be using a hand-held phone and this will he helpful. One half second of silence is the minimum with one second being the suggested amount for the application to not sound rushed. If the length of the caller's input is variable, such as for a dollar amount, and the system must wait to determine if the caller is finished, this may not be necessary. But if the length of the input will always be the same, silence is a must if the caller is to hear the beginning of the next sentence. What you don't say can be as important as what you say:

1. **Allow a half to a full second of silence before each prompt** since many callers have phones with the keypad integrated into the handset and must pull the phone away from there ear to enter tones.

2. **In order to closely mimic the rhythm of the English language, put a half second pause between sentences** and be sure your IVR vendor can down inflect the last word of a sentence under any voicing scenario (i.e., date, dollar amount, spelling of a name, etc.).

Less Is Best. Limit the number of options at a prompt to 5 or 6 max. Most callers get impatient after 3 or 4 though. Always use the word "press" not "touch" as in press 3 or press the pound sign. It is an unspoken industry standard that pressing 0 at any prompt will transfer the caller to a customer service rep.

Prompt With Silence Attached. The two greatest deterrents to designing an efficient and flowing IVR application for your corporation are: 1) a designer who is weak in logic skills and the legal department. Once the lawyers get their hands on your application, you're in trouble. You see, lawyers are not trained to be concise and efficient only thorough, and thorough usually means wordy. We haven't met a caller yet that likes a machine which babbles.

(TALX Corporation)

Prompt Succinctly

Prompts refer to the recorded words that are spoken to the caller by a computer telephony system. Prompts consist of prerecorded words and phrases that are concatenated together to deliver specific information to the caller.

All prompts must be succinct and to the point. Here are some examples of prompts that are too verbose:

"If you want to hear information about the price of a part, press 1."

If you want to hear information about the number of parts in inventory, press 2."

If you want to hear information about shipping lead times, press 3."

At a minimum, the phrase "If you want to hear information about" should be replaced with the word "For." The prompt could be further whittled to say: "For pricing, press 1, for inventory, press 2, for lead times, press 3." The more frequently the caller uses the system, the shorter the prompts can be.

Use language that is familiar to the caller, not internal company jargon. For example, employees at a company might use the term "Product Reference Number," or "PRN" to refer to part numbers. In this case, you should prompt callers for a "Part Number" instead of a "PRN." Have someone outside the company -- who represents a typical caller -- review the final script. It can be an eye-opening experience.

Before recording the prompts, you must decide on the type of voice to use. Female voices are considered less intimidating and are generally preferred when information is being requested f4rom the caller. Male voices are used when most of the application is delivering information, such as movie theater listings. The same voice should be used throughout the application. It is confusing for callers to hear multiple voices in a single phone call.

Prompts should be in a normal conversational tome and pace. There should be sufficient pausing so that the caller does not feel rushed. They should be worded politely, never accusatory. Say "I'm sorry, that is not a valid operation" rather than "You pressed an incorrect key."

Prompts are used both to request and deliver information. Prompts that request information must ask only one piece of information at a time. For example, if the application requires an account number and PIN, ask the caller for the account number, allow him to enter it, then ask for the PIN and then accept the entry This is easier to follow than "Enter your account number and PIN."

It should be clear to the caller how the system expects them to respond. "Please enter your account number, followed by the pound key" lets the caller know that the system is expecting touch-tone input, as opposed to "What is your account number?"

Prompts can deliver either general or caller-specific information. General information contains variables that must be assigned values as the application proceeds. The variable values can be gathered from the caller, form the database or from internal system calculations. Consider the prompt: "Part number 12345 costs $50. The total for five parts, with tax, is $267.50." This prompt is used in an application where the caller can order a part by specifying part number and quantity. The system looks up the unit price in the database, then calculates the final total based on quantity and 7% tax.

The prompt is constructed from four prerecorded phrases, concatenated together with four variables. The first phrase is "Part Number" and is followed by the part number the caller entered. The next phrase is the single word "costs" followed by the unit price form the database. The third phase is "The total for" followed by the quantity entered by the caller. The fourth phrase is "parts, with tax, is" followed by the total as calculated by the system.

When constructing a prompt such as this, remember to use normal conversational flow. Don't try to twist around the words simply to avoid concatenation. For example, use the prompt "There are three parts in stock" instead of "The quantity in stock is three" even though the variable comes in the middle. It sounds more natural. Make sure that the pieces of data that the caller is listening for are properly prefaced. "Your account balance is $500" instead of "$500 is your account balance."

(Voysys Corporation)

Prompt To Touch - Not Push

Avoid the term "push-button" phone when giving recorded instructions to a caller. Some pulse dial phones have push buttons. Use the term "touch-tone" phone.

(Parity Software)

Prompts And The Human Interface

Sounds are restricted to a narrow bandwidth over the phone line. When designing prompts, try to stay away from words that are more difficult for callers to hear, especially S's, M's and N's. Instead, use words with sounds that are easy to hear, such as T's and K's.

Prompts confirm and reinforce. Use prompts not only to confirm a command but to reinforce the types of responses that are more acceptable recognition responses for example, if a user says "erase message" and recognition prefers "delete", have the prompt repeat back, "Did you say delete message?"

You can train users on correct "responses" to improve recognition accuracy by using prompts. The way you word your prompt will have a big impact on how callers respond. Take advantage of the 'parroting effect' -- people tend to repeat words and grammar from the prompt when they respond.

Match Screen with Voice Prompts. If the speech application involves a Graphical User Interface (GUI) as well as a Voice User Interface (VUI), words on the GUI should match the words you want the user to say (e.g., Put "Delete" not "Erase" on GUI buttons if "delete" is what you want people to say).

Sounds are restricted to a narrow bandwidth over the phone line. When designing prompts, try to stay away from words that are more difficult for callers to hear, especially S's, M's and N's. Instead, use words with sounds that are easy to hear, such as T's and K's.

Prompt To Parrot. When using speech recognition, the way you word your prompt will have a big impact on how callers respond. Take advantage of the 'parroting effect' -- people tend to repeat words and grammar from the prompt when they respond.

(PureSpeech)

Prompt Recording Tipfest

When was the last time you saw a really bad commercial on TV, or perhaps heard a poorly-produced radio spot? If you are like most, it really stuck in your mind and reflected poorly on the company. Unfortunately for computer telephony systems, the very same thing can (and often does) happen.

One of the top deficiencies of CT application developers is failure to recognize the importance of using high-quality speech prompts. Without adequate quality, callers can be less willing to have their calls handled by an automated system. Nothing destroys the perceived quality of a voice system more quickly than hearing prompts that sound too loud, distorted, or unrehearsed. To do the job right, scripting, recording and editing high-quality speech prompts takes real effort and an understanding of what affects their quality.

Creating the Scripts. The first step in this process is to write down each and every word that is to be spoken during the recording session. What you want is for every prompt to be spoken in a clear, unhalting manner. The only way to ensure this is to prepare a script beforehand. Since each "prompt" will be recorded as a discrete message, you will need to clearly identify the prompt name or number, then the prompt text, depending on how your prompt recording and editing program identifies each individual message. For example:

Message 100: Thank you for calling XYZ Corporation.

Message 105: If you know the extension of the person you would like to reach, you may enter it now.

Message 150: That is not a valid selection, please try again.

The absolute best time to create these scripts is during the creation of the voice application itself. This is when you are constructing the application steps and when it is most obvious what each prompt should say. A few hints for scripting applications:

1. When announcing menu choices to callers, always mention the digit to press at the end of the phrase. For example: "To hear a list of departments, press 5." This makes it easier for the caller to remember what to press.

2. Always refer to special keys on the telephone keypad in the same way. (for example, "star key", "pound key", etc.).

3. Try to always use the same error message prompts for similar errors. For example, whenever a caller makes an incorrect selection, always use the same prompt (e.g. "That is not a valid selection, please try again").

Prompts And Choosing Voice Talent. When it comes time to record the prompts, the best thing to do is locate talent with solid experience in broadcasting to record them for you. These experts have learned to speak quickly, enunciate clearly, and adapt their voices to the style you are after. And because of their training, it is very likely that they will be able to record what you want quickly and without many retakes. Locating this talent can be quite easy. Make a few calls to local radio stations and inquire about the type of talent you are looking for (male, female, straight voice, animated style, etc.). Chances are you will find more than a few who will be quite willing to record your scripts. Now the next question: what kind of voices should you be looking for to match specific applications? Here are a few pointers:

1. **Go for clarity.** Clarity and intelligibility should be your #1 goals when searching for the appropriate voices to record your speech prompts. Make sure that the voices you choose can enunciate words clearly and distinctly.

2. **Males usually sound better.** With most computer telephony hardware, male speakers sound better because of the data compression method commonly used. That is not to say that female speakers always sound bad, but most male speakers sound less "scratchy."

3. **Avoid too much bass.** Avoid male speakers with too much bass in their voice. Such voices can sound wonderful on the tape but usually sound awful when encoded and played back on computer telephony systems.

4. **Avoid "hyped" style.** Speakers that sound like "hyped-up" top-40 disc jockeys don't make for popular CT systems. Unless the nature of the application will specifically benefit from this style of speaking, it usually grates on most callers in a very short period of time.

5. **Test several voices.** Some voices that sound absolutely great on the tape may come across sounding raspy or rough when they are encoded. Test several different voices you think may work.

The Recording Session. Having created your scripts and selected the voices to use, the next step is to record the prompt messages. Here are a few points that should be considered:

1. **Record "everything."** One of the marks of a good computer telephony system is to have all prompts matching each other - that is, prompts should all have the same volume level, tone, and speech characteristics for a given voice. For this reason, it makes good sense to record prompts that you think are likely to be needed in future revisions of the application.

2. **Recording room noise.** If you are not conducting the recording session in a real recording studio, find a room that is free from noises like ringing phones, noisy air conditioning vents, etc. In particular, make sure the microphone being used is a good distance away from any PCs: hard disks and cooling fans in these units produce sound that can increase the noise level of any recordings you produce if the mike is too close.

3. **Services are available.** If you have prepared scripts but do not have the time or resources to do the recordings yourself, recording service bureaus can turn your scripts into professionally-recorded and edited prompts.

Recording Direct to Disk. The easiest way to record speech prompts for your computer telephony application is to run a program or utility that allows the voice talent to record directly to the system hard disk. With this method, the recording and encoding of prompts can be handled in one step. There are three basic classes of programs that allow direct-to-disk prompt recording:

- **Integrated development environments**

- **Record/edit Utilities, software only**

- **Record/edit Utilities requiring special PC hardware**

With the first two choices, the program is designed to run in a PC with at least one speech card installed of the same brand that will be used to run the application. With the final choice, a special PC card must be installed that interfaces with an external microphone and includes a monitor jack for headphones or speaker amplifier. Each method has its strengths and weaknesses. Figure 3.2 shows a typical screen layout for the many sound editing packages on the market for PC-based components.

Figure 3.2 - Screen From PC-based Sound Editor Application

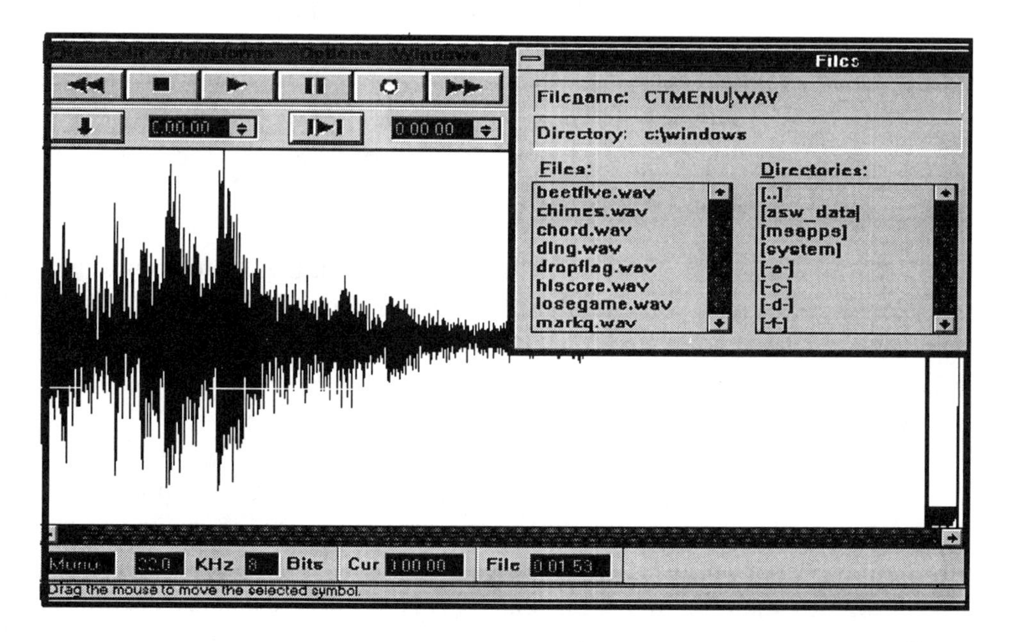

Integrated Development Environments. These "Application Generator" products often include a built-in set of features that enables the application developer to maintain scripts and record them from within a single screen. The advantage of this method is that, because it is fully integrated with the application creation tools, it is relatively easy to jump back and forth between the creation of application steps and the creation of the prompts that each step needs. For an experienced developer this can save significant time.

Record/Edit Utilities. If you have a very large number of prompts to record and edit, need exceptionally high speech quality, need to generate special tones or special effects, or have special editing needs (e.g., cut-and-paste operations), then you will probably want to use one of several products that address these needs.

Bit Works Audio Works Station. This product has earned the reputation in the computer telephony industry as the best set of hardware and software tools to create and edit the absolute highest quality speech prompts. The package includes a PC card with very low-noise high quality circuitry that allows direct recording to disk in a variety of formats. Jacks are provided that allow you to plug in a microphone and headphones directly. The on-card digital signal processing chip works in conjunction with the software to provide real-time filtering, level control, multi-band equalization, and encoding based on settings you choose.

Voice Information Systems VFEdit. This product is a graphical Windows application that gives you the ability to record and edit speech messages using the native speech card that will be running your application. This means you do not have to purchase special hardware to use all the supported operations.

VFEdit supports the ability to record and review discrete prompt files as well as Macro File prompts, with thousands of individual prompts in a single file yielding better run-time performance. It is possible to easily edit prompts by "cropping" their beginning and ending regions; or you can use the mouse to grab a segment of a prompt and cut it out and/or paste it into another message.

Recording to Tape. When application scripts are recorded to tape, an additional step must be executed as compared with recording direct to disk. This step is needed to create the actual prompt files that the computer telephony system needs. It is important to record to tape first for several reasons:

Use of different encoding methods. Let's say you recorded the entire application script direct to disk at a particular sampling rate and using a specific speech card or encoding method (e.g., ADPCM, PCM, WAV, VOC, etc.). If you later desire to import the exact same series of prompts to another format, you may not be able to convert the prompt files without causing a significant decrease in speech quality and/or increase in noise. Prompts recorded on tape can easily be used to generate a new set of encoded prompt files using the exact same recorded script source tape.

Original tape is a natural archive If for any reason you need to regenerate a prompt file, the tape of the recorded script is a convenient archive that you can easily access and use.

Here are a few hints to help you get the most out of your tape recorded scripts:

1. **Choosing the microphone.** Any microphone with a relatively flat frequency response between 100 and 10 kHz should be more than acceptable.

2. **Perform silence test first.** Each time before you start a recording session, at the start of a fresh tape record:

10 seconds of silence with the microphone turned off

10 seconds of silence with the microphone turned on

Play the results back and be sure the difference between the two segments is noticeable but not too pronounced.

3. **Keep levels high enough.** Be sure you record at sufficiently high levels on the tape. If you do, tape noise will not be a significant percentage of the recorded audio.

(Telephone Response Technologies, Inc.)

Rotary Pulse Conversion Snags

The inherent snag -- part one. One problem often encountered during pulse conversion really has nothing to do with the pulse dialing at all. Rotary dialing is much slower than tone dialing.

The application will have to be set with longer inter-digit time-outs so that it does not time out on the caller when they are using a rotary phone. Some applications have no option to modify the inter-digit time-out and will assume the caller has stopped entering digits while the dial is spinning back on the next digit. You have to watch out for this one.

The inherent snag -- part two. Since rotary callers don't have # and * on their dials, the application will have to offer an alternative to these commonly used keys. Some converters translate 11 into * and 22 into #, or the application could accept two digit input and handle one or two digit input itself.

Training. Some pulse to tone converters require a training digit to be entered by the caller to "train" the converter to all the variable parameters in the connection.

This is usually an easy option to add in most applications since it asks anyone with a rotary dial phone to "dial 9 now." From then on, the system will have a much more accurate image of what to look for to match the rotary dial pulses.

(Computer Telephony Magazine)

Scheduling Development

Create the Development Schedule. It is now time for the final step: creating the development schedule. No matter whether the application is quite small or very ambitious, this step is an important one. Why do a schedule? Obviously, it will help you determine the duration of the project if it is done properly. This is important for two reasons. First, it gives you the ability to estimate your development costs and hence what you should charge the customer. Secondly, it gives you a benchmark to compare your progress against. If things start slipping, you can quickly alert the customer.

Create the List of Tasks. The first step in creating a development schedule is producing an outline of the tasks that each developer will be working on, in the order they will work on them. If there is only one developer (you), then there would be one such outline. Then estimate how long each separate task will take. An example is detailed in figure 3.3 below:

Figure 3.3 - Application Task Time Estimate

Program Section	Task Item	Hours
Main menu & Basics	Error handling, transfers, etc	4 hours
Auto-attendant section	Transfer section	2 hours
Auto-attendant section	Audio directory	1 hour
Auto-attendant section	Leave recorded message	6 hours
Order Entry section	Menus, error handling	4 hours
Order Entry section	Database functions	16 hours
Product Info Hotline	Menus, etc	2 hours
Database Host Transfer	Build sample databases	2 hours
Database Host Transfer	Write host simulate program	8 hours
Database Host Transfer	Add functions to application	4 hours
Speech Messages	Create scripts	6 hours
Speech Messages	Record messages	8 hours
Speech Messages	Edit messages	3 hours
Pre-Launch	Preliminary tests	2 days
Pre-Launch	Rework and retest time	2 days
	Total development time	13 days

Of course, in this example many more steps could be included. Notice the last two items: "Preliminary tests" and "Rework and retest time."

After you have finished developing all the application sections, you will need to set time aside to test everything to be sure it works according to the application specification. During tests, you will undoubtedly discover problems that will need to be fixed, so the "rework" estimate is intended to cover that.

Scheduling Accuracy - What to Expect. If you have never produced a schedule for developing a voice application before, do not expect perfection with the first one. It is almost guaranteed that you will underestimate one or more tasks. Don't feel bad - even large established development firms miss on their estimates, and sometimes they miss by wide margins.

One thing you can do to improve your estimates is to assume you have left something out and compensate for it by adding a "fudge factor." Depending on the relative uncertainty of the project, it is not unreasonable to increase your overall time estimates by 25% or more. As you do more application projects, you should get better and better at estimating what it will take to get the job done. With this knowledge, let's apply a reasonable correction factor to our previous estimates:

- **Estimated development time - budget 13 days**
- **Estimate correction factor (30%) - budget 4 days**
- **Probable development time - budget 17 days**

Produce a Time Line Chart. The final step in producing the development schedule is to take the individual task estimates and plot them on a time line chart, sometimes referred to as a Gantt chart. This chart is a useful visual tool for you and your customer - completion dates (called milestones) are easy to see as well as all the tasks that lead up to each milestone. The final milestone is when the project gets delivered to the customer and is installed. This chart should take into account working versus non-working days and show you the actual date when each task starts and finishes. It can be manually drawn on a large piece of paper and should not take too much time to do.

If you prefer, there are several off-the-shelf programs (Microsoft Project is one) that make it possible for you to create this chart in a manner that makes it easy to make changes. You will find that any task changes made will cause all tasks that follow it to shift to the right, which means lots of erasing and redrawing every time a task slips or whenever a new task is added. Figure 3.4 is an example of a possible Gantt chart that you would create for a CT project.

Figure 3.4 - CT Application Development Gantt Chart

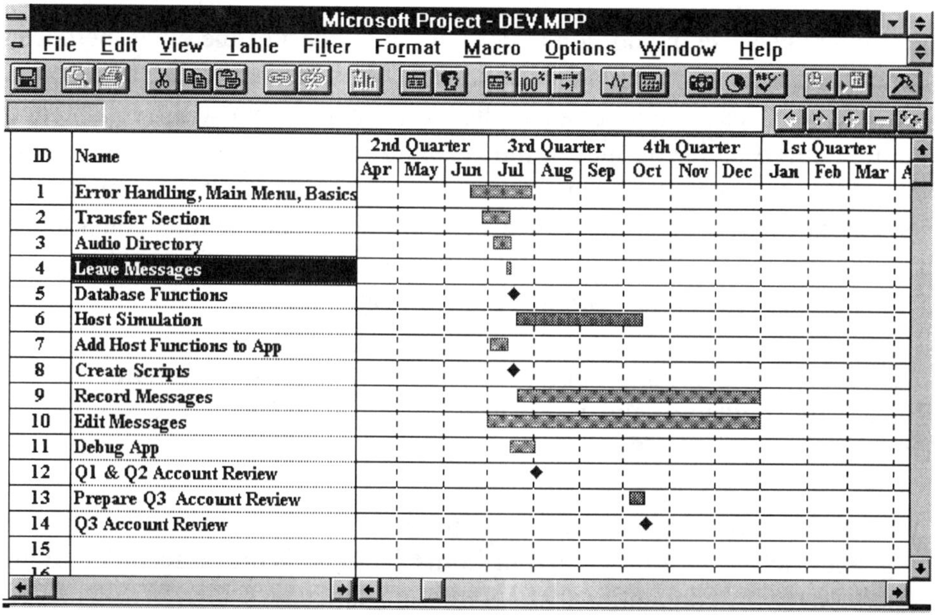

Microsoft Project - DEV.MPP

File Edit View Table Filter Format Macro Options Window Help

ID	Name
1	Error Handling, Main Menu, Basics
2	Transfer Section
3	Audio Directory
4	Leave Messages
5	Database Functions
6	Host Simulation
7	Add Host Functions to App
8	Create Scripts
9	Record Messages
10	Edit Messages
11	Debug App
12	Q1 & Q2 Account Review
13	Prepare Q3 Account Review
14	Q3 Account Review
15	

2nd Quarter: Apr May Jun; 3rd Quarter: Jul Aug Sep; 4th Quarter: Oct Nov Dec; 1st Quarter: Jan Feb Mar

After completing the development schedule, present a copy to the customer and see if the milestones are scheduled to their satisfaction. It is always better to revise things at this stage (before the actual implementation begins) than to wait until development is already underway.

(Telephone Response Technologies, Inc.)

Script-Based Application Development

Computer telephony script languages were designed to bridge the gap between the flexibility programmers are used to in typical programming languages (e.g., C, C++, Pascal, etc.) and the ease of development afforded by Application Generators. Because of this, script languages offer a viable alternative to programmers who do not relish the thought of such a major undertaking in C, and to non-programmers who consider themselves power users of non-voice software tools (e.g., Lotus 1-2-3).

Skill Level Required. Script languages are not a panacea for all application developers because in addition to combining the best of the other methods, script languages also combine the worst of them. Using a script language you will gain flexibility over an application generator but will face a higher learning curve until you master the syntax of the language. Similarly, you will trade the integrated development environment of most application generators for the barrenness of a programmer's text editor. You should possess at least the following skills, or acquire them, before undertaking application development with a script language:

Familiarity with common operating system functions. You should not be filled with trepidation and fear whenever you type : COPY C:\TEST*.* D:\TEST

Minimal experience with either a programming language or a macro language such as those included in Lotus 1-2-3 or WordPerfect. Although experience with conventional programming languages like BASIC, C or Pascal is ideal, experience with macro languages is equally suitable experience.

Familiarity with, or access to, an ASCII text editor (not a word processor such as WordPerfect, Microsoft Word or WordStar

Script Language Fundamentals. As with all development methods before you actually begin writing code you should outline the access paths through your application, preferably with a clearly-drawn block diagram.

Most script languages are built upon blocks of code generally called functions, procedures or occasionally routines. Regardless of the term used the meaning is the same: these are fragments of code used to perform specific, usually discrete operations. For example a typical application may have functions to hang up, act as a menu, transfer a call, and accumulate statistics. Functions are generally discernible through some syntactic convention and in this example functions are enclosed by curly braces:

hangup play(HANGUP); set_hookstatus(ON);

} main_menu {get_response(); get_digits(1); }

Figure 3.5 - Sample Script Language Code Segment

```
main_menu {
 play (MAIN_MENU_MSG, GET_DIG_TONE);
 get_digs(tt, 1d, 6s, call too_long);
 val_single() {
  case 1:
   goto trans_phone
  case 2:
   goto play_product
  case 3:
   goto leave_voice
  case bad:
   call bad_entry
 }
}
```

The example above shows two functions, "hangup" and "main_menu." Each of these functions makes calls to two other functions; "hangup ", for example, calls the "play" and "set_hookstatus" functions. Each function called by another function in your script must either be written by you (as "hangup" and "mainmenu" are in this example) or provided by the company from which you purchased the script language. The power of a particular script language is influenced by two crucial factors:

1. **The ease with which you can perform low-level operations** (e.g., getting a single digit, playing a single message, determining the status of a board, etc.) and

2. **The ease with which you can perform high-level operations** (e.g., create complex menus, fax a set of requested documents to the caller, etc.). It is important that the set of functions provided by your script language include enough higher-level functions so that it is easy to create common application concepts such as menus.

Rather than build a large number of low-level support functions into their packages, some script language manufacturers choose to provide the low-level support through the ability to access code written in C. Although this approach has its advantages you should weigh it against the fact that you may end up writing some C functions to achieve your goals.

From Design to Code. After you have designed the application, you are ready to begin implementing the application. The first step in doing so is to determine the relationship between the blocks in the block diagram and the functions you will write in the script language. To begin with you should assume that you will have a minimum of one function per box in your block diagram. One of the keys to implementing successful voice applications is the design of these functions and how they will interact with one another.

When moving from a design to actual script statements there are a number of guidelines you can follow which will both speed you through the process and give you a more desirable process when you are finished. The guidelines can be summarized as follows:

Write Modular Code. The functions which make up an application should be written in such a manner that each function provides a single useful, well-defined service to the application. For example, in the typical voice mail application a caller will at some point be asked to enter his ID number and a database will be searched with that number to determine if the caller has any waiting messages.

It would be reasonable to expect such an application to contain a function called "ask_for_id," which would play a prompt to the caller and gather tones from the caller. It should not perform the actual database search. The searching of the database should be performed by a second function, presumably called "search_database." The reason for this is that the "search_database" function will be able to provide a useful service to other branches of the program.

Structure the Flow of Control. Early software was often written in what was called "spaghetti code," which was so convoluted with "goto" statements and a general lack of structure that it was almost impossible to follow. The advent of structured programming languages and the lessons learned from them have carried on into most computer telephony script languages.

In large measure spaghetti code can be avoided if the functions created follow the modularity principle outlined earlier. However, you should additionally make use of the loop and control constructs provided by your script language. Looping constructs, such as "while" loops, "do-while" loops and "for" loops allow you to write script functions without using the often confusing "goto" statement. For example the following script statement will play a series of messages (presumably from a voice mail database):

```
not_done = TRUE; while(not_done) {play(current_msg); if (current_msg = last_msg) not_done = FALSE; else current_msg = current_msg + 1; }
```

Using "goto" statements, this same function appears as:

```
not_done = TRUE; start: if(not_done) { play(current_msg); if (current_msg = last_msg) not_done = FALSE; else { current_msg = current_msg + 1; goto start; } }
```

Although the improvement is minor in such a simple function, it should be apparent how the structure imposed on a program through the use of improved control mechanisms like loops will increase the readability of the code.

Write Reusable Code. Never assume that the application you are currently writing will be your last. Even if you are developing an application solely for use within your own company and have no plans for computer telephony beyond the current application you should always try to write reusable code. If you can reuse a function or other code fragment why would you want to waste the time reinventing the wheel?

Keep Maintenance In Mind. You should always write script functions with the idea that you will need to maintain the application, possibly for a number of years. Because of this, you should do what you can while first writing the code to ease that later maintenance burden. All script languages provide some method for adding comments to the actual script statements. You should make liberal use of comments, commenting sufficiently so that anyone reading the script will be able to fully understand it by reading the comments.

Test Frequently. The best way to develop voice applications using a script language is to test frequently. It is not uncommon to test each individual function as it is written. Doing so allows you to detect bugs or problems as soon as they are introduced.

Use the Tools. Hopefully you have selected a script language which provides more than just a language and the compiler to create a program or application from the script files you write; hopefully, the script language vendor has also provided you with tools to record and edit speech messages and to generate reports on the application you have written. It is also desirable to have reports showing if you have any unused (not referenced) functions in your application. Another report which should be considered a must is one which will verify that required speech messages have actually been recorded and are accessible to the runtime system. Without such a report you must manually verify this or risk running a system which may be missing crucial speech messages.

Strengths and Weaknesses. To many users, application development using a script language offers the best of both worlds. You gain the freedom to use a standard text editor with which you can cut-and-paste text and can comment the script code to your heart's content.

However, you lose the protective environment most application generators provide which through their fill-in-the-blanks, form-based approach. In general, there is no reason to expect a script-based language to perform more or less efficiently at runtime than will an application generator-based application. In fact, some script language vendors like TRT provide compatible application generators allowing you to customize your development approach to a suitable mix of the two approaches.

Compared to development in a general-purpose language such as C, script development will provide a much-faster time-to-market, improved maintainability, and greater reliability at the expense of a potential loss of freedom. However, because most script language vendors provide a means of accessing functions written in C this loss of freedom is generally not a factor.

(Telephone Response Technologies, Inc.)

Scripts and Recording Tips

Scripts - Talk like a customer. Use words and terms that your customers use. Department titles and job descriptions are fine for employees, but your customers shouldn't have to learn your company structure to do business with you. Avoid using company jargon. Don't expect the caller to figure out that he needs "Collections", just offer "past due bills." To find out what terms are confusing to customers, ask them in focus groups or get input from your customer contact people. The customer contact people can tell you what the people ask for when they call in. Terms you take for granted can cause confusion. One company used the term "Customer Service" to describe billing functions, while the customers thought it referred to trouble shooting and repair.

Keep it simple. Balance the need to adequately describe each option with the need to be brief. Be succinct.

Make it easy to use. To avoid long descriptions, offer specific options first, then a "none of the above" choice, which can lead to more options.

An example would be: billing questions, installation questions, all other questions.

Voice-Choose your voice carefully. The voice your callers will hear is very important. Don't let the voice distract from the message. A clear pleasant voice without a heavy accent allows the callers to concentrate on the information being provided.

Record in pieces. By recording the script in small sections, changes or corrections can be made without having to re-record the whole menu. The small sections allow you to move items to different menus or reorder the options within a menu.

Record everything! To fully utilize your session, record everything you can think of at the same time. Future options, backup contingencies, holiday greetings all can be done ahead of time to eliminate future recording sessions.

Speak with one voice. Anything that the customers will hear should be in the same voice. That includes default messages, numbers, and error messages. Messages supplied with the voice processing equipment can be re-recorded to use your chosen voice.

(Enterprise Integration Group)

Synchronous Callback Mode Under SR4.1 for NT

Make sure the event handlers for the given channel are enabled from within the thread that is to do the voice processing for that channel. If you do not do this, the event handlers will not being called or are called at the wrong time in a multi-threaded WinNT application that uses the synchronous/callback model.

Under SR4.1 for WinNT, the SRL(Standard Runtime Library) functions sr_enbhdlr() and sr_dishdlr() can be implemented in a multi-threaded application. The limitation is that the event handlers for a channel can only be called in the context of the thread from which they were enabled.

For example, in the main thread of your application, you enable an event handler for channel 1 as part of the system initialization. After initialization is complete, the application starts a thread that is to play a prompt on channel 1. During the play, an event occurs that is to be handled by the previously mentioned event handler. At this point the event handler is not invoked because it was not enabled from within the thread that is performing the play. It is likely that any event handlers enabled in the main thread, will not be invoked until all other threads in that process have completed.

(Dialogic Corporation)

Talk-Off And How To Beat It

Why did the voice mail system cut me off when I was trying to record a message? There is a phenomenon called "talk-off" where the frequencies in a human voice match those of a touch-tone digit for a brief moment, fooling the computer into believing that the caller dialed a digit. This happens most often to women and children, since their voices tend to be higher pitched and therefore closer to the frequencies in touch-tones. A similar phenomenon, called "play-off", can happen when a call processing system plays a menu or informational prompt where the voice recorded has touch-tone frequencies in it. Talk-off and play-off can be reduced by restricting the touch-tones which will interrupt a play or record, most voice cards allow an application to specify exactly which digits will be recognized - for example, if a menu allows only 1, 2 or 3 in response, disable detection of all the remaining digits.

(Parity Software)

TSAPI-Based Applications

During TSAPI technical tutorials, the folks at Novell proffer god advice on how to create dazzling, useful apps for Novell Telephony Services:

1. If there is a name or phone number on your screen -- **point and click** to dial .

2. If you get Caller ID or ANI, use it -- if you don't get it, **get it or complain!**

3. **If you already know the "customer" -- pop a screen.**

4. **Small is beautiful** -- don't take up screen space you don't really need.

5. **Don't overload the user** with call status information just because you can get it.

6. **Let your application (not the user) worry** about the access codes, special dialing sequences, etc.

7. **Think workgroup (not just individual) productivity** -- networked applications can share resources (address books, white/yellow pages...).

8. **Please don't create yet another (corporate) directory** to administer -- integrate and/or import .

9. **Think client-server** -- some applications work better or are more cost effective with server components doing the work on behalf of many clients.

10. **Not all PBXs have the same features** or work exactly the same way. If you are trying to work across many different types, plan on doing some testing.

(Novell)

Type-Ahead – Allow It

Always allow "digit type-ahead" to allow an experienced user to go straight to the menu option they want. There is nothing more frustrating than wasting time listening to prompts you know well because the system doesn't let you interrupt the message by dialing a digit.

(Parity Software)

UNIX Operating System Kernal Tuning

Loading device drivers under UNIX often requires tuning of multiple system parameters within the UNIX kernal. In addition, user application software may also require additional parameter tuning in order to achieve acceptable levels of system performance. UNIX kernal parameters are not always well documented or easily understood, and are not always consistent between different flavors of UNIX.

Here are some general tips on tuning UNIX kernel parameters:

1. **Don't change any unless you really have to**.

2. **Keep track of the settings** you found before you changed them.

3. **Be careful about the interactions** of some of the parameters.

4. **Small changes can make big differences** (usually negative).

5. **Create system activity baselines** with sar and timex before you change anything, and then only change one thing at a time.

6. **mtune contains the default system settings and ranges**. Don't change these. Change parameters by putting them into the stune file, and use the idtune command instead of vi.

7. **The stune file contains two columns**, the parameter external name and its current setting. The mtune file contains the external name, the default setting, and the minimum and maximum settings allowable. The stune entry, if made, will override the default setting in mtune.

8. **If you don't know, off the top of your head**, where and what the stune and mtune files are, you should probably not get involved in editing them.

9. For "AppNote" or simple page number references, refer to the *"UNIX Kernal Performance Tuning"* Application Note for further detail.

10. For "SCO 3.2 notes", the ref. pages are from the *"SCO UNIX System V/386 OS SystemAdministrator's Guide"*, version 3.2v2.0E, of 17 May, 1990, (sag).

11. For "ATT 4.0 notes", the ref. pages are from the *"ATT USL UNIX System V/386 OS System Administrator's Guide"*, version 4, of 1991, (sag).

12. For "IF 4.2 notes" annotations, the ref. pages are from the Information Foundation Administrator *"Advanced System Administration"*, from Prentice Hall.

13. Interactive 3.2 V4.1 from SunSoft, mtune 2.33 - 94/05/16.

(Dialogic Corporation)

UNIX Application SRL Parameters

Setting the SR_INTERPOLLID for SRL parameters is complex. It relates to the specific programming model and the application architecture. SR_INTERPOLLID determines how often SRL will wake up to service non-stream based devices. All the stream messages get serviced immediately regardless of the setting for SR_INTERPOLLID. If you set SR_INTERPOLLID to a very low number (20), SRL will get maximum time and it will be polling non-stream devices continuously. Setting it to a large number (INT_MAX) will yield to other processes more often and that's why it is a good idea to set it to INT_MAX in a synchronous application.

Here is a guideline to set SR_INTERPOLLID:

1. If your application is synchronous or asynchronous dealing with only a single voice channel per process and has only voice devices, then SR_INTERPOLLID should be set to INT_MAX.

2. If your application is synchronous with a single voice and a single DTI device in a process, then SR_INTERPOLLID should be set to a high number up to INT_MAX.

3. If your application is asynchronous with multiple voice devices and multiple DTI devices then SR_INTERPOLLID should be set to a lower number as low as 20. Keep in mind that setting SR_INTERPOLLID to 20 will give SRL maximum time and if you have any other compute intensive process running, it may suffer.

(Dialogic Corporation)

Upper Memory - Running D40DRV And Other TSRs

EMM386.EXE provides access to upper memory space normally left unused. In most PC's, there are blocks of unused upper memory *address spaces*. EMM386.EXE identifies these unused spaces and makes the blocks available for TSR and device driver installation. Normally, D40DRV.EXE will terminate and stay resident in low conventional memory. However, with HIMEM.SYS and EMM386.EXE loaded, D40DRV may be loaded in upper memory instead. D40DRV.EXE will actually reside in extended memory, but will be mapped to an unused upper memory address space. Use a DOS utility like **mem** to check if there is enough unused upper memory for your TSR. To allow TSRs to run in upper memory address space, we include the following line to CONFIG.SYS:

dos=umb

This line allows DOS to control the unused upper memory area. We must also add the "ram" switch to the EMM386.EXE line in CONFIG.SYS to allow access to upper memory:

device=emm386.exe ram

We can then use the DOS command **loadhigh** to load D40DRV.EXE into upper memory. For example:

C:\loadhigh d40drv -sa000 -e192

The **loadhigh** attempts to load a TSR into upper memory. However, if no upper memory blocks are found, the TSR is loaded in conventional memory. **loadhigh** does not report where the TSR was loaded.

EMM386.EXE determines all the unused address space in the upper memory. By default, it only looks in the C000H and D000H segments for unused space. Switches may be added to include and/or exclude any range within upper memory space. For example,

device=c:\emm386.exe i=a000-afff x=c800-cfff

will cause EMM386.EXE to search the A000 segment and not to search C800 to CFFF for possible unused space. Because PC's in general differ in their utilization of upper memory (depending on type of video, network cards installed, etc.) it is important to use these two switches to best utilize the vacant upper memory in your system. The ***I*** and ***X*** switches may be used more than once on a single line.

In many cases, EMM386.EXE will find more than one contiguous blocks of unused address space in upper memory. These blocks are called UMBs (Upper Memory Blocks). In some cases, although the total amount of unused upper memory space may be adequate for a TSR, it may be in the form of multiple blocks, where no single contiguous block is of sufficient size for the TSR. If so, the TSR cannot be loaded in upper memory.

A TSR, once loaded, may perform dynamic memory allocation to obtain more memory for its use. If there are no free memory blocks of adequate size above the location of the TSR, the TSR may at this point fail. Loading the TSR at a low conventional memory address - i.e., loading it without **loadhigh** - will usually solve the problem in such cases.

Note: The driver for the network interface of D/240SC-T1 and D/320SC-E1 boards (NETDRV.EXE) is the only TSR with performs dynamic memory allocation. It may consequently generate, when run with **loadhigh**, a run-time error if it is unable to procure a sufficiently large high memory block.

It is currently in its first beta version, and its dynamic memory allocation will be replaced by an alternative method in a future release, so as to remove the possibility of such an error.

The DOS manual does warn that there is no guarantee that all TSRs and device drivers will run in upper memory properly. The only way to determine is by actually running them.

(Dialogic Corporation)

Upper Memory Installable DOS Device Drivers

Installable DOS device drivers may be loaded in upper memory by using the **devicehigh** command. Unlike the **loadhigh** command, **devicehigh** is only valid inside the CONFIG.SYS. For example,

devicehigh=c:\dos5\ansi.sys

will load ANSI.SYS into a UMB.

(Dialogic Corporation)

Upper Memory And QEMM-386

Quarterdeck's memory manager QEMM-386 provides everything HIMEM.SYS and EMM386.EXE does and a whole lot more. For instance, QEMM-386 can effectively move system ROMs out of upper memory to create more and/or larger UMBs. Although QEMM-386 has more features than DOS 5.0's own memory manager, the basic idea is the same: both attempt to move as much code as possible from conventional memory to UMBs.

(Dialogic Corporation)

V&H Routing

If you want to use the Caller ID information for routing of calls to agents covering different geographical areas, use the Bellcore V&H database to cross reference a phone number and prefix to a location. The Bellcore V&H table is a list of all the area codes and exchanges in North America. It also includes the vertical and horizontal coordinates of the general area covered by the area code and exchange. It is a rather handy database to have.

There are several CD-ROM vendors that offer importable national telephone number databases. All the CD-ROM databases have one failing, they cannot be changed. One way around this problem is to build a local hard disk database that is searched before the CD-ROM is searched. This allows changes to be made to
correct the CD-ROM database changes.

(Computer Telephony Magazine)

Voice Files From RAM Disk

When playing voice files from a RAM disk, it is not necessary to specify the PM_RDISK mode in xplayf(). Using PM_RDISK causes unnecessary memory usage from D40DRV. Do not specify PM_RDISK mode for xplayf() and do not use the -V option for D40DRV.

The PM_RDISK mode for xplayf() was originally added to deal with some limitations of early memory managers. With our latest System Releases and the improvements of current memory managers this mode is no longer needed.

Some early memory managers had problems performing EMS to EMS transfers. Since a RAM disk is just EMS memory and the channel buffers are in EMS memory, it was necessary to create the PM_RDISK mode. When specifying this mode it is required that D40DRV has been installed using the -e option (to put the channel buffers in EMS memory) and the -V option(to establish an intermediate DOS transfer buffer).

With the -V option, to set up a DOS transfer buffer, the transfer now becomes EMS to DOS buffer to EMS. Most current EMS managers do not have this problem, so not using PM_RDISK does not present a problem. In this situation the -V option can be omitted , thus saving you some conventional memory.

(Dialogic Corporation)

Voice Processing Menus

Give the application first. Then the number to enter, e.g. "For information on roses, dial 1." Don't say "Dial 1 for roses." If say dial 1 first, they'll forget which number gives them which option.

(Soft-Com)

Voice Storage

The biggest voice processing problem is over-estimating your storage needs for your voice prompts. The key is to get your voice prompts into as few words as possible. Less disk space, fewer bored listeners.

(Telephone Response Technologies, Inc.)

Voice Won't Stac

Don't use disk compression such as DOS Double-Space or Stacker from Stac Electronics if your hard disk is used mainly for storing messages in ADPCM digitized voice files (most voice boards use ADPCM by default). ADPCM files do not compress more than a few percent at most (they're already highly compressed), so all you do is waste memory and resources and add a complex element with the potential to fail, but for little or no benefit. If you use PCM, you can get compression up to 50%, so it might be worthwhile.

(Parity Software)

Warning Messages

Try to interrupt recordings with a "warning message" a good 15 seconds before the end of the maximum recording time (assuming the maximum is at least one minute). This gives callers a chance to wrap-up their message before ending the recording.

(Computer Telephony Magazine))

Watchdog For PCs

When your 24-hour a day PC locks at 3 AM, what now? Lose the revenue? Get your customers annoyed? Wake a service technician to go and hit the reset button?

Berkshire Product's (Duluth, GA -- 404-271-0088) PC Watchdog card ($145) is an 8-bit PC card that watches the activity on your PC bus and can reset (i.e. warm reboot) your computer if no activity has occurred for a specified period of time -- two seconds to two hours.

The PC Watchdog card with temperature sensor option costs $160. Cooling fan failure is one of the more frequent failures of computer telephony systems. Once a fan fails, the voice hardware dies from overheating, especially the hotter, higher powered DSP-based voice cards.

The PC Watchdog has an adjustable time delay before the board becomes active to give your application time to get rolling. This prevents the card from resetting the PC if your system takes a few minutes to get started. It can either be set to activate two minutes after bootup or only after the first activity on the hardware polling ports. One LED on the back edge of the card indicates whether the card has had to perform a reset already. The other one is a heartbeat indicating that the board is armed. One of the board's default addresses is 300h (hex address) which is a common voice-card port address. This means that as long as the voice-card driver or voice application talks to the board every few seconds, the board won't restart the system.

The port address can be changed to one of several options. By selecting the network-card address or the standard hard-disk controller address or the address of the serial ports, the card can remain active without having to modify the application.

Software to control the PC Watchdog is easy. It can be added to one of your programs with two lines of C source code. You can chose to disable the monitoring of network, disk and comm port activity. Instead you can configure the card to watch the I/O port at 270h and add a line of code to input from that port every few seconds. As long as your application is running properly, everything is okay. But if your app dies, the board will wait out the timeout period and then reset the system. To check the temperature, your program could input a status byte from port 271h. If the high temp bit is set, the system could send out a page alerting the administrator of the high temp condition.

The PC Watchdog card with the temperature sensor causes an audible alarm and triggers an output line that can be monitored by your application or external hardware. It can also keep the PC in reset state after a 120 degree high temperature alarm in the hope of preventing further damage.

You could use the external relay outputs to completely power down your system if your PC got hot. All you would need to do is poll one of the status ports of the PC Watchdog card and handle the 110 degree temperature alert if it occurred. With a voice application, it could call a emergency number and play a high temperature warning message. You don't get the inside temperature from the card, but you do get a bit that goes high when the board sensor is above 110 degrees.

The PC Watchdog will reset the PC if the bus voltage drops below 3.5 volts DC. This can assure a clean reboot from a brownout condition, where many CPUs simply go to sleep and never wake up.

If your system uses a PC Watchdog card, your AUTOEXEC.BAT file should be able to check the database integrity of your data files and be able to re-index them if necessary. Once the data files are stable, it would then restart your application.

The card can also be configured to cause a Non-Maskable Interrupt (NMI) that can cause your application to write out information about the configuration and the current state of each task to help discover what caused the crash. If the NMI handler code determines there is nothing wrong with the program and computer, it can make the board skip the reset and continue normally.

Anyone building systems that allow user configurable situations that can lock up the system really needs this card.

(Computer Telephony Magazine)

Windows 95 And EMM386

In Windows 95, using EMM386 can cause problems in the installation of Dialogic voice boards. Even after an open address is assigned by Device Manager, the board will not download with a memory conflict error. Remove EMM386 from your config.sys.

A user is not required to have a Config.sys or Autoexec.bat in Windows 95. The services are provided automatically. If a user had EMM386 in his Config.sys and upgraded to Win95, it will conflict with the operation of Device Manager which is used to configure Dialogic boards.

(Dialogic Corporation)

Windows CT Performance Tips

Here are some guidelines to follow if you want to prevent CT system performance problems under Windows:

Do not attempt run-time operation while other "resource hog" applications are running (like Microsoft Word). Run live tests with lots of phone lines active while you are loading and fully using any programs you want to run concurrently in a system (especially Windows 3.1).

Watch for efficient use of SQL queries in your CT application while running under Windows. When you access a database in your application, be aware that the way you construct your database query can have a significant effect on performance.

Make sure you have plenty of ram memory in your PC. You do not want Windows 3.1 to begin swapping program space to disk when it has to switch from one loaded program task to another. This will absolutely kill performance.

Don't skimp on the speed of the hard disk where the speech prompts and database files will be located. Faster is always better. Same holds true for disk controllers. Move commonly-accessed speech messages to ram-disks.

Don't skimp on processor speed. For all but the smallest systems, always buy toward the high end of the speed scale.

Watch for file fragmentation on your system's hard disk. The MSDOS (and Windows 3.1) file system performance degrades as the disk fills up and as files are stored in "fragmented" multiple areas of the disk. This can be a problem, especially if you record multiple speech messages simultaneously-- this is virtually guaranteed to cause fragmentation. Performance drops since the disk drive head has to travel longer distances to access a message. If recordings are made to LAN-based file servers, fragmentation is managed by the LAN software itself (e.g. Novell 3.x or 4.x). Another solution: auto-run a defragmentation program as an automated off-line process once a day on your system.

Another potential performance problem: MSDOS (and Windows 3.1) suffers from poor file access performance, especially when more than 200 speech message files are located in a single subdirectory. The solution: use "packed" message files ("VAP" files) that contain speech information and scripts for thousands of speech messages. If your application needs lots of on-line prompts, VAP files provide much better performance.

(Telephone Response Technologies, Inc.)

Windows 3.1 And Windows NT Server

Two lines you must add to System.Ini in its [Network] section if you are running Windows for Workgroups on a network with a Windows NT server:

MaintainServerList=NO
LMAnnounce=No

Without these two lines, your workstation will compete with NT's domain controller for the role of "Master Browser," (whatever that is) which, in turn, will fill your NT Server Event Log with error messages, which it dumps every 15 minutes.

(Computer Telephony Magazine)

Windows INI Files

Today, make a copy of all your INI files. Put the files away. The day will comes when one of your Windows programs, e.g. File Manager, starts acting funny. It will then be time to swap in the old, but good INI file.

(Computer Telephony Magazine)

Windows Memory - Is There Enough?

This tip is short but... it's very important. If you are deploying a Windows-based CT application, and you want to ensure the absolute best performance, you must take care to install an adequate amount of ram memory. If you run a Windows CT application with less than an adequate amount, you will run into a situation where Windows starts going to disk for more "virtual storage."

When that happens, your system performance goes way down because disk accesses are no match for direct ram accesses. In fact, a multi-line system will seem dead to a caller for as long as several seconds while memory things are being swapped in and out of disk "memory."

Unfortunately, there are no hard-and-fast rules for how much memory to use since each application uses computer resources differently. You are left with the end-all: test it thoroughly. If memory is a resource that you want to save money on (say you are a manufacturer and every dollar of margin helps), then you must exercise every possible function *with all lines active* to ensure proper operation. If you are shaving memory to the bone, make sure your speech prompts and other time-critical operations run smooth over an extended period of time.

(Telephone Response Technologies, Inc.)

Windows NT - Compiling With MS Visual C++ V. 2.0

MS Visual C++ users must use the Microsoft MSVCRT.LIB run time library when linking a Dialogic application and set the /MD (Multi-threaded with DLL) compiler option, in spite of whether the Dialogic application is single or multi-threaded mode.

When using MS Visual C++ V 2.0 to compile a Dialogic application, follow the next steps to set up the project correctly:

1. Open your new project and add the source code and Dialogic libraries to it

2. Choose settings from the Project Menu

3. Click on C/C++ box

4. Go to Code Generation category and choose "Multi-Threaded using DLL", being the third option.

5. Set up all other options from the various menus, as needed

Failure in doing this will cause problems when the application is run. A dx_play() will fail with IOTT structure error and a system error for lseek. Problem goes away with the correct settings as indicated above.

Note that this option must always be set, even for single-threaded applications. Projects for voice applications using the single-threaded model should include LIBDXXX.LIB & LIBSRL.LIB. Projects for voice applications using the multi-threaded model, on the other hand, need to include LIBDXXMT.LIB & LIBSRLMT.LIB.

(Dialogic Corporation)

Windows NT A Good Choice

Windows NT: A Definition. Windows NT is a full 32-bit, preemptive multi-tasking operating system with a huge capacity to enable power users to fully exploit line-of-business and personal productivity applications. It also supports the latest generations of microprocessors including Intel and RISC systems and symmetric multi-processing systems.

Cross-Platform Portability. This means that software written for Windows NT can work on a variety of computing platforms. For example, the software will run on an Intel-based PC or a RISC-based system like the DEC Alpha. The ability for the software to work an "small" and "large" computing platforms makes your software investment more durable over time. This also implies a higher degree of flexibility in your vendor selection.

Multi-Tasking and Multi-Threading. Windows NT allows for multiple processes to happen simultaneously. You can processes voice calls, handle security, search foreign databases and back-up files all at the same time. This is not a new concept - UNIX has been doing it for years. The advantage in NT is its popularity and worldwide acceptance.

Multi-Processing and Scalability. Windows NT supports computing platforms that use more than one processor. This is helpful for transaction-intensive applications or applications that are mission-critical. A lot of resiliency can be built-in to a computing platform that has more than one processor. If one processor fails, the system can continue to work. Having multiple processor design allows you to "scale" or grow your application to handle more and more over time.

Distributed Computing - Built-In Networking. NT is designed for client-server implementations. It is easy to hook-up many computers in an NT environment and make them share files and services with one another. This built-in networking capability is especially useful for call center applications where you need to hook-up file servers, client workstations, routers and remote access points.

Windows NT vs. MS-DOS Windows. The similarities are mouse and keyboard techniques for working with windows, menus and icons, desktop tools such as Program Manager, File Manager and Print Manager, Accessories, games and other components. Windows NT has an extended set of these. There are also similarities in the complete object linking and embedding (OLE) capabilities supported by applications created for versions of Windows for MS-DOS.

Windows NT and MS-DOS Windows differences. First, Windows NT is a complete operating system - with inherent support for networking, remote access and client-server use. Native DOS Windows falls short of this. Then there's the multi-tasking. DOS is not inherently multi-tasking. You have to "fake-out" the system to make it think it can do this (usually with proprietary tinkering and poor results).

Windows NT uses 32-bit addressing which provides faster memory and file access. This (WIN32) standard does not come with DOS. In addition, the NT file system (NTFS) sports security and automatic error-correction capabilities and can handle long file names. Windows NT & Computer Telephony. The most notable attributes of Windows NT versus DOS are the ones that are important for computer telephony applications. They are:

- **Reliability and robustness for mission critical applications**
- **Preemptive scheduling for improved "real-time" data handling**
- **32-bit addressing for better memory management**

- **Scalability for higher-density applications**
- **Portability for increased flexibility**
- **Built-in networking for client-server applications**

(APEX Voice Communications)

Windows Programming Preamble

Before installing new Windows programs, back up your INI and GRP files. A crash or lockup can wipe out one or more GRP file.

(Computer Telephony Magazine)

Chapter 4

Successful Call Center Strategies

Call centers are rapidly being re-defined. They are no longer just a collection of ACD (Automatic Call Distributor) lines and agent terminals. Now, there are small (informal) call centers made up of a handful of workers in a department. Maybe they're working on a one-time campaign together. Call centers can spring-up overnight to handle emergency situations, product recalls and special sales.

The Internet is even getting into the act with call centers. With the right software, an Internet-based transaction can spawn an outbound telephone call to a customer. And even the SOHO (Small Office / Home Office) market is beginning to take advantage of computer telephony. With a decent PC, a voice modem and some new software, you can collect Caller ID, rout calls and do screen pops in a two-man office for as little as $3,000.

This chapter represents a collection of the best ideas from the top consultants and vendors of call center technology. They include application tool vendors, ACD manufacturers and some leading-edge users. What with PC-based switches now bursting onto the market – even the ACD manufacturers are re-thinking how to attack this ($4 Billion) market.

But the fundamental issues of call center management remain the same. Issue like agent productivity, ergonomics and alternate routing for switches. If you are growing your existing call center or starting a new one, these tips will help you stay ahead of the confusion.

ACDs And What They Do

Consider the benefits of an ACD in both traditional and non-traditional niches (i.e. departments) of your company. Consider both the obvious and the not-so-obvious benefits of an ACD. The payback comes in both hard and soft returns.

Looking Beyond The Obvious. Your cal call center is more than a bunch of people manning terminals and telephone. Look at it as a strategic weapon with which to differentiate your self from your competitors. Your call center is also an economic catalyst and a personnel management tool. It's really a socio-political differentiator in the office of tomorrow

Call Center as Strategic Tool. You can provide unique services with your staff and your ACD. You can use it for surveying, routing calls and special campaigns. You can implement innovative marketing plans and put ACD groups together in an ad-hoc fashion. Virtually every department in your company can play a part in this.

You can also use your call center as a means to achieve improved operating results. This includes economies on your phone bill and increased staff productivity which equates to higher profit margins. You also enjoy a better and proper image to customers and prospects.

Mission Control. Every company has a mission, so your call center can provide the tools to control many details of the mission. You can manage the quality of customer handling. You can be accountable for and understand the cost per call. You can get to know your customer better - even those who hang up.

Know the Unknown. Start thinking about the power of information - not just call distribution. Collect information on the randomness of certain call patterns so you can better predict call occurrences. Knowing what your service level should be equates directly to staffing levels. Force yourself to consider you caller's delay tolerance (this changes depending on the time of day and the purpose of the call). Armed with this information, you can make precise staffing decisions and know where your true system costs reside.

There are tools available to catalogue and report on all of this information from the better ACD and ACD software companies. Plan on spending between $10,000 to $50,000 for customized reporting packages to automate this reporting. In the long run, the better decisions that come out of this information mean long term customer satisfaction. The payback period is usually less than six months. You also do a better job of retaining revenue grows for customers and making then even more loyal. Remember, it costs 5 times as much to get vs. retain a customer. See figure 4.1 for an example.

Figure 4.1 - Value of a Customer Service Call

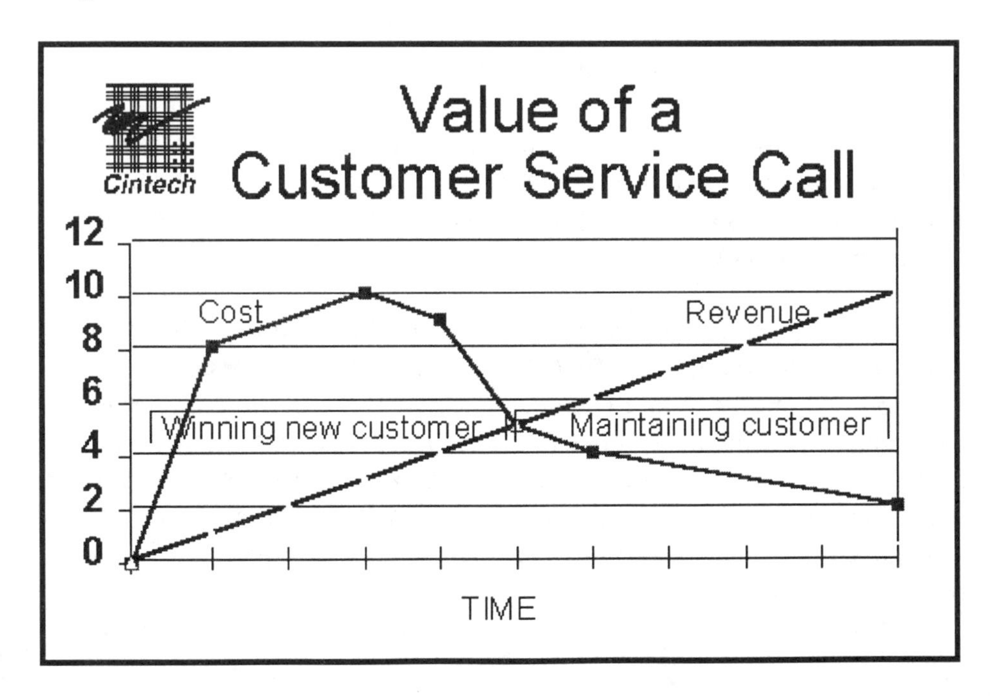

Customer satisfaction allows a business to realize a return on the initial cost investment of gaining a new customer--good call centers help provide that service.

Implementing the "Important" in VIP Customer Service. Send calls from VIP customers directly to the "front of the line" (Priority queuing, DNIS, DID) Make sure the most capable employees handle VIP customer calls (skills-based routing).

A Quiet Workplace is a Productive Workplace. Bring organization to your call center by minimizing or eliminating noise from calls ringing to numerous sets. Use displays or PC programs to eliminate agent-to-agent communication on the status of calls and lines. This will also help to eliminate unnecessary calls from the receptionist as to status of agents. Use acoustical tiles on the ceiling, walls and furniture components to cut down on noise. Customers don't like to hear chatter in the background, and your agents will be more productive without it.

(Cintech Tele-Management Systems)

ACDs And Web-Initiated Callback

Consumers now have the ability to request a live agent callback in order to provide security, ask a question, or inquire about a nonstandard item. The consumer does this by initiating a request from a Home Page which is sent to the page's owner.

Figure 4.2 illustrates how a Web-initiated Callback works. This example is based on Rockwell's *Internet ACD*. The Internet browser (consumer) lands on a Home Page and becomes interested in a product or service on the screen. The browser "clicks" on an icon or hypertext requesting a callback. A second screen is activated which asks questions such as name, phone number, and time requested for callback. It is the consumer who schedules the callback. The browser then sends the request by "clicking" on another button. It's that simple.

Figure 4.2 - ACD Meets The Internet

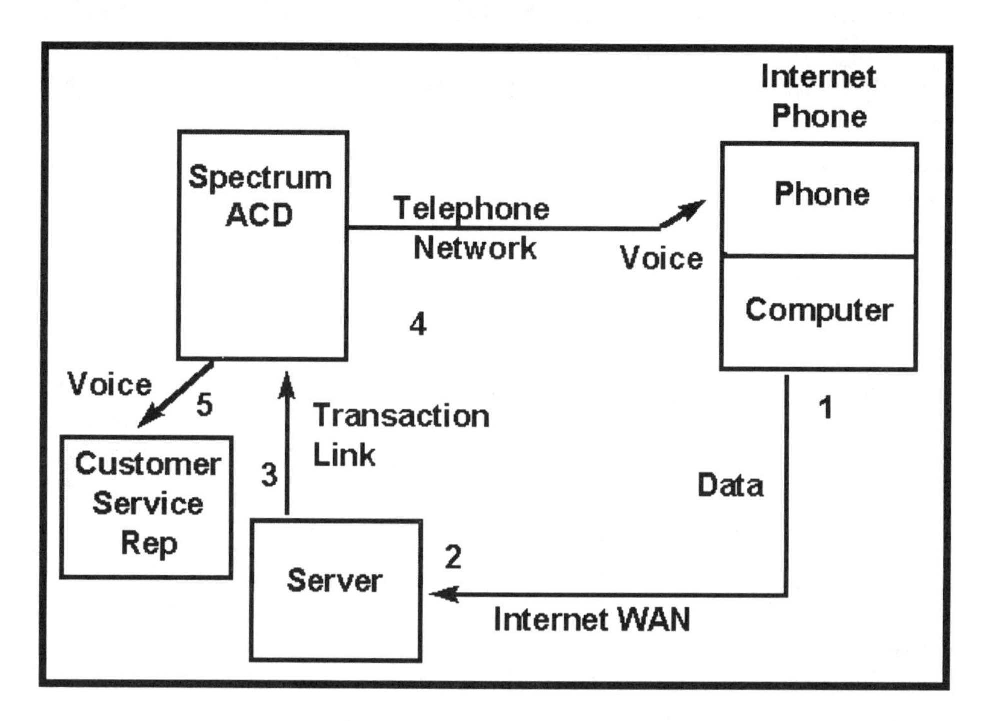

After the browser initiates a callback, the request is routed back to the Home Page owners Internet Server via e-mail/TCP-IP, which in turn sends the request to an Electronic Response Unit (ERU). The ERU is a PC with special software that translates the Internet message and customer information into a predictive dial message to the ACD.

What the figure cannot illustrate is the speed at which a callback can occur. Given an open phone line at the customer's end, callback occurs in less than 10 seconds. That's right, the time from browser initiation of the call to actual ring is under 10 seconds. This costs the customer nothing because the call is initiated by the Home Page owner.

In the case of ISDN or multiple phone lines, the callback can be immediate. This allows a customer service representative and customer to be looking at the same Home Page. Alternatively, the return call can be scheduled at an appointed time or after the customer logs out of the Internet.

Web-initiated Callback Benefits. Web-initiated Callback is a method of providing *efficient service* to customers. This can help reduce time and actually increase customer service. Instead of a customer calling on a higher priced 800 line, the call center can make the call on a lower priced regular long distance line. The call center can efficiently capture customer information for free. Think of the "request callback" screen as a VRU with the full graphic imagery of the Internet -- with information being supplied on the customer's time.

Travel agencies can use the potential power of the Internet to capture customers' interests in destinations, dates and travel packages. Customer Service Representatives can assemble the requested information before the callback is made, thus cutting time and calling charges.

The customer data captured is incredibly beneficial to the Home Page owner. Companies using the Internet haven't really been sure who's checking out their wares and what, if anything, they can do to improve selling rates. Current technology may allow a company to see that 50,000 people logged onto its Home Page. It doesn't allow simultaneous live questioning of consumers, and it cant capture the sales of potential, security-conscious buyers. Nor does it allow the ability to cross-sell or up-sell related merchandise.

Couple current ACD abilities with Web-initiated Callback and a much more complete picture of Internet marketing unfolds. You can now answer the marketing questions of who are the buyers, what products do they purchase, how often, when and why. The amount of information available is almost endless. Creating a marketing strategy for the Internet has ceased being hit or miss.

Providing personal contact and secure transactions improve customer service, making a customer more comfortable doing business with a company. Also, with Internet ACD, browser and service representative can interact much like they do in retail stores, taking into account individual needs and requests.

In addition to the obvious potential for sales and service, Web-initiated callbacks can help a company manage their image on and off the Internet. Most Home Pages have a way to e-mail questions or suggestions which fall into two types: either the simple, "send a question and get a private answer" or the online chat board where anyone logging into the system can access others' questions and the company's answers. Here are some Web-initiated Callback Applications:

1. **Customer priority.** Part of good service is the ability to track merchandise that has been shipped. Web-initiated Callback could be used in order to provide different information pages about shipping and offer the customer the ability to check the status of an order. By clicking on hypertext, the customer is prompted not only for a name and address, but for an order or shipper number as well. The callback time is reduced because the service representative has the information about the package in hand before making the call.

2. **Customer service centers.** Users can utilize the "request call" screen to ask for a brief description of problems so they can be investigated before the customer service representative calls the customer. A sure way to escalate a consumer's anger is to make that customer feel ignored. Today's e-mail may as well be "snail mail" to a customer who is unhappy. It's too easy for an unhappy customer to start or join a chat room for the sole purpose of telling the world about the poor service they received. And as we all know, nothing speaks louder than a testimonial --good or bad. Web-initiated Callback removes the time barrier and allows customer service centers to immediately solve problems.

(Rockwell International Switching Systems Division)

ANI Considerations

Advantages Of ANI. Automatic Number Identification is useful in identifying incoming callers before they are connected to the appropriate resource. You are also able to design your applications to track the sources of incoming calls (this can be helpful in building caller demographic information).

ANI is also helpful in allowing you to return calls from hang-ups (recovery from abandoned calls). Combined with DNIS, user requests and connections can be easily defined without human intervention.

Figure 4.3 - ANI In The Call Center

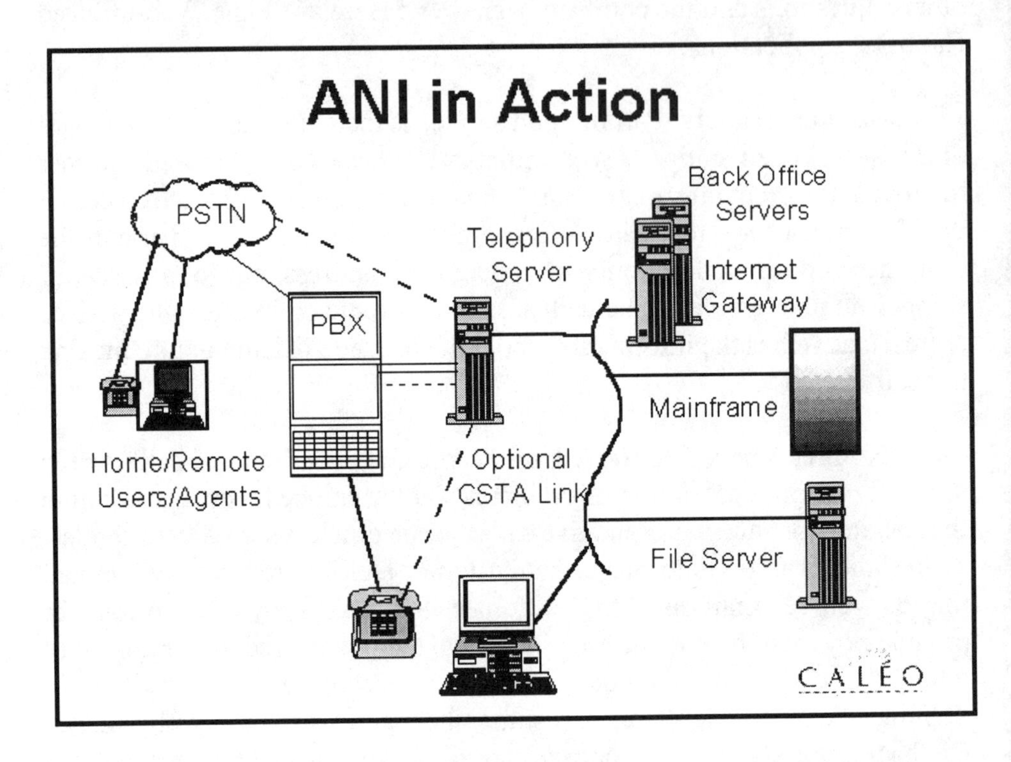

As pictured in figure 4.3, the basic call center topology remains the same when you use ANI. You still need your switch, your computer system and your agents. ANI is delivered on either T-1 or ISDN PRI trunks. You can also get DNIS (Dialed Number Identification Service) along with ANI. DNIS is like a network-level DID (Direct Inward Dial Service): the last four digits dialed by the caller are captured. This helps to route the call for certain products or services. A block of DNIS digits can be used by a call center to distinguish between support calls for different products or orders for different services. When with over one hundred DNIS numbers, you can have only 48 lines, let's say.

This bank of virtual trunks makes a lot of economic sense. So the ANI tells you who is calling, and the DNIS can tell you what they're calling for. Figure 4.4 shows the basic steps involved in processing an ANI-based call. What makes ANI calls different from regular calls is the link between your switch and your database. After the ANI digits are collected from the telephone network, they are processed either by a computer telephony server or sent directly to your host computer system (or file server). This is done over a data link or LAN connection. The digits are further processed so a look-up of a customer record associated with the ANI digits can be found. An agent station is then signaled, and the customer record appears on his or her screen. At the same time, a command is sent to the switch to route the call to that agent telephone. In some cases, it is the telephone system that simply routes the call to an available agent – whereupon the computer system catches-up by "popping" the screen to the receiving agent.

Figure 4.4 - Steps Of AN ANI-based Call

Disadvantages of ANI. Often, the billing number is ambiguous, so you have to design your application for these cases. For example, the lead trunk numbers of PBXs, payphone numbers and cellular phone numbers yield dubious results. In addition, most PBX's drop ANI or do not pass it to external telephony servers. For these reasons, using ANI and DNIS requires a well designed rule set.

(Caléo Software, Inc.)

Advanced Auto Attendant Features

Most companies use auto attendant transaction boxes to record digital messages that give callers DTMF options towards a destination, which is usually a phone extension and that's it. We do that too, but we took it a step further to solve a problem.

Here's the problem: We get hundreds of phone calls from customers who want our the latest version of our software drivers, our BBS number or our web site's URL or our fax-on-demand number. These simple calls flooded our ACD hunt queues every day. It's silly to have these rudimentary calls tying up our very technical and overwhelmed support reps. Worst of all, customers could only get answers to these questions when we were open for business which was 9:00 to 6:00, Monday through Friday. I figured out a way to get callers answers to these questions 24 hours a day and seven days a week.

I set up transaction boxes that played back a message offering DTMF options for other transaction boxes that played back messages about our latest software drivers, new products, web site URL, BBS number and fax-on-demand number. So customers hear, To learn about our latest drivers PRESS 1, to learn about our on-line services PRESS 2, to learn about new products PRESS 3 and so on. Our daytime message includes an option to go to a CSR. Our nighttime message doesn't. It works great and off-loads a significant number of calls that we would usually get during the day.

(ENSONIQ)

Customer Relationship Sensitive Service

In order to supply great service, you need to look way beyond computer telephony and integration. Good customer service starts with sound work-flow and process management. Look at your call center systems as an opportunity to improve customer relations.

Remember that most businesses are demand-driven. You should challenge your staff to provide performance guarantees in order to achieve customer service improvement. Your basic philosophy should be to *improve call content* -- not simply automate connections. There's little room to improve switch productivity (call connection), but major room for agent and management productivity (style). The biggest opportunity you have is to improve customer information delivery (Content). Here's some tips for developing a project plan to improve your information (content) delivery:

1. **Buy in by all client elements.** There are both internal and external clients. This means your call center staff actually serves fellow employees and also outside customers. Don't forget to include other departments in your buy-in process. Remember that your users should be king. This may be the most political project ever attempted – but it's well worth the effort.

2. **Have a detailed plan.** Everything should be in the open. Involve your project managers, business analysts, salespeople and customer service departments. Everyone should work together on a detailed, written plan. Write down everything from everyone and compile everyone's ideas for review. Everyone's input should go into the plan.

3. **Define actual project goals.** Turn all of the desires from internal and external customers into well-defined goals. Prioritize the goals and mimic them back to your constituency. Make sure they still agree at every step.

4. **Develop a detailed design specification.** Based on these goals, get your technical people and MIS department to design a technical specification. Remember that MIS people are drawn to standards. Make sure you develop a specification based on standards that are easy to implement. Don't re-invent the wheel.

5. **Client validation.** You should put every step of feature development into a validation loop with the user community. There should be no surprises when your new service goes on-line.

6. **Embrace "change" management from the start.** Change is good. Make sure everyone involved in the process knows that you accept and anticipate changes in the way you will do things. By designing a new service, you are inviting change. Deal with it in a straightforward and honest fashion.

7. **Appoint project management on both sides of your team.** This means you should get active involvement from users, agents, managers – everyone who will be affected by the new service.

8. **Phase and stage development.** Establish regular status meetings. This includes project development as well as training meetings. Train and re-train. Ask for feedback on your training.

(Teknekron Infoswitch Corporation)

DMI - Call Center Desktop Management Interface

The PC industry is growing at a phenomenal rate, encompassing an ever-larger number of users, more sophisticated technologies and increasing levels of complexity. The emphasis on networking has made the PC environment even more complex. As a result, the gains in the reduced cost of hardware and software have been offset by the increased cost of support. The bulk of this can be attributed directly to the burden of managing PCs and PC networks.

The need for managing information, computing resources and networked systems has never been greater. Network managers are seeing their LANs grow at alarming rates, both in the number of workstations attached to the network and in the types of applications being deployed. At the same time, low-cost gateways and bridges have spurred rapid growth in LAN/WAN connectivity. Since these branch offices rarely have their own administrators, support for them falls back on the host site administrator.

Consider the following questions in designing your call center:

1. Do you plan on supporting callers who have PCs, LANs, Servers and other computer-based equipment?

2. Will you provide help desk services for troubleshooting nd technical support of PCs?

3. Do you or will you send flash ROM, software drivers, and software updates to your customers.

If you answered yes to any of these questions, then you need to plan on learning more about DMI. This is a snapshot of how it works and how it will help every call center.

Figure 4.5 - Enterprise-Wide Administration Duties

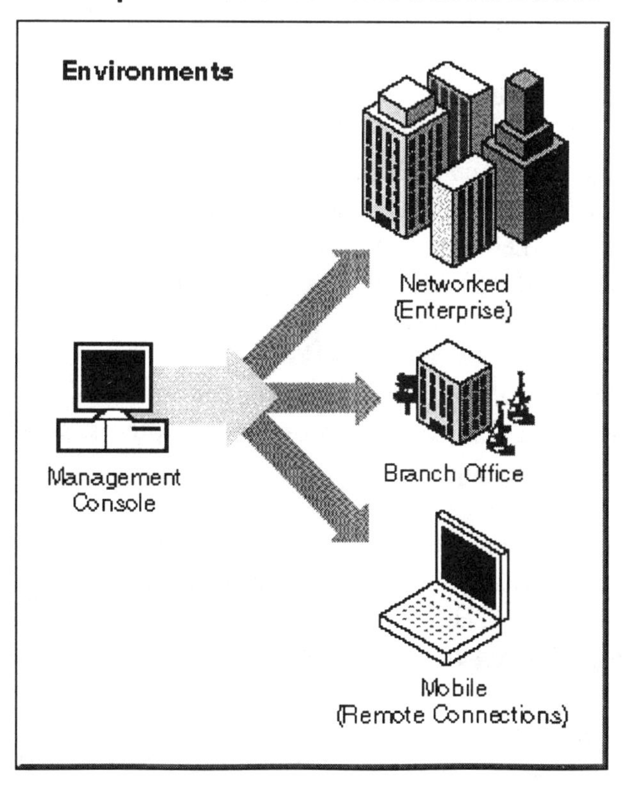

Figure 4.5 illustrates how the support for managing computing resources across the enterprise falls on the host-site administrator. A new breed of network management is needed to deal with this crisis. LAN administrators need hardware and software products that integrate seamlessly and work as manageable, cohesive units. They need systems that can self-configure, self-adjust and communicate with the user.

Integrated Management Needs. Despite different structures and configurations of local networks, the need for PC LAN management remains consistent. These environments need a way to integrate desktop-level information, such as resources and configuration, into the network foundation. Necessary functionality includes centralized management, remote control support, and asset management.

When the PC was designed, no standard was founded for uniformly identifying, installing, or integrating hardware and software into the platform. Consequently, the PC is essentially non-exclusive, open and flexible.

Because the corporate world lacks a common, comprehensive method for managing PCs, and peripherals, LAN administrators have had to bear the burden of managing the network, the desktop systems it links and the assets of those systems. They've also found it extremely difficult to configure and troubleshoot networked PCs, especially remotely. With most management systems available today, even knowing what software and hardware resides on a given system is difficult.

What is DMI? The Desktop Management Interface (DMI) is an open architecture standard for managing desktop computers, servers, hardware and software products, and peripherals.

The DMI standard was developed by the Desktop Management Task Force (DMTF), a consortium of computing industry leaders that includes Digital Equipment Corporation, Hewlett-Packard Company, IBM Corp., Intel Corp., Microsoft Corp., Novell Inc., SunConnect and Bay Networks. In addition to those eight charter members, the DMTF also includes more than 300 participating members.

The DMI can be implemented in hardware, software, or a peripheral attached to a desktop computer or network server, enabling that product to become manageable and intelligent. A DMI-enabled component or application can communicate its system resource requirements with a DMI-management application and coexist in a manageable network and PC system.

Because it's an open standard, the DMI also guarantees vendors a common strategy for designing supportability and manageability into all of their products. Vendors can take advantage of the DMI technology as a reliable mechanism for building easy to use, manageable and interoperable products, including: hard disks, word processors, CD-ROM drivers, printers, motherboards, operating systems, graphics cards, sound cards, modems and network adapters.

By making the DMI available for implementation, the DMTF provides the industry with a flexible and comprehensive structure to manage systems and products throughout the intrinsically open PC LAN environment.

What Are MIFs?. Various multi-vendor groups are cooperatively defining standard DMI Management Information Format (MIF) files, which facilitate implementation by describing manageable attributes for certain product types. MIFs contain two types of information about the computer and its hardware and software components:

1. **Static or resident** data, like a component's name and version data (resides in MIF database)

2. **Dynamic or instrumented** data generated by instrumentation code that examines the current status of various hardware and software components.

3. **The DMI also provides network managers real-time information** for managing assets across the entire enterprise as well as the ability to become more proactive concerning network re-appropriations and the future. With the manageability enabled by the DMI and products that implement it, network managers can plan for growth and change.

What's the Structure of DMI?. The DMI is structured with multiple communication levels, which provide the inherent manageability of all DMI-enabled products. These communication levels include the Service Layer, the Management Interface and the Component Interface.

Figure 4.6 - The Structure Of The DMI

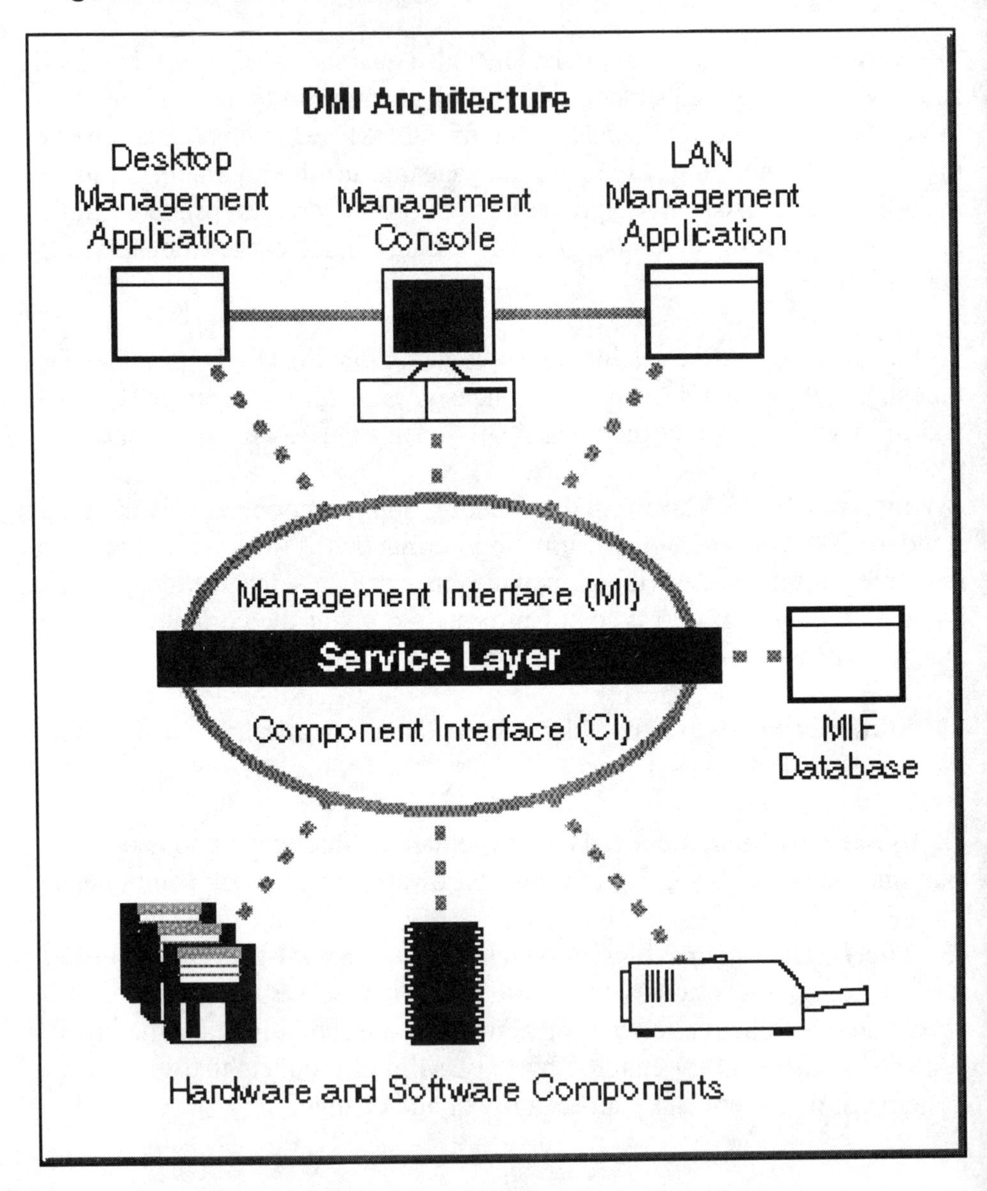

The Service Layer is a local program that gathers information from DMI-enabled products. It then manipulates that information in a MIF database and transfers the information to independent management applications as requested. The Service Layer handles communication between itself and management applications by means of the Management Interface (MI), and handles communication between itself and manageable products via the Component Interface (CI). This is pictured in figure 4.6.

The MI allows a management application to query for lists of manageable products, access specific components and obtain and set specific attributes. Additionally, the MI allows a management application to tell the Service Layer to send information about details from DMI-enabled products.

Simply put, the MI provides for management applications to access, manage and control desktop systems, components and peripherals.

The CI provides the ability for components to be seen and managed by applications that interface with the DMI in a standard way. This saves component vendors from having to make decisions about management applications, protocols and operating systems, and allows them to focus on providing competitive management for their products.

The cornerstone of the DMI is the MIF database, an accumulation of all available MIF files located on the system or network. Each manageable product provides information to the MIF database by means of a MIF file that contains the pertinent management information for that product. Once installed, DMI-enabled products communicate with the Service Layer through the CI. They receive management commands from the Service Layer and return information about their status to the Service Layer.

When a manageable product is initially installed on the system or network, the information in its MIF file is added to the MIF database. That information is then available to the Service Layer, and thus to management applications. Once a DMI-enabled product is installed, its information becomes available to management applications, such as Intel's LANDesk Management Suite.

The suite's DMI Control Panel helps LAN administrators identify and manage activities at PC nodes and peripherals, making it easier for them to integrate all the pieces into a cohesive management strategy.

With a click of a mouse, the LAN administrator can inventory, diagnose, troubleshoot, configure or test any DMI-enabled component on the network, including motherboards, chips, LAN adapters and printers. The DMI provides a standard way for all enabled products on the market - now and in the future- to be integrated, centrally managed and controlled. This has special meaning for help desk applications.

(Intel Corporation)

DMI In Action: Help Desk Applications

The DMI enables manageability by allowing applications that call the DMI to access MIF information. This paves the way for the development of new kinds of powerful help desk software.

How can DMI reduce the OEM's cost of providing maintenance?

DMI can help the OEM's call center staff to:

1. **Reduce the number of calls**. DMI cuts down on the need for calls to technical support centers because DMI-based utilities make it easier for home users to diagnose, fix and configure their PCs. Installation procedures can access and rest configuration parameters.

2. **Deflect calls.** Automated agents can access the user's MIF database to obtain data about a user's problem, diagnose and perhaps even resolve the problem without help from a technical support professional.

3. **Handle calls.** Call center staff can remotely browse the MIFs in the user's PC to identify problems and reset parameters.

As a result, you can resolve problems faster and with less expense.

How do I get started with DMI? DMTF Corporate Secretary, 2111 SE 25th Ave., Mail Station JF2-51, Hillsboro, OR 97124. www.dmtf.org

(Intel Corporation)

DSVD For Technical Support

Most PCs are now bundled with modems, so the idea of downloading support tips, updated drivers – and even making regular telephone calls with your modem is commonplace. This is enabled by DSVD (Digital Simultaneous Voice and Data).

Voice Calls. A voice call can be initiated from a regular telephone, or from the computer itself if it has an attached modem with speaker phone or headset and microphone. During voice-only calls, technical support agents are essentially "blind and handicapped" because they can't see the user's monitor or directly manipulate the mouse or keyboard.

The support agent must verbally instruct the user to perform sequences of keyboard and mouse manipulations, and the user must report the results of each step. The disadvantages of this approach include:

Communications between the user and the support agent are often tedious and error-prone:

- **The support agent can't directly manipulate the user's PC.**
- **The support agent can't download software fixes and upgrades.**

Data Calls. If the technical support agent needs to directly manipulate the user's C or download software fixes and upgrades, the original call must be terminated so the support agent can call the user's PXC and establish a data connection. This enables the support agent to directly manipulate the user's PC, including the following capabilities:

Upload and Download Files. The technical support agent can upload files such as AUTOEXEC.BAT, SYS.INI and CONFIG.SYS for review. The support agent can edit the upload files and then download updated files and device drivers back onto the user's PC.

Remote Execution. The support agent can write a script, download it to the user's PC, and remotely execute the script to change parameters and configurations on the user's PC.

Remote control. The support agent can remotely initiate and control the user's PC. Using tools such as Reach Out, CoSession, and PC-Anywhere, the support agent can see the user's entire display and use the mouse and keyboard on the agent's PC to drive the application on the user's PC. Remote control tools may be slow because the transfer bitmaps form the user's PC to the technical support agent's PC. Most remote control tools do not allow the user to drive the PC.

Application sharing. The support agent can share the windows of a single application executing on the user's PC. The user can demonstrate the problem and the support agent can demonstrate how to resolve the problem. Application sharing usually works faster than remote control because GDI commands rather than bitmaps are exchanged between the user's PC and the technical support agents' PC.

Disadvantages of the data call. The user has to hang up form the original voice call and switch the modem to data mode. The user and the support agent can't communicate verbally during the data call.

Voice/Data Calls. In voice / data calls, technical support agents and users can exchange both data and verbal messages during a single phone call. This can be accomplished by using one telephone line for a voice call and another for data transfer, but few users have two lines available at the PC's location for this purpose. Voice and data can be merged into a single call on a single line using either of two new technologies:

VoiceView's RAD. Rapid Alternating Data protocol enables switching between voice and data. The user and the support agent can communicate verbally except during short periods of time when data is transferred. The RAD protocol is available on most 14.4 Kbps modems.

Digital Simultaneous Voice and Data. Digital SVD modems can transmit both voice and data at the same time on the same line. Digital SVD modems can simultaneously support data at 19.2 Kbps and voice at 9.6 Kbps. Digital SVD works on a single Plain Old Telephone Service (POTS) line. The DSVD standard was created through a cooperative effort between Creative Labs Inc., Hayes Microcomputer Products, Intel Corporation, Rockwell International, and U.S. Robotics Inc.

While RAD protocols are currently more widely available, the DSVD protocols transfer data at a higher data rate (19.2 Kbps or better). DSVD is also more convenient because the user ant the support agent's conversations are not interrupted by a data transfer. The combination of DSVD with application sharing provides a powerful tool for technical support and just-in-time training.

Switching Among Information Transfer Modes. There are situations during which each mode is desirable:

Voice only. This is the traditional mode for users to describe problems and technical support agents to describe solutions. Voice mode is necessary for VRU (Voice Response Unit) software to generate verbal instructions and menus to the caller ("If you have a software problem, press 1. If you have a hardware problem, press 2. ..." ACD (Automatic Call Distribution) computers can only route voice calls to technical support agents. A call can be routed from one technical support agent to another only if it is a voice call.

Data mode. This is necessary to transmit data between the technical support agent and the user's PC. This is a requirement for large file transfers. Technical support agents traditionally switch to data mode by hanging up and calling the user's PC modem.

Both voice and data are desirable since they enable the technical support agent to greatly speed up the diagnosis and resolution of the user's problem. Users and technical support agents can interactively talk with each other while data is being transferred. The technical support agent can diagnose the problem quicker. Because data can be transferred, the problem can be resolved faster. RAD is currently available on many modems today. DSVD provides improved data rate and usability.

To be truly effective and efficient, technical support agents must be able to switch among voice, data, and voice/data modes, and all three modes must be available during a single phone call. When support agents can easily switch among voice, data, and voice/data modes, agents and users can talk with each other as the agent diagnoses and resolves the user's problem. This increases user satisfaction, preserves the existing call center infrastructure, and saves time and money for technical support organizations. This is all possible today on a single phone line between the user and the technical support agent.

Digital SVD Application Examples. Technical support questions can be greatly assisted with Digital SVD. The customer calls the vendor using the voice channel. The vendor discusses the issue with the customer until the issue is understood and informs the customer that a driver update is needed. Since both parties have DSVD modems, the update is transmitted directly to the customer over the data channel while they remain on the voice channel. The customer continues to receive phone support while installing the new driver and making sure it works. They hang up after the customer's issue is resolved.

These examples show how easy it is to use DSVD modems and how game play is enhanced with voice. The only thing that the developer provides is an initialization string, just like any other modem. Since voice is completely handled through the RJ-11 jack on the back of the modem, nothing needs to be done to the software to add voice.

(Intel Corporation)

Ergonomics For The Call Center

What is ergonomics? Ergonomics matches the task, the tool and the environment to fit the needs of the worker. According to Websters Dictionary, it's a science that seeks to adapt working conditions to the worker.

Why is ergonomics important to call center management? Ergonomics play a large part in encouraging worker productivity, preventing worker injuries and reducing workers' compensation costs or potential lawsuits.

Office workplace injuries are rising. According to the *Center for Ergonomic Research*, more than 8 million new cases of back injuries are reported each year. Backache is second only to colds and flu as a cause of lost work time in the U.S. In fact, the *U.S. Dept. of Labor* says that *repetitive stress Injuries* or *cumulative stress disorders* are 10 times more prevalent now than 10 years ago.

Figure 4.7 - Dollars Saved With Ergonomic Planning

Return on:
Ergonomic Investment

Productivity Increase For One Shift		2%	5%
A. Average hourly labor rate	$	8.00	8.00
Labor = 20% of output	X	5	5
B. Average hourly production			
Output	$	40.00	40.00
Rate of productivity			
Increase	X	.02	.05
C. Productivity increase			
Per hour	$	0.80	0.80
Per month (173 hours)	$	138.40	346.00

CenterCore

What do these injuries cost your business? The *Center for Ergonomic Research* estimates that lost wages and medical costs from white-collar injuries total $27 billion a year. *OSHA* says motion and back injuries account for 30% of worker compensation claims. *OSHA* projects 50% by 2,000.

What do these injuries cost in lost work time? Almost 100 million work days are lost each year due to back pain says the *Center for Ergonomic Research.* The group's studies show the average time for returning to work after a back injury is 14 days.

How does call center work cause injuries? Call center agents sit for long periods, use their keyboard for hours and stare at a VDT screen all day long. In addition, their workplace is often noisy and may have "bad office air." In short, this amounts to a stressful work environment. Add to this the pace with which call center agents are made to work, and ergonomics become very important.

What have I got to lose if I don't invest in ergonomics? Employers will pay approximately $30,000 per employee for direct medical expenses to treat an injury such as *carpal tunnel syndrome* (STS). Add to that legal fees from workers' compensation claims and lost productivity, and one injury could cost a company as much as $100,000. According to a **Risk and Insurance** magazine survey of 260 risk management executives, more than 82% of those responding reported dealing with workers' compensation claims for *carpal tunnel syndrome*. Yet only 51.5% are purchasing or are planning to purchase ergonomic equipment. Ergonomic measures pay dividends by reducing workers' compensation premiums and health costs. You also increase the productivity of your staff and their morale.

Ergonomic Workplace Creation

How do sitting, keyboarding, reading, noise and air affect workers? Sitting for long periods can cause chronic backaches. Keyboarding for long periods contributes to repetitive stress injuries. VDT's (CRT's, monitors, etc.) cause a great amount of eye strain.

Noise at certain sustained levels can cause hearing damage. Respiratory problems can be aggravated or even caused by bad office air. The fast pace of the call center work itself can cause a great deal of overall stress.

Take Breaks. Short periodic breaks are one of the best and most basic ergonomic measures. They alleviate back, eye, and muscle strain. They reduce stress, especially if workers do simple ergonomic exercises. The trade off is a few lost seconds for exercise. This, however, will result in more productivity.

Seating. You should look for seating that offers a sturdy five-blade base for stability. The seat should be adjustable -- both seat and back to conform to the individual user. Look for seating with lumbar support, because it conforms to curve of spine. Height adjustment is another simple, but important feature. A seat with a back angle and back height adjustment is also preferable (seat tilt adjustment is a plus). A waterfall seat edge is also comfortable and less stressful when sitting for long periods.

Best way to sit. 1. Angle the seat so the feet support the legs; 2. Keep feet flat on the floor with thighs parallel to the floor and circulation is not constricted; 3. Keep chair height at about 90 degrees at hip and knee for proper weight distribution; Use lumbar support at belt level to support curve of spine and decrease force on back.

How NOT to sit. Slouching puts too much upper body weight on a small area near the base of the spine instead of distributing it over the seat pan. Sitting forward prevents you from using the chair for back support. Sitting too low causes pressure in the abdominal area, base of spine, and upper legs.

Use Adjustable Workstations. It costs less to adjust a workstation to the worker than to build a customized workstation for every worker. Adjustable ergonomic designs allow people on different shifts to share workstations. People who work more comfortably also work more efficiently.

Use Energy-Safe Non-Glare Monitors. Avoid unnecessary glare, emissions and eye stress by investing in a new, safe monitor. They cost less to operate and they are easier on the eyes.

Use Contoured Keyboards. There are plenty of good split and/or contoured keyboard on the market that cost less than $100. The difference in ease of use and stress reduction on the hands, back and arms is amazing. Try one for a few days. It takes a while to get used to the new layout – but they are better all the way around. In fact, most users claim they can type faster and longer with these new ergonomic keyboards.

(CenterCore, Inc.)

Headset Considerations

In today's call center environments there is a growing preference for headset-equipped PCs vs. Traditional switch-based telsets. This goes for large call centers with thousands of people all the way down to small support desks with a dozen people and individuals who work-at-home.

Favored uses include computer telephony integration (CTI), personal conferencing, business audio and multimedia applications

A Santa Clara Valley Medical Study found up to 41% reduction in muscle tension in the neck, shoulder and back by using headsets. The first clinical proof of headset superiority over regular telephone handsets was conducted by the Institute for Treatment and Prevention of Repetitive Motion Injuries. The group studied 62 subjects including a variety of office workers. All subjects used the phone at least 3 hours a day, but were not considered "dependent" users. The study measured muscle tension of reading, writing, typing while on phone. Another study by H. B. Maynard & Co. cites up to a 43% increase in user productivity. This time and motion study at four firms compared headset to handset use. The scope of work included 2,000 telephone transactions over an eight hour period. This included sales, travel agents and technical field sales support.

Maynard & Co. says that companies using headsets claim improved morale and a work environment with fewer mechanical distractions. This means a reduction in employee downtime by allowing people to concentrate on the conversation, rather than the "pain in the neck" of cradling a handset. You should also consider the user of headsets instead of speakerphones. Most callers enjoy the personal privacy of a headset and consider the use of a speakerphone to be an invasion of privacy. Headsets also offer better sound quality. All of this adds up to improved worker productivity.

(Plantronics)

Help Desk Management - Handling Support Calls

How can PC vendors meet the rising requirement for technical support cost effectively? Calls to PC technical support centers are on the rise — and for good reasons. More PCs are being sold than ever before, so there's a larger population of users to support. More than fifty per cent of PCs sold are for home use, and home users don't have access to the computer expertise often available in businesses. In addition, PCs are becoming more complex, especially with the new multimedia enhancements, so more things have to play together for the system to work successfully. All of this adds up record numbers of technical support calls and to calls that require more expertise to resolve.

Installing more phone lines and hiring more technical support staff provides only a partial solution. It meets user needs, but it adds expense that inevitably is passed on to the user in higher PC costs or charges for technical support call support.

Intel Architecture Labs has investigated ways of using technology to help solve the technical support call dilemma. Their research has identified three ways to resolve the call center dilemma:

Avoiding calls to technical support centers by detecting and resolving problems locally, often before the user is even aware of them.

Deflecting calls to remote automated agents.

Use new tools that support the technical support staff.

1. Avoiding Calls. The ideal place to tackle technical support calls is at their source, before they even become technical support calls. With the appropriate amount of information and intelligence, the PC can help users avoid technical support calls by detecting and resolving problems, often before the user is aware of them.

Information about the PC's hardware and software modules is contained in the Microsoft registry and in individual MIFs (Management Information Files) associated with each hardware and software module. MIFs can also contain data collected by triggers, alerts, and other instrumentation. Since MIFs contain data about the current status of PC components, they can be a valuable resource to help users identify potential problems.

Users can review selected information from the registry and MIFs to help them solve their own problems before calling Technical Support. Examples of the information that can be presented to PC users include "the printer is out of paper," "the modem is not connected to the PC," and "there is no disk in the floppy disk drive."

Three types of technology can supply the intelligence needed to reduce the need for technical support calls:

Wizards can help novice users reconfigure a PC by applying intelligence in the form of carefully crafted algorithms.

Expert systems can apply rules gleaned from human technical support agents to detect and resolve PC problems.

Software agents can monitor the PC to detect and automatically resolve several types of PC problems.

These technologies can enable a variety of software that avoids technical support calls, including PC health monitoring, virus scanning, memory utilization, hardware testing, and PC tune-up applications.

2. Deflecting Calls to Remote Automated Agents. Wizards, expert systems, and software agents are limited to the types of problems their programmers had the foresight to anticipate. Given the many software and hardware components available, not all problems can be anticipated. Problems happen, and when they do, users make technical support calls.

Technical support agents (the humans) at the call center often detect trends in the technical support calls they receive and develop strategies for analyzing and resolving these calls. Some of these strategies can be formalized into scripts that solicit information from the user and then instruct the user in resolving the problem.

Many call centers use Voice Response Units (VRUs) that present voice menus to users, who respond by pressing keys on their telephone keypads. VRUs can be programmed with hierarchically oriented scripts to identify many of the frequent types of technical support calls and advise users about how to correct those problems without help from a technical support agent. These extended VRUs act as remote automated agents.

If the user has a modem that can transfer both voice and data, the VRUs can be extended to transfer data as well as voice messages, allowing users to select options from screen menus and use their PCs to fill information into forms. Modems supporting the VoiceView protocol can switch between voice and data modes. Modems supporting the Digital Simultaneous Voice and Data (DSVD) can transmit voice and data simultaneously.

VRUs and automated agents can present a menu of frequent problems to the user, and instruct the user in how to correct many of those problems, thus deflecting calls from the technical support agent. This savings in technical support agent time translates directly into cost savings for the PC OEM.

3. Improved Tools for the Technical Support Agent. Technology can help the technical support agent to handle technical support calls faster and easier. There are several tools at the disposal of the technical support agent:

The VRU can solicit information about the problem from the user. If the VRU can't solve the problem, an Automatic Call Distribution (ACD) switch can route the call to the technical support agent with the skills or tools for solving that particular type of problem. During peak times, the ACD can route calls to auxiliary technical support agents, possibly at different sites. This approach can also make it possible to outsource technical support calls.

Information solicited by the VRU is presented to the technical support agent with pop-up windows containing information solicited from the user or uploaded from the user's PC. This saves time by enabling the technical support agent to "hit the road running" and avoid asking the same questions previously posed to the user.

Browsers can enable the technical support agent to remotely scan the Microsoft Registry, configuration files, and the various MIFs. The technical support agent can examine critical parameter values and identify problems not resolved by call avoidance or call deflection techniques.

Procedures for uploading, editing, and downloading files let the technical support agent scan and update configuration files.

Procedures for creating, downloading, and executing scripts allow the technical support agent to remotely repair problems on the PC.

Application sharing allows both the user and the technical support agent to see the windows generated by the application and allows either person to use the keyboard or mouse to drive the application. In this way, users can demonstrate to the technical support agent what the problems is, rather than trying to describe it, and technical support agents can show how to use the application rather than having to verbalize what may be fairly complex instructions.

Many of these approaches, including application sharing, require that the technical support agent have remote access to the user's PC, which means at a minimum that the user's PC must have a data modem connected. First the user places a traditional voice call to the call center. After talking with the technical support agent, the user hangs up and the technical support agent makes a data call to the user's PC. During the data call, the user can't communicate verbally with the technical support agent.

If the user's PC has a voice-data modem, the agent and the user can talk and transfer data simultaneously. First the user places a voice call to the call center by clicking a button in a help menu on the PC. When the VRU and automated agent comes on-line, the call is switched to a data call so that the automated agent can pop menus and forms on the user's PC. If necessary, the call can be transferred to a technical support agent. At this point, the DSVD protocol is needed so the technical support agent and user can communicate verbally as they both view and drive the application. DSVD and application sharing together enable "just in time training. "

Customer Care Center Example. Let's look at a hypothetical example of how these new tools might work in action: the Customer Care Center (CCC).

When users need encounter a problem, they click a button that displays the Customer Care Center window. The CCC provides three levels of technical support: local (call avoidance), remote automated agent (call deflection), and personal help from a remote technical support agent (improved tools).

Local Support. The Customer Care Center attempts to avoid technical support calls by resolving the problem locally. When users enter the CCC, they complete a form describing their problem. This triggers the display of several hints and suggestions for fixing the problem. For example, if the user describes the problem as "Can't print," then various status messages about the printer such as "Printer not turned on" or "Printer out of paper" are displayed. Users can invoke various diagnostic routines that may return additional advice.

Remote Automated Agent. If users are unable to resolve the problem, the Customer Care Center connects the user to the call center's VRU. The user's problem description and results of any diagnostic routines are uploaded to an automated agent, which attempts to deflect calls from the technical support agent by automatically diagnosing and repairing the problem. The automated agent is programmed to ask the user for additional symptoms, identify simple problems and, instruct the user to perform sequences actions to resolve the problem.

Personal Help from a Technical Support Agent. If the VRU is unable to resolve the problem, the call is transferred to the technical support agent, who uses a variety of tools to handle the call. Information collected from the user is displayed on the technical support agent's PC, enabling the agent to review the data without asking questions the user has already answered.

If the technical support agent can't resolve the problem, he or she can escalate the call to another staff member, transferring the call and all associated information and history to the new agent. Calls can even be transferred to another help center that is better prepared to handle the problem. For example, problems dealing with a specific application may be transferred from the PC OEM's call center to the call center of the ISV who developed the application.

Security. It is natural for users to be concerned about the security of their system during technical support calls. We recommend the following security policies while handling technical support calls.

Maintain a log of all changes made to the user's PC. Make this log available on the user's PC so the user or other technical support staff can review what changes were previously made to the PC.

Ask permission before performing each access to the user's PC. Usually this question is asked by presenting the user with a dialog box explaining the nature of the access, and having the user click the OK button before the access can take place.

Some users may tire of clicking the OK button, so it's wise to allow the user to override this policy by granting the technical support agent permission for performing various classes of operations.

Ensure that the user can see what the technical support agent is accessing and stop the technical support agent at any time.

These policies are superior to the situation that occurs when the user leaves a PC at the repair shop with no security protection whatsoever.

Technical support calls are on the rise, often forcing users to wait long periods on the phone or hang up frustrated. In addition to adding phone lines and increasing the number of technical support staff available, three approaches - avoiding calls, deflecting calls, and utilizing better support tools -- can help deliver improved service, cost-effectively. Combining all three approaches can help give users the support they need when they need it -- and meet the OEM's budget objectives as well.

(Intel Corporation)

Help Desk Support Roles In Your Call Center

As pictured in figure 4.8 - there are plenty of people you need to involved in your call center support team. Here's three critical positions you should staff for in your call center. They are all big jobs, so don't try to get one person to do all of this. Budget for distinct functions and you will be much more profitable, efficient and your customers more happy:

1. **LAN Administrator.** This person support connectivity to computer applications. This includes personal productivity applications as well as screen pop, CTI and call routing applications. If your LAN is not optimized for your call center applications, you're out of business. This person also provides on-site problem assistance and performs timely backups.

2. **Call Center Analyst.** This person will make sure you are really using your workforce management tools.

This person will use these management information tools and make suggestions on how to improve call flow, reduce call abandonment, etc. You paid a lot for all those reports. Make sure someone is not only reading them – but taking action on them!

3. **Content Administrator.** This person is trained on how to update on-line databases and keep company specific data up-to-date. This includes fax-on-demand, Web page, "Intranet," shared folders, etc. This person is in charge of keeping local specific data up-to-date. You can also put this person in charge of altering workstation configurations and applicators so they have access to the data. Your content administrator should also be in the loop on voice prompt and scripting changes.

(Siemens ROLM Communications, Inc.)

Figure 4.8 - Call Center Support Roles

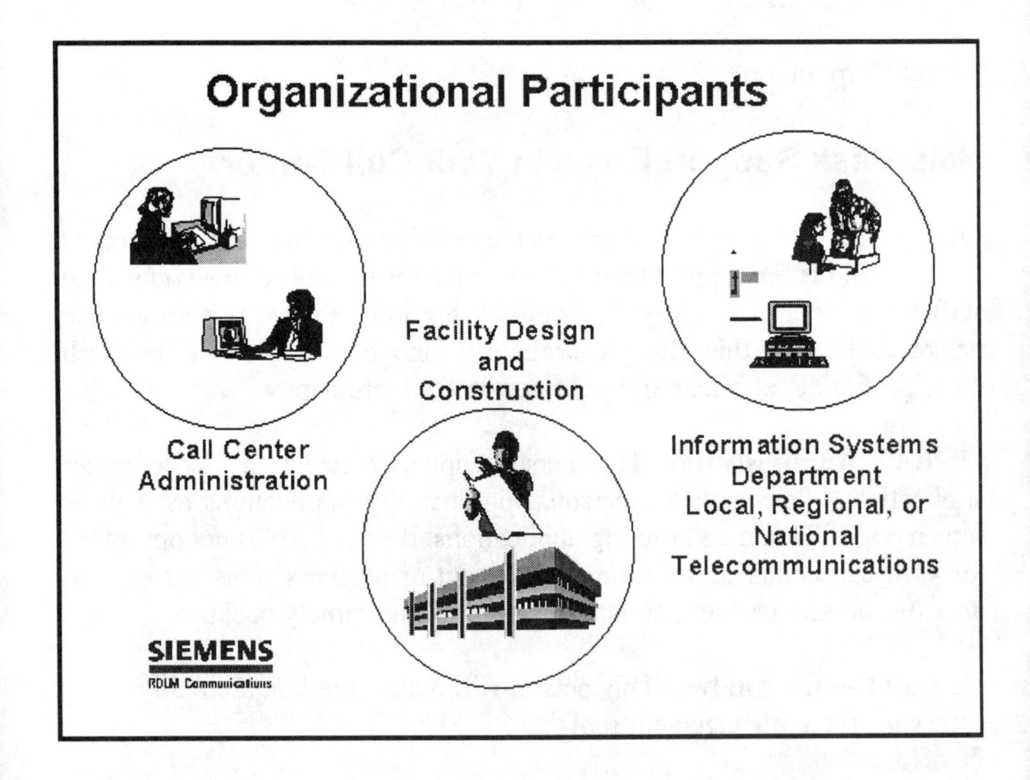

Home Agent Planning - Debunking Myths

Myth 1. At-home teleworking agents are not "real" call center agents. Work-at-home agents can now have a complete complement of call center capabilities including full voice and data feature access, comprehensive system reporting and both public and private telephone network services. They are only as real as you expect them to be. Certainly, form a technology perspective, there is nothing in the way of your call center being geographically dispersed. The bigger challenge is the proper management of these individuals.

Myth 2. Anyone can work at home successfully. Not everyone is cut out to be a good teleworking agent. The key attributes of a successful teleworker is someone who not only can work independently, but someone who is self-motivated and self-disciplined. It is also important to be computer literate, since the technology required for teleworking revolves around computer telephony.

The individuals you choose should also be comfortable working outside the social context of your office. This means the individuals you pick must be able to make decisions without a lot of consultation and be overall effective communicators.

Myth 3. A teleworking program can be implemented overnight. The path to teleworking success is careful planning. You need to have clearly defined goals and criteria for success before you begin. Well-suited applications are also important. Of course, if the job requires physical contact and paperwork, home agents may not really solve your problem. It also take plenty of time to painstakingly set the criteria for participants. Choosing the right people is often mo0re difficult than the technology and information planning. You should also chart the cost-benefit estimates and get your colleagues to fully buy-in before trying to do teleworking.

There are still other considerations that all add up to more time investment. Remember that teleworking touches all areas of your corporate culture. Will it be in keeping with your management philosophy and policies?

How do you account for organizational behavior, ad hoc meetings and social events? How will you have to adapt your company policies (human resources and legal). How will this all impact your telephone bill and your facilities security?

Myth 4. Teleworkers have it easy. Not so. At-home agents must be prepared to solve common problems on their own. They do not have the luxury of walking down the hall to get face-to-face assistance on a problem. They can suffer from a sense of social isolation and stagnation. In fact, cabin fever can set in or the agent can become obsessed with compulsive overwork. This has an impact on their family, as well.

Teleworking tips for at-home agents:

- **Communicate, communicate, communicate**
- **Alert others when help is needed**
- **Ensure home office is ergonomic and suited to work habits**
- **Find creative ways to nurture relations from afar**
- **Take regular breaks**

Myth 5. Teleworkers are difficult to manage. This is not really true. Because of the nature of call center operations, agents are more a natural for teleworking than other employees. Call centers are designed to track and monitor individual performance and do not require face-to-face interaction. Good call center systems provide data on all agent activities, and provide easy access to help (conferencing and monitoring with supervisors).

How to set up managers for success:

- **Insist on proactive, increased communication with telecommuters**
- **Commitment to ongoing monitoring and timely feedback**

- **Establish objectives and manage by results**

- **Think along non-traditional lines; be flexible**

Myth 6. Work is something we all "go to." Work is something we do – anywhere. Work is not a place, but rather a state of being and activity. There are case study examples (see Aspect) of customers who have easily expanded their call center operations by creating "virtual" call centers using at-home and satellite agents. This enables flexible working conditions and superior productivity for the right individuals.

(Aspect Telecommunications)

On-Line - Call Center Strategies

If disaster strikes your call center, you can be prepared for the worst. Consider the critical links in your enterprise and make sure that the most important components are "safe." Consider:

1. **Power supplies for desktops and server.** Make sure your file and communications servers have UPSs. They cost little and they are worth their weight in gold if you lose power. Don't forget peripherals such as LAN switches and other control elements attached to the LAN. If you're concerned about cost, pick-out some of the most important desktops and make sure they at least have a few hours of battery back-up.

2. **Traffic rerouting.** Talk to your local phone company and long distance carrier about alternate routing. In some cases, you may wish to consider splitting your trunks between several carriers. You can also make arrangements for the forwarding of calls to another call center if you are in multiple locations.

3. **DC switch with four hour minimum backup.** Ever notice how you phone still works in a power outage at home? That's because phone companies use DC power to their switches (fed by wet cell batteries).

You can do the same with your ACD or PBX. Talk to your supplier about DC options for the switch.

4. **Multiple servers.** If your enterprise data is important, plan on having more than one server. You can back-up the standby server manually or with software routines. A RAID (Redundant Array of Inexpensive Disks) subsystem is a good idea for your server, but if the server is down, no amount of RAID is going to keep you on line.

5. **Multiple CTI links.** Computer Telephone links between your computer system and your switch are simple physical connections. They can break or get severed by mistake. You can have more than one. You can route the wiring along separate paths. It's inexpensive to do – you just have to plan for it.

6. **Hot standby units.** This goes for Desktop PCs, Severs and Switches. It's expensive, but so is the loss of business you incur when you are down. At the very least, consider having critical spare parts on hand (at your site).

7. **A way to make trunks busy in case of shut down.** Make sure you can busy-out your trunks in case all else fails. If you have an analog system with regular tip and ring circuits, you can physically busy-out the trunks by breaking the line. You can also call your phone company and have them do it in the central office. It's better to give a busy signal to callers then having them hear a "ring no answer."

(Harris Digital Telephone Systems)

Screen Pop Test

Many organizations are deploying Caller ID applications that display the customer's data just as the voice call rings on the agent's station set. This is known as Coordinated Screen and Call Answer -- "screen pops" to the rest of us. It the most popular way to use Caller ID information. Do you need your screen popped today? Here are 10 questions to consider. If you answer yes to any of them, you need screen pops:

1. Should I be treating some callers differently?

2. Are my regional campaigns effective?

3. Should I route some calls to specific agents?

4. Do I want a closer bond between specific callers and specific agents?

5. Can I verify the caller's security before I answer?

6. Should I redirect this call to an IVR system?

7. Do I have customers who should never be answered by a machine?

8. Would I like to call back callers who abandon?

9. Would reducing average call length by 15 to 25% be good?

10. Can I up-sell the caller? Would shorter calls allow that?

(TKM Communications)

Agent Scripting

When agents handle several types of products, you need something to display accurate scripts for each call. Most scripts are interactive with the agent and they enter the caller response into the script program.

The script should change based on several criteria. Some call centers are set up using DNIS. This lets a service bureau cater to the needs of multiple clients or products or campaigns. The system picks up the number the caller dialed to get into the system and queues up the call based on that number. Inbound call routing can also be performed using the line or channel number the call arrives on.

(Computer Telephony Magazine)

Staffing Levels For Call Centers

Matching staffing levels to call volumes is an imperfect science, even with the assistance of complex Automatic Call Distribution (ACD) scheduling packages. One inevitably encounters the dread Call Center Paradox: you can maintain high service quality with a lot of agents, but then you have low agent productivity. Lower the number of agents to boost individual productivity and your queue grows too long and you diminish service quality. What to do?

Blending. Perhaps the most innovative idea to resolve the Call Center Paradox is *blending*, which actually combines inbound customer service with outbound telemarketing or collections.

Let's say you've hired lots of agents but there's a lull in inbound activity around the traditional 3 PM "slack-off" time. If your agents are also trained to handle outbound telemarketing or collections, you can keep them busy with a system that continually monitors and evaluates incoming call volumes and transitions agents to "outbound mode" when necessary.

If your system permits it, you can even have your outbound agents leave messages on any answering machines they encounter, since incoming calls can now be handled.

Thus you can keep a lot of agents on staff (maintaining service level quality) yet maintain agent productivity. Agents actually tend to like blending since they aren't restricted to performing the same repetitive, boring task all the time. Call blending comes in two forms. Reactive blending automatically switches pools when there is an overflow. Predictive blending will automatically switch pools during pre-defined business hours. If your system can predict patterns, go with predictive blending. If there's no rhyme or reason to your calls, go with reactive. There is a distinction between "Dialer-centric" blending (where the Dialer takes over for the ACD as inbound calls are routed to the dialer to be handled by a dialer agent), and "ACD-centric" blending where an agent is dynamically reassigned from the dialer to the ACD or from the ACD to the dialer as required by inbound / outbound call volumes.

ACD-centric is better from the standpoint of better queuing, better accessible ACD functions (like DNIS and ANI) and better CT integration. The only expense of call blending is training agents in both areas, unless you have a system that can include / exclude agents from groups depending upon their attributes.

The only problem with blending appears to be one of office politics -- managers of outbound services dislike their agents' outbound activities being usurped by the specific goals of inbound services when the incoming call volume increases. Ultimately, however, the bottom line will also force a "blending" of management styles.

(Davox International)

Wireless Phone Considerations

How come most of us have wireless phones at home, on the road and in the friendly skies, but we're still stuck with constricting wired-up phones at work? A look at the obstacles and today's solutions.

Capacity. You need to support many conversations within the system and behind the switch.

Interference. Interference among multiple base stations.

Cost. Wireless phones have been more expensive than office phones.

Integration. Working with the existing office communication system is key, but in ways you haven't thought of yet.

Having access to all of the switch's basic phone features is one thing (a very important thing). But mobile workers also need to be integrated with CT-type apps. Most people who don't move around a lot have the luxury of the integrated phone and computer telephony desktop. Mobile workers don't.

Features like voice-mail and e-mail message indications are critical for roaming users. And the next step could be integration with emerging media processing technologies like text to speech, so that you can read text message headers on wireless LCD screens.

Site planning. Installing wireless so it reaches all nooks and crannies means site preparation and inspections. Certain building materials, like metals, will interfere with wireless. After testing wireless systems over the years, we know first-hand that Manhattan, for one, with its clogged airwaves, can be a real drag for some cordless technologies.

Security. This has always been a major concern. Since the early cordless telephones, there have been considerable problems with people listening in on others' conversations or using other people's frequencies. In business, encryption is essential.

The number one benefit is mobile employees won't get mired in voice-mail hell. It's bad for the callers. And it's bad for the users, who have to rifle through 35 messages at the end of every day.

With Caller ID and usable displays on the phones, you can also make real-time decisions on who you want to talk to, just like you can with many of the emerging screen-pop and unified messaging products. Another really nice benefit is for people who work in multiple offices. Executives could move from domain to domain and use the same phone sets.

MIS and IS staffers who are constantly on the move need it. Hospital workers like nurses and technicians need it. Paging isn't always enough. Having a supply of wireless phones for VIP visitors is a great way to treat customers.

So far, only a small percentage of phone-system vendors offer wireless solutions, though some outside vendors, most notably Uniden, are making office wireless a reality by bringing it to market in a non-proprietary way.

They all work pretty much the same way schematically -- Phone systems connect to base stations (or cells) which in turn provide the wireless call coverage for the phones. When there are multiple cells (which is usually the case) some sort of controller generally added to the mix. Good news: Since we're talking a private network-type thing, there are no $1 a minute for airtime.

(Harris Digital Telephone Systems)

Workforce Management With Skill Based Routing

Traditional Call Handling. With most ACDs, calls are routed to a split or gate and then answered by an agent group as pictured in figure 4.9. Calls may overflow among groups, however, no consideration is given to the specific agent's skill. This ties-in to the traditional approach to call center staffing. Most of the time, call center managers determine the workload based on the combined work volume generated by the gates/splits. Managers typically staff to meet a service goal using Erlang-C queuing theory. Unfortunately, this does not take into account the increased holding times and abandoned calls due to unskilled handling of callers.

Figure 4.9 - Basic ACD Call Routing

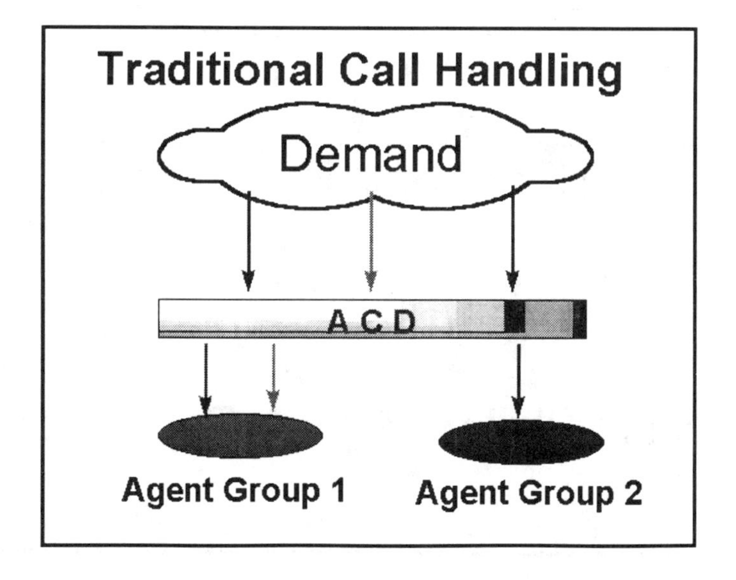

For example, a call could come in for a Ginsu knife only to be handled by someone who has no knowledge of the Ginsu knife. The caller either muddles through with the agent or asks to be transferred to someone with the appropriate knowledge.

Skill Based Routing Environment. Intelligent call switching technology is required for true skills-based routing. When calls are presented to the ACD, the calls are routed to agents based on a skills routing table for that call type. Call types are identified with DNIS or dedicated trunk identifiers. Some agents may be skilled to handle multiple call types. See figure 4.10.

This is more efficient than traditional routing methods and much more service-minded. With this approach, callers are matched more closely to the agent or agents that can handle the call in a more efficient and predictable manner. After all, predictability is the whole key to good call center staffing.

Figure 4.10 - Multi-Skilled Workforce Model

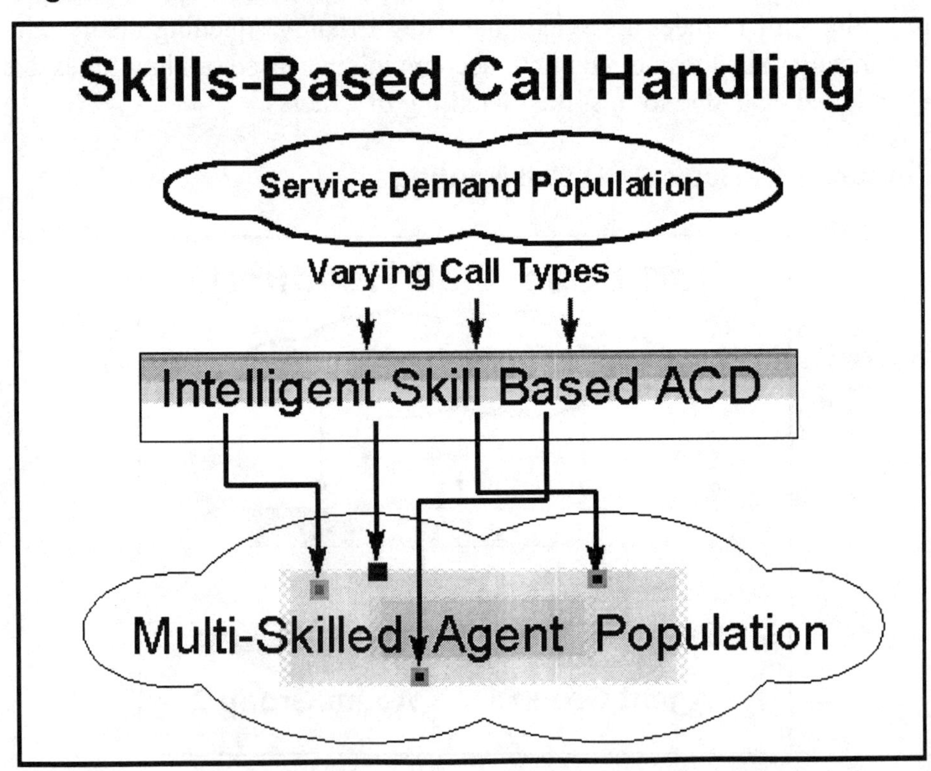

There are several good vendors of workforce management software. But before you pick a vendor, prepare your a knowledge base first. The Workforce Management approach to skill based scheduling involves taking an employee skill inventory an then pooling multi skilled employees. First, you need to assess the skills of your team and catalogue the following:

1. **Multi-lingual Agents.** Find out how fluent your staff is in speaking and understanding multiple languages. Bilingual or multi-lingual agents should be identified in your skills database.

2. **Problem Solvers.** Every call center has agents who are better at problem-solving than others. Make sure you account for this in your database.

3. **Subject matter expertise.** Make sure that every service and product your agents are representing is mirrored in the database. Their skill level on each of these should be reviewed monthly in order to update the database.

4. **The Quick and the Friendly.** Each person has their own style. Some are good at quick, informational calls. Others are well-suited to the "consultative" call. Make sure you identify who these people are so you can match their "people skills" with the job of helping callers.

Skill based forecasting. Workforce Management software helps you to collect data by call type and then estimates the work volume required for each call type. It helps you to determine the occupancy for each call type and combined call types. Once this data is collected, you can establish staffing bounds (minimum and maximum bounds). Your staffing bounds are dynamic. Some of the things causing the changes in staffing bounds include time of day, nature of campaigns, and skilld of your agent pool.

A good skills-based routing package will use a scheduling algorithm that considers your workforce database meets combined skills requirements while matching the distribution of requirements by skill. The best packages optimize based on a user-specified priority scheme. The best vendors will help you to simulate and forecast staffing requirements given a pre-set schedule and routing configuration that you specify.

You can use simulation software to determine expected results based on the schedule and ACD routing configurations.

Uses of simulation techniques. Simulation software helps to calculate your overall staffing and scheduling requirements. You can determine the total employees required for the call center (accounting for skill mix). Good software packages also handle current day schedule management, because of the effects of schedule changes on service. ACD configuration planning is also important, because switch upgrades (and when they are done) will have a profound effect on your schedule and service to customers.

(Cybernetics Systems International)

Chapter 5

Platforms For Computer Telephony

The operative idea in choosing the right computer telephony platform is to "go rugged." And why not? Folks easily spend $35,000 on a rugged off-road vehicle. Why? Because it's made to handle worse-than-average conditions. The same goes for industrially-hardened, ruggedized, heavy-duty, fault-tolerant, fail-safe and high availability PCs. If you are ready to launch a computer telephony-based service - don't waste your bright, new shiny application on a regular PC. Look into the "rugged" ones for the sake of your business.

Let's face it: your computer telephony system could be either save you hundreds or even thousands of dollars a day in reduced operating costs and a lower phone bill. It could also be generating lots of revenue. Why would you relegate such a function to that of a lowly, $1,500 PC? No matter how you describe them, industrial-grade computers offer huge advantages over rinky-dink desktop clones. They offer NEBS compliance (for CO-based enhanced services), easier servicing and upgrading, oodles of slots and much better reliability.

PCs have made their way into every corner of business today - and most importantly to those of you reading this book - into telecommunications. Let's look at the essentials in selecting a PC system. Let's also take a look at the basic difference between desk top and industrial / specialized PCs.

PCs have become popular because they are easy to use. They are also inexpensive, flexible and there are plenty of applications from multiple vendors. This includes custom applications for everything from call routing to Internet servers with call center links to unified messaging.

Rule of thumb: An industrial grade computer is five times as reliable under "stressful" conditions as a regular PC. No wonder that more and more smart developers are building their mission critical computer telephony applications on these beasts. CT demands reliability. It's a natural fit.

Buying Industrial PCs

According to the April, 1995 issue of **Measurements & Control** magazine, Value Added Resellers and systems integrators account for about 75% of industrial PC sales. The other 25% are intrepid individuals who buy system components from here and there and create (or attempt to create) their own systems. If you want things like NEBS (Network Equipment Building Standard -- an intense telco equipment standard) compliance so your PC can go through an earthquake and a fire and keep answering phone calls, you've got to pay for it.

Says Bruce McGrath of Industrial Computer Source: "Our PCs are used as a central office platform. The key to NEBS compliance is the flame retardance. You've got to find just the right materials." The main reason people don't have a lot of NEBS machines is because of the cost of testing, not the cost of construction.

Once you get into construction there is certainly some extra cost, but by the time you get through NEBs testing, depending upon how well you designed the unit the first time through, it takes about three or four months and can cost $100,000. That's a real barrier to market entry for PC makers.

Michael Vaughn of ICS, told us: "As far as mean time between failure (MTBF) goes, a fully ruggedized industrial PC system should be good for 100,000 to 150,000 power-on hours (POH) versus 15,000 to 30,000 POH for a good quality clone. This is due to more robust components and extra cooling and filtration which, in dissipating excess heat, cause system components to last significantly longer." Here are some other things to consider:

Servicing. When a conventional PC really breaks down badly, it can take two hours to replace the motherboard, since you have to take the whole system apart. High availability PCs have passive backplanes and front access doors that allow major system components to be swapped out in a few minutes.

Upgrading. The same gizmos that let you quickly replace system components also let you quickly upgrade the system. Need a 133MHz Pentium card plugged in right away? No problem.

Lots of Slots. A typical PC clone motherboard has about eight slots, whereas an industrial grade chassis has up to 22 on its passive backplane. You could in theory have 30 slots, but impedance problems and signal interference would cause difficulties. Even 20 slots lets you run big CT applications with lots of voice/fax cards without resorting to a room full of PCs.

Keep it Cool. Few people realize that the quickest way to shorten the life of any PC's electronics is to let heat build up in the enclosure. Heavy duty PCs can have up to six cooling fans and washable filters to keep dust (and heat) from building up.

(Computer Telephony Magazine)

As you can now imagine, the biggest mistake most commonly seen is that businesses treat PCs as if they were all the same. Thousands of man-hours can be spent on specialized applications that generate good revenue, and still PCs are too often an afterthought. The truth is that all PCs are *not* the same.

Consider the fact that your application *is* your service. Consider also that failures reduce your effectiveness in delivering service. Since PCs are part of your application (most likely), it would only seem logical that your PCs should be:

1. **As stable as your application.**

2. **Tailored for your application.**

3. **Planned into your application.**

The basic essentials to look for when making a PC purchase decision are hardware flexibility, space efficiency, serviceability and reliability. All of these thing taken into consideration will allow you to minimize costs and down-time and thus maximize service to your customers.

Desk top PCs were built for desk top applications. They are built to handle a typical eight-hour work day typified by the use of spread sheets and word processors. These units either sit on a desk or on the floor next to it. But when you consider the use of standard desk top PCs for commercial or revenue generating applications, they fall short in a number of ways. You'll notice instantly that they are not as flexible. Many of them are actually proprietary or have only a few expansion slots. Some desktop (and laptop) makers are constantly changing their specifications. Many of these machines are bulky and hard to rack or stack.

You will also notice that issues of reliability crop-up almost instantly. Most desktop PC manufacturers use the lowest cost components and their designs are not "fault tolerant." This is because it is very competitive in the desktop market and the demands on these PCs are less stringent. Desk tops were made for desk top applications, and they serve that purpose well.

But revenue generating applications need Industrial PCs that are especially tailored for high-demand use. Lets review each of these essentials as they relate to industrial PCs.

Hardware flexibility. Industrial PCs in most cases have standard passive backplane architecture (ISA, PCI, EISA). These are non proprietary architectures that are easy to find components for and allow for lower cost upgrades from a variety of vendors.

High space efficiency. Industrial PCs also have more slots / more flexibility. You can cram in as many as 20 boards into one unit. This is done routinely for high density applications. Many industrial grade PCs allow you to customize for your application. For example, you can get units with split backplanes (separate hosts in one box) and they are also available with different rack-mount kits (slots in 19” rack space). If you will be using multiple PCs in your application, space efficiency is a major consideration. Don’t make space an afterthought - ask yourself these questions ahead of time:

1. **How many PCs will we need?**

2. **How much space can we afford?**

3. **What happens if we grow faster?**

The most popular & efficient form of storage for PCs are rack systems. Some of the most common configurations are 19” or 23” x 7’ racks. You can also get custom racks. It’s common to fit as many as six to 32 PCs in one rack depending on what platform you are using. You know what your computer floor space costs you -- do the math - its part of your application’s cost. A real life space problem with standard desk top PCs that Crystal was involved in was at MCI Telecommunications. See figure 5.1.

MCI’s needs expanded, so this forced the issue of space being unavailable for their expansion. Their solution was to use small industrial PCs or multiple PCs in 1 box from industrial PC vendors. They used crystal’s PCs and got a 4 to 1 space savings. They didn’t need to move. They didn’t need more space -- they just made better use of it. Only industrial PCs give you this kind of space efficient capability.

Figure 5.1. - How Big Companies Use PCs

Crystal Group Installation
at MCI Telecommunications

High serviceability and predictability. With industrial PCs you can get better product consistency. Components change less often, so your applications require fewer modifications. Industrial grade manufacturers also pay more attention to "rev" control and are more likely to give you prior notification of changes.

This allows you to make proactive development decisions which, in turn, help you to save money. This predictability also reduces unanticipated interruptions and additional development costs.

High serviceability is critical to maximizing up-time --and because you've chosen PCs as your platform, you are likely to have some problems... so when they happen, how fast can you get them fixed? Consider how easy it is to install or remove components. How will you rack-mounted PC attach to the rack? What is the weight and size of the PCs? What about the cable connectors?

Figure 5.2 - Quick Disconnect Connectors

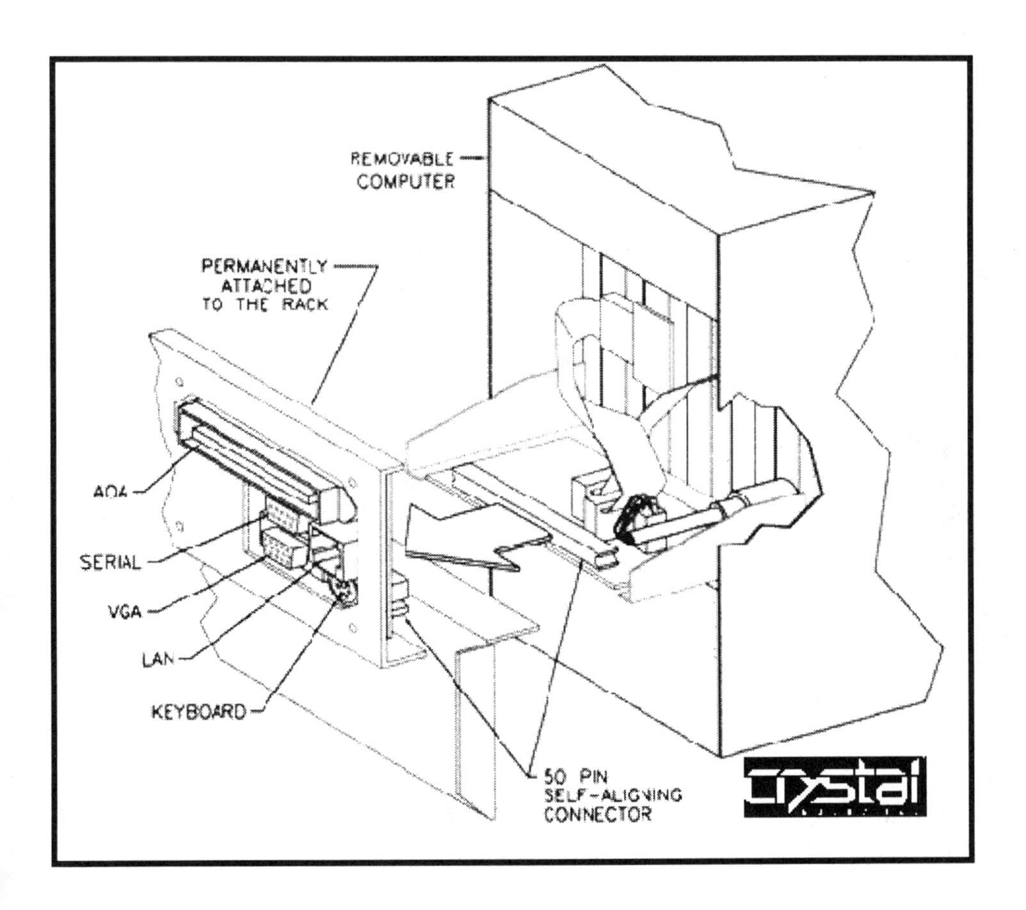

The connector shown in figure 5.2 has all cables rigged into one "quick disconnect" connector. The unit can be bolted or shelved. The example here can be configured at 4.5" to 19" wide. It is optimal to slide the PC in the rack or pull PC entirely out of the rack with ease (reduce installation & service time). It's also best if there are no cables to connect or disconnect (eliminates cable failures).

Once the PC is out, how easily can it be serviced? Can you easily get to the individual components? Industrial PCs have easy access to - cards, hard drives and power supplies. They are designed to be easy to field service. This minimizes service time / cost, and helps to maximize up-time.

High reliability. With industrial PCs you get power options. Besides the standard 120/240VAC, you get options for -48 VDC & +24 VDC. DC power is common in central office environments. You get better and more reliable power efficiency and uninterrupted power flexibility with DC power. That's why telephone companies use it.

Industrial PCs have up-time designed in. They are made with industrial grade components and are therefore more stress tolerant. They receive a greater amount of testing. Some are also equipped with watchdog timers so alarms are tripped if there is a failure. Of course, they are made to run longer without failure, as well.

Look for units that look like they are packaged in a rugged way. Look for steel or aluminum "skin." Look for greater cooling capacity. A good industrial grade PC should have extra fans, vents, filters, etc. to handle more heat. All of these attribute ad-up to units that are made to run around the clock.

Some industrial grade PCs have redundant components, like power supplies & hard drives. Redundancy improves up-time. Basically two ways to go about PC redundancy or back-up:

Partial back-up (back-up power supplies; back-up or "hot swap" hard drives) or full back-up (complete redundant PCs).

If you want to insure that your app is up all the time, and you want to use PCs, then you must provide for full, multiple PC back-up or "multiple redundancy." In summary, industrial grade PCs are more flexible, more space efficient, easier to service and more reliable.

(Crystal Group, Inc.)

Mission Critical PC Platforms

Is your application mission-critical? Here are some characteristics that identify applications as mission-critical:

High port count. Higher density boards enable more ports in a single system. Suddenly a single PC can provide unified messaging to 500 or more workers. Scary? It is if the app is running on a commercial or even an industrial grade PC.

Multiple applications. Few of today's applications are restricted to a single function. A PC-based single system may be used for auto attendant, voice mail, IVR and FAX-on-demand. The more essential functions provided by a single application, the more high availability becomes essential, even if there are relatively few users.

PC-based switching. The trend is for the PC to become a telephony server in a client server architecture. Increasingly, the PBX or other switches are being incorporated into the server as line cards.

In such cases, the PC becomes an integral part of the telephone system. Does an organization tolerate and expect an occasional LAN crash? Of course. Does the same operation tolerate and expect the phone system to go down? Absolutely not (unless you live in certain parts of Europe or South America)! Large Service Bureaus and Central Office applications. Service bureaus and CO applications (like residential voice mail) have the highest port counts and often service thousands of users. Here the need for fault tolerance is great and widely acknowledged.

The PC is already making headway there, largely due to the availability of specialized fault tolerant PCs. If your application encompasses one or more of the above characteristics, you should consider a heavy-duty, higher availability PC platform.

(I-Bus PC Technologies)

Mission Critical PC Benefits

High availability PCs are distinguished by a range of features that include: Hot-swappable power supplies, RAID and hot-swappable disks, system monitoring, live insertion of I/O boards, and special chassis designs. Each one either tolerates, prevents, or predicts failures in the system. High availability PCs incorporate some or sometimes all of these features.

Redundant and Hot-swappable Power Supplies. The power supply subsystem is often the most likely source of failure in a PC. Some PCs are equipped with redundant power supplies where one supply takes over from the other in the event of failure, allowing the system to continue operating.

However, once a supply fails, the system needs to be brought down in an orderly fashion and the chassis cover removed to replace the failed supply. Hot-swappable power supplies let you replace the failed supply while the system is running, increasing uptime and simplifying replacement.

Hot-swappable power supplies also come in N+1 (need + 1) configurations. N+1 architectures share the load across two or more power modules and enable the output power to be scaled up (by adding modules). These are found in higher end platforms.

Load-sharing hot-swappable power supplies extend the life of the power supplies by running each individual supply below its maximum output. The electronics of load-sharing power supplies are relatively complex, representing a potential single point of failure and adding to system cost.

Non load-sharing hot-swappable power supplies have one power supply standing by while the other supplies the majority of power to the system. The electronics are simpler and their cost lower. Some PCs offer both AC and DC versions as well as modules with different output wattages.

RAID and Hot-swappable Disks. The drive subsystem is another area where conventional PCs have been relatively weak. RAID (Redundant Array of Independent Disks) is a proven, relatively mature technology available for off-the- shelf, commercial IDE and SCSI hard drives.

RAID eliminates downtime and protects data by being able to hot swap drives and reconstruct any data contained on a failed drive. In some cases it can improve performance. Most applications can be upgraded to RAID without having to change the operating system or modify the application software.

A typical RAID solution in a high availability PC contains a set (two to five or more) of drives and an ISA or PCI-based controller. Some operating systems (like OS/2 and NT) can do simple mirroring, eliminating the need for a controller.

The 3 1/2" one-third height drives are usually housed in removable 5 1/4" half height canisters. The canisters are also available in hot-swappable versions that manage the data bus and isolate devices from the bus. The hot-swappable versions are not needed with many RAID controllers because the controller performs these functions.

RAID solutions are also available as separate tower or rack-mounted subsystems that are cabled to the computer.

The most common RAID configurations are 1, 3, and 5. RAID 1 (disk mirroring) is the simplest and most common approach. Data is written to two disks that operate in parallel. If either of the disks fail, the other takes over. Once the failed disk is replaced, an identical copy of the data on the functioning disk is reconstructed to return to 100% redundancy of data.

Although providing a high degree of data availability, this is a relatively expensive approach because it requires twice as much storage as the data requires. In contrast, RAID 3 and 5 distribute (stripe) data across a number of disks, often five, using sophisticated redundancy algorithms that utilize parity data.

RAID 3 breaks up the logical record into bytes, requires a dedicated disk for storing parity information, and enables a high data rate because the data can be accessed from different disks in parallel.

RAID 5 stripes blocks or records of data, rotates parity information across all the disks, and enables a high transaction rate because separate I/O requests can be serviced concurrently.

System Monitoring. System monitors are intelligent cards that observe various vital signs of the system and perform related activities such as resetting unattended systems and alarming.

The monitors anticipate and detect failures, enable orderly reboots and shutdowns, and increase the mean-time-to-repair. They increase system availability and prevent many common causes of catastrophic failure -- such as detecting when a system begins to overheat and shutting it down, rather than having it crash and burn-up.

System monitors range widely in their design and functionality. They can be ISA cards, standalone cards or rack modules with their own power connector, and as built-in features on CPU cards. They typically have an LED display integrated in the chassis or module. Some have the display housed in a separate package that fits into a 5 1/4" half height drive bay. The simplest system monitors are enhanced watchdog timers. More sophisticated units also monitor temperature, power, voltages, air flow, and fan rotation as well as send alarm signals to hard relay contacts and/or over a serial connection to a remote terminal. The most sophisticated monitors have their own microprocessor and advanced features such as multiple levels of alarm, user programmability, Network Equipment Building Standard (NEBS) compliance, automatic switchover to a stand-by system, multichassis support, LCD display, and push button user interface.

Live Insertion of I/O Cards. A historical limitation of the PC architecture has been that it does not support live insertion; this is about to change.

It is possible to isolate I/O cards from the backplane using a special purpose board seated between the backplane and the card. This board enables and disables signals and power via a manual switch or software control. This board can also be used to isolate a secondary switching bus like MVIP.

Carriers attached to the cards enable easy insertion and extraction while the system is live. Since the ISA bus doesn't support multiple masters, live insertion of CPU boards is not feasible. Also, live insertion of PCI cards is not feasible either because of the bus specification's strictness.

Chassis Considerations. There is more to the design of a high availability system than having fault tolerant features, system monitoring and live insertion. Chassis design considerations like cooling, segmented backplane support, and accessibility also come into play.

The engineering of the chassis' cooling is critical because of the heat emitted from the high number and density of boards in many systems. Adequate pressurized airflow is important and many designs benefit from multiple fans. Some designs incorporate hot-swappable fans or fan trays.

So-called segmented backplanes allows for multiple systems in a single chassis. This not only distributes processing but also saves on cost and rack space.

Power-sequenced segmented backplanes are ideal in systems where two or more systems reside on the backplane and need to powered up or down independently, perfect for applications requiring multiple servers or redundant systems.

Whether a rack mounted chassis is accessed from the front or top affects the ease of repair and maintenance. Conventional top-loading chassis save vertical rack space, but are difficult to service because they have to be pulled out of the rack and the top cover unscrewed.

Front-loading chassis, built primarily for central office applications, are accessed by opening a hinged door. The cards are oriented vertically and optionally equipped with mechanical carriers.

Distinguishing High Availability PCs. There are several ways of distinguishing high availability PCs. A few are listed below. You may find these characteristics helpful in selecting the right type of PC platform for a given application and in comparing platforms of a given type:

Packaging. PCs with high availability features come in desktop, tower, conventional 19" and 24" rack mounted, and front loading packages. All of these vary in their physical dimensions. Desktop and towers are oriented toward medium to large Customer Premise Equipment (CPE) applications.

Conventional 19" rack mounted systems are suited to larger CPE and service bureau applications, while 24" rack mounts and front loading platforms are typically installed in central offices.

NEBS compatibility. NEBS is a broad range of Bellcore specifications commonly embraced by the telcos. NEBS as it pertains to computers includes standards for: form factor, flammability, system monitoring, cooling, temperature and humidity, grounding, shock and vibration, EMI and ESD.

Chassis from a growing number of vendors claim NEBS compatibility. Their equipment ranges from meeting a small subset of requirements (e.g. system monitoring and flammability) to more or less full compliance (and are thus NEBS certified). The telcos vary greatly in the degree of NEBS compatibility they need for any given application.

The Feature Set. Some high availability PCs only offer redundant power supplies. Others may have hot-swappable power supplies, but no hot-swappable drive option. Many do not even offer system monitoring capability.

Truly full-featured, high availability PCs offer hot-swappable power supplies, hot-swappable drives and system monitoring. Each system then varies as to the capacity and capability of each of these subsystems.

Scalability. Scalability refers to the ability to add fault tolerant features to the system. This enables the platform to be configured or upgraded to the degree of fault tolerance demanded by a wide range of applications. Some systems let you add only a few features. Others come fully loaded. Ideally, the base platform comes with a minimal feature set and the ability to add options.

Modularity. Most high availability systems are monolithic; all the subsystems integrate into a single mechanical package. At the higher end, some systems are modular: each subsystem is a separate module. Modular systems have added flexibility in configuration, more options, and it's easier to upgrade portions of the system.

(I-Bus PC Technologies)

PC Buses, Fans And Slots

EISA VS. PCI Buses. Don't got near EISA cards anymore. PCI is much better. PCI is four times faster. PCI is easier to install. PCI does Plug and Play. PCI is the standard everyone is following.

EISA cards are hard to install. If you move them from one slot to another, you have to reconfigure your PC with -- of all things -- an external diskette utility.

Card makers are ramping-up PCI production, but still have stock of old EISA boards they need to get rid of. Your wait for PCI versions could be four to six months. If you buy EISA now, get a guarantee that EISA card will be replaced by a PCI one later for free, or for cheap.

Fans And Their Height. When choosing a computer for use with voice systems, be aware of the height of CPU cooling fans and the location of local bus cards in the standard system. CPU cooling fans add height to the CPU and often will interfere with a two to three full length card slots. Since many voice products still use full length cards, it could impact system expansion options.

Local Bus And Cables. Many PCs come with local bus combination controller cards that perform several jobs with a single board. The downside to this design is that the local bus slots are almost always on the outer edge of the motherboard and any cables from these slots will have to either go under or over all the cards in between the board and the devices. If your cables are not long enough to twist and lie flat under the middle cards, you may not be able to use the system for full length voice cards.

Number of Slots. Internal PC card converters make sense only if your computer has available slots. If not, you will need an expansion box or an external converter.

(Computer Telephony Magazine)

PC Switching Configuration Tips

Many useful systems can be built using MVIP, H-MVIP, PEB and SCbus architectures for shared resource switching. For medium-density applications, you can use less than the full H-MVIP interface. Boards that implement the full interface are referred to as "compliant" while boards that implement less than the full interface are referred to as "compatible." The difference is time to design and manufacture – and of course: cost. If you see "compatible" MVIP cards, they cost less because the entire chipset is not being used.

Extensive experience with MVIP-90 suggests that even when the complete interface is available in a single integrated circuit, the economics of some applications benefit from implementing a subset of the interface using a minimum of gates within an application-specific IC.

Applications with a total required system capacity less than 512 time-slots or 32 Mbps aggregate bandwidth should use MVIP-90, not H-MVIP. In addition, devices that must work within high capacity systems, but themselves require only limited interface bandwidth, may also use an MVIP-90 interface. Note that devices with MVIP-90 interfaces may be used within an H-MVIP system by operating one or more of the H-MVIP data streams at 2 Mbps.

Overall, the need for H-MVIP was driven by changes in the density and complexity of the computer Telephony market. As end users clamored for more capacity per box, whether for multimedia conferencing, multi-media servers or video servers, the need for a higher capacity integration standard bus in a box has emerged.

From the systems perspective, critical components in the PC have arrived, including higher speed ISA bus follow-ups such as EISA and the PCI Bus. Coupled with H-MVIP, new EIDE drive technology, Intel Pentium processors and newer operating systems like Windows NT, these changes make a formidable platform for the dense and complex of CT's near future.

(Natural MicroSystems)

Chapter 6

Component Magic Tricks

Alas, not all components are "plug and play." In fact, there's plenty of gadgets out there from DTMF pads to ROMs to PBXs – that you'll have to deal with when you install a CT system for yourself or someone else. The industry has made great strides in taking the "black magic" out of boards and software alike, but there's still lots of quirky stuff out there.

The challenge is not so much getting boards to work inside your favorite PC, but rather: how to get *everything* to work together. Perhaps the biggest challenge is to train your automated attendant or IVR system to understand busy or disconnect signals behind a PBX, or how to understand paging terminal tones. Each major computer telephony board maker has software that will help you to "train" your system for these situations.

This chapter takes a look at the most popular fixes for some of the most infamous problems. This includes BIOS conflicts, signaling parameters, ROM upgrades and debugging serial ports. The good news is that many of the component manufactures are now putting software fixes, upgrades and technical notes in their home pages on the World Wide Web.

Many of the "tipsters" in this book have home page URLs listed in Appendix B. It'll take you hours to visit them all, but you'll be impressed with the level of support available.

AEB Signaling on D/41E Boards Under MS-DOS

Software programmability for enabling AEB (Analog Expansion Bus) signaling on D/41E boards is alluded to but not described in System Release 4.1. documentation. AEB signaling is enabled on D/41E boards through the use of a downloadable voice board parameter file.

The standard JP8 on D/41D boards, used to select signaling from the built-in loop start interface or AEB, is not required on the D/41E board. Instead, a downloadable voice board parameter file must be created or modified to enable AEB signaling. The following method can be used to enable AEB signaling on the D/41E board under MS-DOS.

1. **Edit your baseboard configuration file** (DIALOGIC.CFG) in the \DIALOGIC\CONFIG directory.

2. **Find the configuration line beginning with SRB**: TYPE=D41E that configures the D/41E board that you want to enable AEB signaling. Then, check for a PARM=xxxx.PRM statement on that configuration line. This statement specifies a voice board parameter file to download to the D/41E. If the PARM statement is absent, PARM=name.PRM needs to be added to the end of the configuration line (for example, PARM=D41E.PRM). After modifications are complete, save and exit the baseboard configuration file.

3. **Change directory to the \DIALOGIC\DATA directory.**

4. **If the PARM statement was added to the D/41E configuration line** in the baseboard configuration file, the voice board parameter file for the D/41E (for example, D41E.PRM) that was specified in the baseboard configuration file needs to be created.

5. **Edit the voice board parameter file.** Parameter 104 controls AEB signaling on the AEB. By default, parameter 104 is set to 0 to enable the front-end loop start interface. For non-zero values between 1 & 15, the front-end is disabled and AEB signaling is enabled on channels selected through a bitwise value as follows:

1 - AEB signaling enabled on channel 1
3 - AEB signaling enabled on channels 1 & 2
7 - AEB signaling enabled on channels 1 - 3
15 - AEB signaling enabled on channels 1 - 4
Parameter 104 is assigned by adding the following statement to the file:

PARM[104]=x/* 0 x 15 */

At this time, AEB signaling on the D/41E board can only be enabled on channels 1 and 2.

(Dialogic Corporation)

AMI WinBios Configuration Problems

Setting up Shared RAM correctly on PC's using AMI WinBios (usually new Pentium systems with PCI bus) is a little more complicated than it might seem. Preventing Caching in the Shared RAM area to be used by a Dialogic board is not as easy as simply disabling the External Caching menu item.

This relatively new BIOS is a 4 window user interface to modify the BIOS setup information. In the SETUP window is the ADVANCED section in which you will find settings controlling Shadowing for the various Shared Memory Address ranges. As you would expect, make sure Shadowing is disabled for the address range in Shared RAM that will be used by the Dialogic board. Also found in this section is a menu item for configuring EXTERNAL CACHE MEMORY. Setting this menu item to DISABLED will cause the Dialogic boards to fail firmware download with Shared RAM conflict errors.

It is not immediately clear that you must ENABLE this option so that the menu items needed to exclude the area are made accessible. These newly accessible menu items are located in another BIOS window named the CHIPSET window.

In the CHIPSET window, the Shared RAM area to exclude can be setting the Non-Cacheable Area 1 menu item to AT-BUS. The Non-Cacheable Size menu item should be set to the size of SRAM segment to exclude for the Dialogic board in decimal Kbytes. The Non-Cacheable Base menu item should be set to the SRAM base address that you are using in decimal Kbytes (i.e. 832KB for D000).

It is also important to be sure that the PCI slots are not using the same IRQ that the Dialogic boards are going to use. Set the PCI Setup menu item to MANUAL and set the IRQs for each slot to some unused IRQ. If the PCI slots are not being used, they may all share the same IRQ.

(Dialogic Corporation)

B8ZS Line Encoding Detection

The Bipolar 8 Zeros Substitute line encoding is one of the possible coding schemes commonly used in North America and other countries for T-1 digital systems. It is used to prevent the loss of synchronization when long string of zeros are being transmitted on the T-1 line. The scheme would insert bipolar violations to the AMI encoding any time 8 consecutive zeros are being transmitted on the data stream.

Both the CPE and CO should be set for same line encoding, being AMI, B8ZS or bit 7 stuffing (BIT7). The Dialogic T-1 interface cards, as a DTI/211 or D/240SC-T1 can be configured to use either one by means of the following API function calls:

dt_setparm() ***Unix, Windows NT, OS/2***
chgglobal() ***DOS***

The Dialogic DTI/2xx API also allows the board to be notified of bipolar violations and B8ZS line encoding detection. This can be done by adding (enabling) the following bitmask to the T-1 alarm notification mask:

DTEC_B8ZSD (Unix, OS/2, NT) DTECM_B8ZSD (DOS)
DTEC_BPVS (Unix, OS/2, NT) DTECM_BVCS (DOS)

This can be done by using the following API calls:

dt_setevtmsk() with DTG_T1ERREVT *as the event mask in Unix, Windows NT, OS/2*

chgglobal() *with the* **DTG_ENERRC** *parameter in DOS*

The BPVS condition should be treated as an alarm condition on the T-1 data stream. This can be caused by noise; other transmission impairment, as well as a notification of the CO transmitting in some other coding scheme different from the DTI's one.

The B8ZS events, on the other hand should not be considered as an alarm condition, but rather as a notification of bipolar 8-zeros substitution detection. The following rules should be followed by the end user's application to accommodate for this:

Configure the Dialogic card for B8ZS line encoding transmission:

in UNIX, OS/2, NT do:

value = DTSP_B8ZS;
dt_setparm(devh, DTG_CODESUPR, (void *)&value);

in DOS do:

chgglobal(devh, DTG_CODESUPR, DTSP_B8ZS);

Disable the B8ZS bitmask from the event mask, so that to disable B8ZS detect:

in Unix, OS/2, NT do:

dt_setevtmsk(devh, DTG_T1ERREVT, DTEC_B8ZSD, DTA_SUBMSK);

in DOS do:

chgglobal(devh,DTG_DSERRC, DTECM_B8ZSD);

(Dialogic Corporation)

D/41E-SC Under Non-SCbus Releases

The following is a set of rules for properly configuring the PEB Terminator on standalone D/41E-SC boards along with configuration file settings:

1. A PEB Terminator must be placed in the Resource position on the D/41E-SC.

2. The following O/S dependent software configuration modification has to be performed:

For DOS & OS/2:

Add **MODTYPE=NETWORK** to the existing SRB configuration for the D/41E-SC in **DIALOGIC.CFG**

For UNIX:

Specify a Type of **D4XE-PEBN** for the **d4xe** device in **.sbacfg**

(Dialogic Corporation)

Detecting Dial Tone With D/42-NS

The GTD template used on a D/41D board to detect North American dial tone does not work on a D/42-NS. The GTD template for a D/42-NS board needs to be defined as a dual tone template. When building a GTD template on a D/42-NS board use the rules that apply to the D/81A Rev 2, D/121B, and D/41E boards.

The D/42-NS Base Board includes memory enhancements which allow the D/42-NS board to detect dual tones with a resolution of 62.5Hz separation between tones. The D/41D board is limited to detecting dual tones with a resolution of 125Hz. If you wish to detect dial tone you will need to configure a template to look for 350Hz and 440Hz dual tones.

(Dialogic Corporation)

DOS Communications Software

Don't throw away your old DOS communications software (just yet). You'll need it to go off line and talk to your modem using AT commands. The most useful modem commands are ATI3, ATI4, ATI5 and ATI6 according to Mark Larson, MegaHertz's excellent technical support manager. They'll give you information about your modem, what it is, its firmware revision, what its technical specifications are, etc. Useful stuff. Easier to get at through DOS than Windows.

(Computer Telephony Magazine)

DSVD Setup

Digital SVD (Simultaneous Voice and Data) modems are hitting the market. Eventually, we will all use them. A software application is Digital SVD-ready when the user can select the Digital SVD modem type from a setup menu. This occurs using a data file that is accessed through a user interface using the following steps:

1.In the software application's setup options, select the modem setup.

2. Select a Digital SVD modem.

3.The software application issues the correct initialization string to the modem. The initialization string typically goes to the modem each time the software application is run.

The Initialization String. The initialization strings for the U.S. Robotics 28.8 Digital SVD modem and the Rockwell 28.8 Digital SVD modem are (respectively):

DSVDAT&F1-SSE=1
DSVDATQ0E1V1-SSE=1

These initialization strings are only intended for their respective modems. If a user places a different modem in the computer, then a different initialization string must be selected. If the Digital SVD Modem initialization string is used as a default initialization string for another modem, the other modem does not recognize the -SSE=1 Digital SVD Modem command and sends an ERROR result code back to the software.

Connecting to Other Modems. When Digital SVD is enabled on the original modem and Digital SVD is not supported on the answering modem, the modem proceeds with a standard, non SVD, V.34 startup sequence. A standard V.34 connection with V.42bis compression and V.42 error control is established, if possible. If V.34 is not supported by the remote modem, the software tries to connect at lower bit rates such as V.32 and V.32bis.

Voice Connection Tones. When a handset is picked up, a low-high tone is sent across the voice channel to signify the transition from data-only to Digital SVD mode. When a handset is placed on-hook, a high-low tone is sent across the voice channel, to signify the transition from Digital SVD to a data-only connection.

Two users can increase data rates to the maximum bandwidth of 28.8Kbps by simply hanging up one handset. The remaining user hears the high-low tone if their handset remains off-hook. If the handset is picked back up, the user is notified by the low-high tone, to signify the transition back to Digital SVD.

Questions and Answers. Here are the most common DSVD questions that Intel and its DSVD partners have encountered:

Q: Can you output the Digital SVD-decoded voice to the SoundBlaster (digitally)?

A: The only mapping of the Digital SAD Audio is out through the RJ-11 phone jack. No connection to a SoundBlaster is provided.

Q: How tolerant is Digital SVD to line noise?

A: Line noise tolerance is a function of the V.34 protocol. The Digital SVD connection seems no more susceptible to line noise than any other V.34 modem.

Q: Can you output Digital SVD-decoded voice to the hard drive or filter voice data on the host before encoding it on the modem?

A: It is not possible today because current modems map voice to RJ-11 phone jacks. There is nothing in the DSVD specification that prevents this from happening in future implementations. When DSVD is implemented through a software solution, as envisioned in Intel's Native Signal Processing or NSP specification, this will be possible.

Q: Can you control how much of the bandwidth gets allocated to voice or data?

A: The only way is by hanging up the handset. This allocates maximum bandwidth to the data channel.

(Intel Corporation)

DTMF Dialing Levels

Occasionally, an application may need to adjust DTMF dialing levels for a unique testing requirement. By modifying a set of firmware parameters, it is possible to change the amplitude of DTMF tones during dialing. While this is seldom necessary or advisable, there are some unique application needs, such as in production testing environments where this may be necessary. It must be noted that these parameters are not to be changed when connecting a Dialogic board to any PSTN in any country. For each country which Dialogic holds approvals in, there is a set of pre-determined parameters which must not be changed in order to maintain compliance with the PTT regulations of that country. DTMF dialing levels are among those parameters which should not normally be tampered with.

You will need to create a parameter file which is named in your dialogic.cfg file for each voice board entry you have there. Just add the path and filename of the parameter file to the end of each line.

The contents of the file will contain 4 parameters which govern the amplitude of the DTMF tone pairs. They are (including their default values):

parm[73]=28690 /* this is the lsw of the lower tone */
parm[74]=21 /* this is the msw of the lower tone */
parm[75]=30002 /* this is the lsw of the upper tone */
parm[76]=18 /* this is the msw of the upper tone */

These provide approximately -3.5 dbm output level.

Take the default values and convert them to hexadecimal. Then combine the **lsw** and **msw** hexadecimal numbers. Then convert that number back into a single decimal number **n**.

Use the following equation to calculate the new values based upon the number of dbm change you wish to make + or -.

Take **n** from step 3 and plug it in here...

$$y=n(10^{x/20})$$

Now convert **y** into a hex number, split it into **lsw** and **msw** and plug those values into the appropriate places.

REMEMBER! the calculation you are making is based on **x** dbm change from the default provided.

(Dialogic Corporation)

DTMF Generators - Acoustically Coupled

Hand held, battery operated tone generators do not reliably work with computer telephony applications. These devices are placed at the mouthpiece of the handset and buttons are pressed which created DTMF tones. These are usually small calculator sized devices. In effect, there may be no reasonable solution and using a hand held DTMF or tone generator should be discouraged. The qualification template that is used for tone detection by some of the APIs could be "relaxed" and made more tolerant. This has a technical cost and other problems could surface because the system would be more likely to sense a DTMF digit that is not present.

To date, all of the known inexpensive acoustic coupled DTMF generators present several serious problems. The two main problems consist of low levels launched into the mouthpiece and in the relative level between the two tones (TWIST).

Tone Level. When a DTMF button is pressed on a typical phone, the expected tone level is about -5 dBm. This is 5 dB below the maximum allowed full level voice signal of 0 dBm (1 milliwatt). When a hand held acoustic coupled tone generator is pressed against the mouthpiece, there is an acoustic loss between the two transducers. This is about 14 dB below the DTMF produced on a normal phone. The level is now about -19 dBm. If the generator is not pressed close to the mouthpiece or if it is not oriented correctly, the losses will be even greater.

Assuming perfect alignment and consistent pressure against the mouthpiece while pressing buttons (no easy trick), there is an additional loss of about 15 dB through the switched network to the remote end. The level received at the remote location is -34 dBm or about 1/2 of a micro watt. The level where noise and distortion start to become significant contributors to the received signal is about -40 dBm and lower. This margin of 6 dB is very close considering the variations in loss through the switched network and coupling variations at the mouthpiece. The story, unfortunately, gets worse when the effect of TWIST is included in our calculations.

Twist. A DTMF signal consists of two tones. As could be expected, the level of these two tones may be different from each other. Usually a tone at the center of the voice spectrum will have a higher level than a tone near the edge of the voice spectrum. This results from the characteristics of the transducers and transformers used in the phone. The difference in tone levels is called TWIST. Because a DTMF generator adds an additional transducer in creating the tones, even more TWIST is introduced.

Since all DTMF signals are not created equal, there is a variation in TWIST depending on the digit pressed. Tone combinations which use the lower frequencies, tend to produce the largest amount of TWIST. Two of the must troublesome digits would be “3” and “6”. If a keypad has 16 buttons rather than the typical 12, the “A” and “B” DTMF signals would also create a large TWIST. In a typical phone, there is a TWIST of about 3 dB.

When a hand held DTMF generator is used, there will be about 6 dB of additional TWIST for the digit “3”. This is a function of the mouthpiece transducer and the generator transducer. This results in a total of 9 dB of TWIST. If the level calculations which were produced earlier included TWIST, the lower frequency tone will be typically about -43 dBm. This is a low signal where noise and distortion are significant contributors. As a contrast, consider the Bellcore specification of a minimum of -25 dBm for the design of their DTMF receivers.

(Dialogic Corporation)

Fax On Demand System Parts

Although the form may be different, the key components of fax-on-demand systems of each type are the same: a computer, voice hardware, fax hardware and software. Most fax-on-demand (FOD) systems in operation today are microcomputer based. A stand alone micro-computer based FOD system normally consists of the following components:

1. **One computer platform**, usually a PC 486 with a hard disk for image storage.

2. **One or more voice-processing boards**, usually with two or four telephone ports per board.

3. **One or more PC fax boards.**

Until recently these boards usually had one port per board. Currently, several multiple port fax boards are available for FOD applications. Also, the newest boards integrate fax, voice and data functionality so that a multifunction board replaces the separate voice and fax boards. Extensive software to consider includes the following:

1. **An operating system**, usually DOS 3.0 or higher (Windows, PS/2 and UNIX based products are also available).

2. **Voice processing software & firmware** (bundled with voice board).

3. **PC fax software & firmware** (usually provided with fax board).

4. **Configuration/installation/menu construction**. Pre recorded voice messages, Voice message generation, Image enhancement and manipulation and format conversion.

5. **Functional modules and options.**

(ABConsultants and Nuntius Corporation)

Flash ROM Updates For Supra Modems

The SupraFAXModem 288, 288i and 288PB support Flash ROMs. This feature allows for easy updating of the modem's firmware. In order to get the best performance from your modem, it is strongly recommended that you update your modems to the current ROM code. ROM upgrade utilities are posted on all Supra online services including the BBS (541-967-2444), World Wide Web Page (www.supra.com), FTP Site (ftp.supra.com/pub/flashers_current), and a CompuServe forum (GO SUPRA).

Supra's Technical Support Department publishes a newsletter that is sent via Internet email every couple of months. It is filled with announcements, upgrade information and technical tips. Signing up for this newsletter is easy. Simply visit the Support page (www.supra.com) and fill out the form. Supra will keep you informed about new products as well as provide tips to maximize your modems performance.

Supra maintains a fax-on-demand server filled with technical tips documents. (These same documents are available online from the Support page on the World Wide Web page.) All you need to do to order a catalog of these documents is call into the fax-on-demand system with a touch-tone telephone and enter in the fax number where you want the catalog to be sent. Upon receiving the catalog you may select the items about which you want detailed information, then call back and order those documents.

(Diamond Multimedia Systems, Inc. -- Supra Communications Division)

Hard Disks Must Be Fast

Keep in mind if you use a PC that's too slow, callers will hear speech prompts that are broken up. This sounds bad and will result in caller complaints. Worse, they may never call the system again. No matter what, always use fast hard disks -- 12 millisecond access time or less. Consider the purchase of caching disk controllers to speed-up the access time.

(Telephone Response Technologies, Inc.)

Interrupt Dueling

Some Dialogic voice boards ship with a default IRQ of 3. This conflicts with COM 2. Be sure that your PC does not have a COM 2 port, or that it is disabled, or that you change the Dialogic IRQ.

(Parity Software)

LAN Cards A Low Priority

If you have a LAN card, be sure that it is installed at a lower priority IRQ than the Dialogic board (lower IRQ numbers have higher priorities). Otherwise you may find that record and play of network files does not work correctly.

(Parity Software)

Memory Conflicts Go Away

One way to get around memory conflicts is to use Rhetorex boards and the company's new Configuration Wizard for Windows 95 and Windows NT. This wizard suggests dualport addresses, IRQ's and IO Ports which are available according to the plug and play information contained within Win95.

(Rhetorex, Inc.)

Modems And Call Waiting

Call waiting? At hotels, for SOHO applications or at home, that horrible phone "beep" knocks your modem or fax machine off the air. Here's a fix: load up your communications software; get it to **"go local";** then type:

ATS10=20

That will increase your S10 register to two seconds. This register sets the time between loss of carrier (caused by the 1.5 second call waiting signal) and internal modem disconnect.

Modem Fast Dial

Add **S11=50** to its dialing string.

(Teleconnect Magazine)

NEC Mark II PBX Integration With D/x1Ds

D/x1Ds may have difficulty detecting inbound rings from NEC Mark II switches. NEC can provide information on adjusting the ring generator's ring frequency. Have this frequency increased to 22-25Hz.

D/21Ds and D/41Ds, in general, require a ring frequency of about 22Hz or greater to reliably detect an inbound ring. The NEC Mark II switch provides a frequency of about 20Hz, which, in some cases, is not high enough. There is a ring generator module on the switch which can be adjusted to provide a higher frequency ring. Contact NEC Technical Support for more information about this adjustment.

(Dialogic Corporation)

Norstar Busy Signal Detection

Standard templates for detecting central office busy do not work when trying to detect a busy Norstar extension. To detect a busy Norstar extension you must modify you tone template to account for a difference in the cadence of the Norstar busy.

The Norstar uses a dual tone busy signal which consists of 480Hz and 620Hz and has a cadence of 700ms on time and 700ms off time. To properly detect the Norstar busy signal, programmers should code a dual frequency GTD template with cadence.

The deviation for the frequencies should be 60Hz and the deviation for the cadence should be 50ms to 70ms. With the GTD template established with these parameters you will be able to detect a busy extension with the D/42-NS board.

(Dialogic Corporation)

Norstar And Collecting Caller ID

Caller ID is sometimes blank when collecting it with the D/42-NS board. The solution for Caller ID collection is to answer the call on the second ring and lengthen the Caller ID collection time-out.

The Norstar switch sends the Caller ID packet between the first and second rings and depending on the switch traffic load and the application if the line is answered during the first ring you may receive a blank Caller ID packet. If you wait for the second ring to start and then answer the line the switch would have already sent the Caller ID to the D/42-NS and the problem can be avoided.

Under the DOS operating system you collect Caller ID with the function getdtmfs(). Dialogic recommends you set the maxsec variable in the RWB block to 5 seconds. This will give the switch enough time to send you the Caller ID without timing out.

Under UNIX, dx_gtcallid() would be used to collect Caller ID, and under OS/2 dl_getcallid(). If you get a blank Caller ID under UNIX or OS/2 you need to call the function again to retrieve the Caller ID. Since these functions return immediately you can not activate a timeout like the DOS driver. This tip is targeted to applications that do not answer the line on the second ring. If your application is answering on the second ring then the Caller ID will be in the digit buffer, and the function should terminate immediately.

(Dialogic Corporation)

PEB Configuration Tips for HD Series

Here are the rules for properly configuring the PEB Terminator on HD Series boards, with matching configuration file settings. This is for all operating systems.

1. **D/240SC & D/320SC.** The PEB Terminator is *always* in the Resource position.

For DOS & OS/2: **MODTYPE = RESOURCE**
For UNIX: **TYPE = SPAN-PEBR**

2. **The D/240SC-T1 in Drop & Insert.** There must be a cross-over cable to another network device. The PEB Terminator is *always* in the Network position.

For DOS & OS/2: **MODTYPE = PEB_NET_DROP**
For UNIX:
TYPE=SPAN-PEBM for spanB1; TYPE=SPAN-PEBR for spanB2

3. **The D/240SC-T1 in Terminate Mode.** Stand alone or with a straight PEB cable to other resources. The PEB Terminator is *always* in the Resource position.

For DOS & OS/2: **MODTYPE=NETWORK**
For UNIX:
TYPE=SPAN-PEBN for spanB1; TYPE=SPAN-PEBR for spanB2

(Dialogic Corporation)

R2MF High Rate Signaling

In some countries, the Dialogic R2MF signaling rate may not be fast enough for the PTT. If the signaling rate is below the rate required by the PTT, R2MF signaling will not be completed and the call will terminate.

In UNIX, using System V version 4.2, Dialogic has developed a solution that allows a high exchange rate between tones while doing R2MF signaling. This is accomplished by moving the exchange protocol from the application level in the PC down to the firmware on the board.

R2MF signaling is used outside of the United States, most commonly in Europe and the Far East. A tone is sent from the central office (PTT) and a response is expected back from the subscriber. A sequence of tones is exchanged back and forth which conveys information about the call. In most environments, the R2MF exchange rate of 4 to 4.5 tones per second works without any problem. Some countries, however, require tone exchange rates of up to 8 transactions per second. If the central office sends a tone and is not answered fast enough with a return tone, the call is terminated. The delay introduced by asking the application for each new tone limits the rate of signaling.

The request has to pass up through the driver over the PC bus, be processed and return back again over the PC bus through the driver to the network resource. With very high exchange rate requirements, the delay introduced by making requests through the application becomes a limiting factor. Dialogic's UNIX release 4.2 provides an enhanced feature set by using a build template for the signaling exchange.. A macro can be programmed at the firmware level that eliminates the need for the protocol exchange up to the application. This can now be done at the firmware level and the final result of the exchange is passed up to the application. The macro involves setting up a state machine in the firmware which greatly speeds up the signaling process. This new feature is a part of Dialogic's "Global Call".

(Dialogic Corporation)

Reboot Switch For Master Computer

How often have you needed some way of having a master computer reboot other computers? Many larger computer telephony systems are built with several PCs. Often one PC will monitor the others and can signal a failure or lockup on any of the others.

Western Telematic (Irvine CA -- 714-586-9950) has come up with the RPB-115 ($395) rack mounted reboot switch for five separate AC power outlets into which you can plug five PCs.

The reboot switch has a RS-232 link back to the host PC and has a simple command structure to control each port.

You must use a PC or another task under a multi-tasking operating system to keep track of all your Western Telematic-connected machines. If any machine appears to be locked up, it could order that machine to be rebooted via the Western Telematic machine.

If you are running short on serial ports, the RPB-115 could be cabled in a Y configuration. The RPB-115 will ignore the data on the line unless it matches the exact six byte control sequence in front of the command.

The RPB-115 also has a "reboot all" command to allow synchronized starting if desired. You can select either a 2 or 5 second off time delay on the dip switches depending on how long your equipment needs to restart.

The unit is under two inches tall and has five boot in progress LEDs as well as power and data LEDs. All the power connectors are located in the back of the device for cleaner appearance. This unit works on 120vac power but there is an international version available.

They are working on an $100 add-on box to control the switch from a phone line using touch-tone commands.

This present device only turns off the power, waits a 2 to 5 second delay, then turns it on. It cannot turn the power of each outlet off and leave it off. In short, it is a reboot switch, pure and simple.

(Computer Telephony Magazine)

SC2000 Internal Registers

When writing to an internal SC2000 indirectly addressed register, nothing seems to happen... The internal register address is cleared with each read or write operation. If you assumed that the address remains, wrong reads and writes will result. You must always write the address before each read or write with one of the internal registers. This must be done even if you are writing to or reading from the same register time and time again. A good test of your ability to read from one of the internal registers is to read the chip version number. This resides in internal register number 4.

Only four of the registers in the SC2000 can be accessed directly because there are only two address bits. Access to the other 256 internal registers within the SC2000 requires a sequence of read and write operations. For example, when writing to configuration register "2" ; write 0x01H into the internal address register 1, write the clock divider information, etc., and finally write into register 0 with bit 2 (the write bit) set to 1. If you change your mind about the configuration, you must go through the entire sequence again. The address 0x01H has been erased from register 1 when you set the write bit to 1 in register 0.

(Dialogic Corporation)

Serial Port Overruns

Modems are getting faster. Serial ports are not. Bingo, serial port overruns and communications problems. There are explanations and solutions. Here they are from an expert in PC communications

"Overrun" refers to the condition that occurs when a computer cannot keep pace with the data arriving at a serial port. Imagine characters arriving on a conveyor belt with a small holding area at the end and a worker who unloads them. If the worker cannot keep up with the arrival rate, or if he gets distracted for a moment, then the holding area may overflow and some characters fall off and get lost. That's an overrun.

How Can Overruns Be Avoided? There are five possible approaches to avoiding overrun errors:

1. **Upgrade your hardware.** Getting a faster PC, assuming you don't already have the fastest on the market, will reduce the risk of overruns because the processor can run interrupt-service routines more quickly and so keep the COM-port waiting for shorter periods. Upgrading your entire system is, however, not usually an economical solution and it's typically not one that can guarantee to reduce all risk of overruns anyway.

If you are in a position to consider a more modest hardware investment then consider upgrading your serial port. Users of portable PCs can switch to a PCMCIA modem from a serial port one. Thus your new modem serial port will be what's in the PCMCIA modem. In the newer ones, you often find an upgraded "serial port."

The simplest possible upgrade is to change a chip in your existing port. The heart of any COM port is a chip called a UART (Universal Asynchronous Receiver-Transmitter). This is the chip that holds the incoming characters until the processor can take them. You can determine what type of UART your COM ports have by running one of several software utilities designed for the purpose. One candidate is Microsoft's MSD. If you find out that your port is built using an UART identified as an 8250 or 16450 then it's one of those dinosaurs that has only room for one character. If that chip is socketed you can replace it with a 16550A which should cost about $10. The 16550A is the latest and greatest PC-compatible UART and can hold 16 inbound characters!

The 16550A is a great advance over earlier UARTs but it's still far from ideal. For one thing, while capable of storing 16 inbound characters, it only actually does so when software tells it to. By default it acts like a dinosaur and stores just one! Many communications packages these days recognize a 16550A and initialize it to use its 16-character memory but, be warned, some do not. The other consideration about the 16550A is that, in extreme cases, even the elasticity permitted by a 16-character buffer may not be sufficient to totally avoid overruns. In most practical cases it is; but there are no guarantees.

A slightly more expensive step is to replace an entire port. Serial-port cards equipped with 16550As are commonly available and should cost around $35. Especially smart serial ports are available from Digiboard, Hayes, Multitech and Telcor. They are more expensive but often a worthwhile investment, especially if you run Microsoft Windows.

We use and like the Hayes ESP (Enhanced Serial Port) which lists at $99, and also hear good reports of the Telcor T/Port Adapter. (Hayes Microcomputer, Atlanta, GA -- 408-840-9200. Telcor Systems, Natick, MA -- 800-826-2938.)

Many magazines are now noting in comparative reviews of notebook PCs which units are built with 16550A-based COM ports. Sometimes these are flagged as simply having "high-speed" COM ports. We recommend buying only notebooks that are so equipped.

The advent of high-speed PCMCIA modems has proved to be a blessing for notebook users. Such modems have COM ports built into them and, of necessity, these are designed as 16550A-compatible. So, should your notebook manufacturer have decided to save a couple of bucks on every unit by installing cheaper UARTs for the COM port(s), you need not be affected, at least for modeming.

2. **Reduce the COM-port speed.** Short of upgrading the hardware, reducing the speed, say from 9,600 to 4,800 or even 2,400, is the surest method to avoid overruns. You may consider this only a last resort but at least it's a recourse you can depend on.

3. **Upgrade your software.** As mentioned above, even if you have a 16550A-equipped port, you still need software that recognizes that and takes advantage of it. If you use an advanced operating system such as NextStep, NT, OS/2 or UNIX then that job falls to the serial-port driver in the system itself and, in all cases we know of, you can depend on it's being done. If you use DOS, then responsibility falls to each application. If you use Windows then the situation gets rather complicated; this will be covered later.

4. **Use a protocol.** When software detects an overrun it can do one of three things. First, it can ignore the condition. It is not uncommon for terminal emulators to do so, which means that you may observe characters missing from your output. Second, the software can report the condition to you. Third, the software can try to recover from the loss; but only if it is operating a protocol that allows such a move.

When a program is using ZMODEM, say, to transfer a file it may well ignore overruns per se because it can depend on an overrun giving rise to a block-check (CRC) error, just as it would were data corrupted by any other fault. The damaged data block is simply retransmitted until it gets through okay. The trouble with overruns may be that they occur often enough that many blocks have to be retransmitted and performance degrades. In a pathological case, one could see better file-transfer performance at, say, 9600 bps than at 19200.

Modern modems support error correction by operating a protocol such as MNP or V.42. This means that your data is guaranteed reliable passage over most of its journey and is vulnerable to damage only on its way between your modem and software. While other problems can occur here, overruns are the only errors likely to impinge. In a few cases, a remedy is to use software-based MNP rather than a modem-based protocol.

Using your modem's MNP is usually preferable in terms of performance. If you get overruns, however, and your software can do MNP then you may need to trade some performance for reliability. One software package that can operate MNP is MCI Mail Express for PCs, a DOS front-end for MCI Mail.

It's the user's choice: one can configure the program to use its MNP driver or not. A user who lacks an MNP-capable modem should certainly should take advantage of the software MNP; but even with an error-correcting modem, one may properly prefer to use software MNP. It may be necessary to determine by experiment whether you get better performance operating modem-based MNP at a low speed or software-based MNP at a high speed.

Some people who should know better have published statements that it is never right to use software MNP if one has a modem that is MNP-capable. That claim is **wrong**.

Switching on software MNP rather than using your modem for error correction does not actually eliminate overruns but allows the software to recover from them in the manner described above for ZMODEM. When switching to using Software MNP, bear in mind that the modem must be configured to not operate MNP.

5. **Eliminate the Culprit.** It is rare but sometimes the case that, when overruns are being caused by the action of some software keeping interrupts off too long, you can modify or eliminate that software. Disk-caching programs for DOS, especially those that do write caching, may be such culprits.

Special Considerations with Windows. Overrun errors occur more frequently in Windows than in plain DOS systems. The main reason is that Windows itself turns interrupts off for substantial periods while task switching. When considering overruns and Windows, one must consider native Windows applications and DOS programs running in Windows DOS VMs separately.

Windows Applications. Fortunately, the Windows 3.1 COM-port driver (COMM.DRV) has the capability of enabling the FIFOs (those 16-character buffers) in 16550A-equipped ports. Unfortunately it does so only when told to! The operative cue is the COMnFIFO parameter in the [386Enh] section of the SYSTEM.INI file. This should be set to 1, as in this example:

COM1FIFO = 1

Be careful, some documentation tells you to use the form:

COM1FIFO = ON

But this does not work! Unfortunately the driver is hard-wired to set the interrupt threshold to 14. This means that the COM port generates an interrupt only once it has accumulated 14 characters, leaving but two character times (i.e. while the 15th and 16th characters arrive) for the processor to respond. The hardware supports thresholds of 2, 4, and 8 as well, and 8 would have been a much better choice. Still, even with the threshold of 14, having the FIFOs enabled does make a big difference.

We wonder why COMM.DRV demands that COMnFIFO be set. We cannot see any danger in just always enabling the FIFOs when they are present.

Of course, if you do not have suitable COM ports then this facility of Windows cannot help you. It is our experience that, without FIFOs, it is rare that communication at speeds of 9,600 bps and higher can be sustained without overruns. One does, however, have all the options listed above for improving matters. When working with Windows, we particularly recommend considering use of software MNP when available.

DOS Applications under Windows. The Windows COMM.DRV mentioned above serves Windows applications and Windows applications only. DOS applications running within the Windows environment are served by another COM-port driver called CommBuff. This serves to partially virtualize COM ports for DOS applications. Unfortunately, CommBuff is not as clever as COMM.DRV -- it does not support 16550A FIFOs -- and thus makes all COM ports appear to DOS programs as "dinosaur" COM ports.

Therefore, when running under Enhanced Mode Windows, a DOS application cannot take advantage of a COM port's FIFOs even when it knows how! In this situation, upgrading the COM-port hardware does not help and one must often fall back on either software MNP or reducing speed to avoid overruns.

An alternative is to use third-party replacement COM-port software called TurboCom. This product provides replacements for both COMM.DRV and CommBuff, the latter with a driver that does support FIFOs.

TurboCom is available from Bio-Engineering Research Labs, Ashland, OR -- 503-482-2744; CIS 71521,760 and MCI 344-537. While we like and use TurboCom's CommBuff replacement, we are not fans of its alternative COMM.DRV as it seems to cause problems for some Windows applications.

(Computer Telephony Magazine and Pete MaClean)

Shared Memory - Enabling ISA (Pentium PCs)

Dialogic boards require the use of shared memory in the upper memory area of a PC's system memory to communicate with the CPU. This area of memory may not be available by default on certain systems. Certain segments of the system memory can be reserved for use as ISA Shared Memory. This can be done through the system's BIOS settings.

On PC's with certain BIOS's, the shared memory segment of the system may not, by default, be available for use by other devices. This area of memory is instead used for shadowing, a process whereby a block of memory from an adapter card's ROM is copied to the same address in system memory to allow faster access to the code. This process of shadowing, causes problems with adapter cards, like Dialogic's, which use the shared memory area to communicate with the host processor. For Dialogic boards, this type of memory problem will reveal itself when an attempt to download the firmware is made (e.g. under DOS, genload would return "WSB0008: Error configuring memory at address D0000"). Typically this problem can be avoided by setting a BIOS option to disable the shadowing of RAM.

Recently, however, it has been observed that an additional or different step may be necessary to allow the Dialogic board to be installed in the PC. A block of system memory may actually have to be reserved for use by an adapter card. Reserving this space of memory will actually set up a block of system memory that will not be shadowed. This has been found to be the case with more recent versions of AMI BIOS found on Pentium processor PC's.

How this is done will vary with different types of BIOS, but will typically require that an ISA Shared memory base address field be specified along with the memory size to be reserved. An example of how to set this for AMI BIOS version 1.00.10.AX1 is given below. Do the following under the Plug and Play Configuration Sub-Menu:

1. For Configuration Mode, choose Use Setup Utility.

2. Now set the ISA Shared Memory Size field to 16, 32, 48, or 64 KB.

3. Specify the starting address for this block of shared memory in the ISA Shared Memory Base Address fields.

(Dialogic Corporation)

Shared RAM Voice Card Exclude

If you are installing a voice card under Windows 3.1, make sure that the EMMExclude setting in the [386Enh] section of SYSTEM.INI is set to exclude the range of shared RAM used by the voice card (Windows is too dumb to detect the card automatically). For example, if the voice card uses 1000 (hex) bytes at address D000, then set EMMExclude=D000-D0FF.

(Parity Software)

Software Interrupt 6D

The Dialogic driver for MS-DOS (D40DRV) uses software interrupt 6D by default. Many clone PCs use 6D in the VGA BIOS, which causes a conflict. Utilities such as QEMM's Manifest, or Parity Software's FREEVECT, allow you to check whether a vector is free before installing the driver.

(Parity Software)

Chapter 7

Testing And Installation Tips

What a great thrill it is to install your first computer telephony system. The contributors to this chapter have collectively installed thousands of them. Even with all of this experience - there are still situations that arise that have never been seen before.

No one relishes the idea of stumbling upon "something never seen." In most cases the unknown will hurt you in this business. What most readers of this book want is a good measure of predictability in their installation and maintenance routines. That's what this chapter helps you to achieve. You'll get some good advise on how to test, install and troubleshoot computer telephony systems.

There's a good measure of "mistakes to avoid" in this chapter, too. For more on this subject, you should invest in a copy of Steve Gladstone's book: *Testing Computer Telephony Systems and Networks* (Flatiron Publishing -- 1-800-LIBRARY). He's made a thriving business out of helping telephone companies, switch manufacturers and enhanced platform providers to test and debug their equipment and applications.

BRI ISDN Installation Tips

Here are the best tips for getting ISDN BRI up fast from Nynex's best ISDN installer, Matt McGuire:

1. **Cross your Ts and dot your SPIDS.** The Service Profile Identifier (SPID) is a ten-digit "phone number and suffix" which defines the characteristics of the circuit to the switch. Most BRI lines being installed today have two SPIDS -- one for each of the B channels. Most Basic Rate ISDN user-configurable terminal software asks you to poke-in the type of switch your line is connected to.

If you mess this up, it won't work. Most packages prompt you to enter a digit indicating one of three switch types: NI-1 (National ISDN Standard Switch), or DMS-Custom (Northern Telecom) or #5ESS Custom (AT&T).

2. **National ISDN switches need two different SPIDs.** If you have a choice, get National ISDN service. #5ESS Custom doesn't need a SPID, but you often get one and you can enter the same number twice. There aren't many DMS-Custom lines.

At Computer Telephony Magazine, one of the BRI line SPIDs is 212-366-6670-0000. At the Computer Telephony Expo 96 in Los Angeles, Matt saw an equivalent SPID that was 213-693-0097-00. Note that the later is shorter. If you don't have your two SPIDs, you might not be able to bond the two 64 Kbps lines into one 128 Kbps. That happened to one exhibitor at CT Expo. Solution: Check the SPIDs on both channels. One customer ended his with 01. He should have ended it with 00. When Matt changed the SPID, the channels bonded perfectly.

Make sure your initial installer writes down all pertinent information for each ISDN circuit at time of installation. This includes SPIDs, circuit numbers, and DN (Directory) numbers. The best piece of information you can get is a full listing of your central office ISDN configuration information.

This can be many pages long. But make sure you get it, because ISDN CPE equipment is not standard and sometimes you need to know if it's your CPE equipment or your ISDN line that's the problem. Read the stuff to the vendor. He'll know instantly what the problem is.

Preferably, this information should be affixed to the jack (demarc) so you don't loose it. When you call for service, have the number assignments ready.

3. **So it looks like a regular tip and ring twisted pair.** But it's not. BRI ISDN lines have no analog dial tone. You won't get dial tone if you plug into a digital line. Don't assume the line is broken if you don't hear anything. You can plug an analog phone into certain NT-1 terminal adapters and pull digital dial tone.

You check for "dial tone" on an ISDN line by using ISDN test equipment. Matt uses Tektronix Craftek CT-100. He loves it. He says it's the best hand-held ISDN tester. It can cost you $4,000.

4. **You can get ISDN lines in quickly.** At Computer Telephony Expo 96 in Los Angeles, PacBell's installation crew, which was really super, had some lines in and working in under eight hours. From order to installation in eight hours. It's possible.

5. **Some ISDN networks are built for 56 Kbps only** and some are built for the faster and more standard 64 Kbps. PacBell's handoff to the long distance network for ISDN has traditionally been 56 Kbps. You can get 64 Kbps "clear channel" out of PacBell but you have to ask for it and they have to rework some equipment. Suffice, ask when you order. Are you getting 56 or 64? 64 is better. Make sure you get it.

6. **BRI ISDN lines cannot be installed like regular analog lines** from an outside plant or installation perspective. Not only does significant load balancing have to occur in the C.O., but you can't run BRI copper through load coils and pass it through all kinds of splices and bridges. Distance can be a major problem. ISDN needs a nice, clean and noise-free circuit to work well.

7. **Connectors.** Most NT-1s need a RJ-45 jack. Make sure you have extra connectors, wires, etc. -- so you can rig the correct physical connection in a pinch. In some cases, the phone company delivered the ISDN circuit with a standard RJ-11 jack, which didn't fit well into the RJ-45 plug which virtually all ISDN equipment has. In one case recently, Matt replaced the RJ-11 with an RJ-45 and it worked perfectly.

8. **Pre-Assign a Long Distance Carrier.** If there is no long distance carrier assigned to the line, you will be automatically switched to the PSTN (voice) part of the network when you try to dial a long distance call. This is particularly annoying if you are making a data call and you keep getting "no carrier" messages back.

Some customers do not specify a long distance carrier, and therefore are forced to dial an extra sequence of equal access (10-xxx) numbers in order to reach the data network. Some software won't accept these extra dialing digits. You can avoid this problem by telling the long distance telco you want to use when you order your ISDN service. They can't assign one for you, so if you don't tell them who you want -- you'll get none, and hence the problem some people have.

9. **Do a loop-back test.** If your NT-1 has loopback test capability, go ahead and try to use it as soon as your line is installed. Many terminal adapters come with software that lets you do this. A loopback test determines if you have a 64K clear channel on each virtual circuit on your line. The test software will send a data stream down the channel in order to read a "bounce back" image from the C.O. If you don't have a loopback telephone number to call in order to run your test, call 212-240-9967. Only 64 Kbps. Not 128 Kbps. Doing a loopback test first is a great idea, because 99 times out of 100 -- your product will work if the loopback test does.

10. **Trouble reporting.** Many customers call for service and fail to tell the telco trouble number that it's an **ISDN** line they are reporting trouble on.

(Computer Telephony Magazine and Nynex's Matt McGuire)

Cables Are Always Bad

If in doubt, assume the cable is bad, the connections are bad, you're connected to the wrong plug or your jumper settings are wrong. Electronics and chips are rarely bad.

(Teleconnect Magazine)

Cool Your Phone System

Phone systems work better when they're kept cool. Sadly, most companies hide them away in cramped closets and rooms. To air condition a phone system costs more than air conditioning a normal office. However, there are some hugely economical solutions for air conditioning a phone system -- e.g. a device called a spot cooler. It's a small air conditioner with a couple of vacuum cleaner type tubes. Push them under your PBX and blow the cool air through it and out the top. Most PBXs now come with holes in their tops. (If yours doesn't, drill some. Check your warranty first.) Don't believe your PBX vendor when he says his equipment will work perfectly in a confined, warm place.

(Teleconnect Magazine)

DID Mistake to Avoid

On analog DID trunks: Make sure that when you take your IVR system down for maintenance that you leave the 48V DC power on the DID trunks. If you unplug the power supply, an alarm will go off in the C.O. and they'll disable your trunks. You then have to call the telco to get them turned back on...

(Computer Telephony Magazine)

DTI/1xx T-1 card Installation Tips

The term DTI/1xx refers to Dialogic's PC-compatible Digital Telephony Interface expansion boards. The DTI/100, DTI/124, and DTI/101 are specific models of this board. The DTI/1xx is a time division, byte-interleaved multiplexer that multiplexes 24 DS-0 subchannels of voice onto a 4-wire, 1.544 Mb/s T-1 line (DS-1). Each subchannel is sampled at 8 kHz (with 8-bit samples, for a total of 64 kilobits per second). The DTI/1xx is designed to connect a T-1 line to Dialogic voice processing resources.

Multiplexing equipment in the voice communications field is commonly referred to as a channel bank. In the data communications field, similar equipment traditionally is called a T-1 multiplexer or simply a MUX. Typically, the channel bank multiplexer in a T-1 voice application connects T-1 channels to various kinds of local analog telephone lines which are connected, ultimately, to a telephone.

The DTI/1xx replaces the channel bank in this type of configuration and typically terminates the individual channel at a Dialogic voice channel instead of a telephone. This voice channel automates several call functions. Besides multiplexing, the DTI/1xx performs several other functions on the outgoing (transmit) bit stream. These functions include providing framing, bipolar encoding, inband signaling, and synchronization with the network for the outgoing multiplexed stream. The DTI/1xx interprets signaling and framing, converts bipolar format to unipolar format, and recovers clocking from the network for the incoming (receive) stream.

Supervision. E&M signaling is the default protocol supported by Dialogic firmware. The E&M voltage level is not important, since the demuxed channel will not terminate at a real E&M line. The Dialogic firmware will interpret the signaling bits and control the D/xxx channel accordingly.

Timing. Most DTI/1xx applications that connect to the Public Switched Telephone Network will use loop timing. Timing slips will occur if internal timing is used when connecting the DTI/1xx to the network. Timing slips are typical of new installations.

T-1 Line-to-CSU Cable. T-1 connectors vary from carrier to carrier, but the cable provided by the T-1 line carrier usually consists of two pairs of individually shielded, twisted 22-gauge wires (one transmit pair and one receive pair) terminated in either a DB-15 or modular RJ48-C plug.

The industry trend has been to standardize on the RJ48-C. Make sure when you order service that the type of plug the carrier provides is compatible with the type of plug your CSU accepts. Older CSUs may require that you connect the wires directly to a terminal strip or similar arrangement. Refer to your CSU documentation for more information.

Wet or Dry T-1 lines. Circuits that carry a voltage to power regenerative repeaters are called wet T-1. This voltage traditionally has also been used to power the CSU at a site. Since 1987, the Federal Communications Commission no longer requires carriers to provide line power, and the trend has been to discontinue its use. T1 lines without power are called dry T-1. If your carrier supplies dry T-1, then your CSU must have an AC power source. Check with your carrier to see which kind of line they supply so that you can obtain an appropriate CSU. Many of the latest CSUs can handle either kind of line.

Building a CSU-to-DTI/1xx Cable. If you are using a Dialogic DTI/1xx board, you must build your own CSU-to-DTI/1xx cable. The recommended cable type should be shielded twisted-pair cable (ABAM 600 or equivalent) in which each of the two pairs is shielded and the two pairs have a common shield as well. This kind of shielding helps to avoid crosstalk problems on the line. Choose an appropriate cable length for your installation. Eighty-five feet is the maximum standard cable length between the CSU and the DTI/1xx for standard T-1 signals (6 volts peak to peak). In general, you should keep the cable length as short as possible.

Refer to your CSU documentation for guidelines concerning cable lengths greater than 85 feet. Use a 15-pin D-subminiature type MALE connector (AMPHENOL part number 205205-2 or equivalent) for the end of the cable attaching to the DTI/1xx. Use this connector with snap-in solder cup contacts (AMPHENOL part number 66570-3 or equivalent).

The connector should also include a slide latch assembly (AMPHENOL part number 745583-5 or equivalent) for attaching to the locking posts of the female connector on the rear bracket of the DTI/1xx. Connect the cable wires to the proper pinouts on the DB-15 male connector on the DTI/1xx end of the cable. Pinout designations are detailed in the DTI/1xx documentation.

Consult your CSU documentation for the proper pinout designations and connector type for the CSU end of the cable. Follow your CSU documentation concerning grounding the cable.

IMPORTANT! When you are building a CSU-to-DTI/1xx cable, be sure you understand how the CSU documentation is labeling CSU pinouts for transmit and receive to local equipment (in this case the DTI/1xx). For example, the transmit pins 1 (tip) and 9 (ring) on the DTI/1xx are OUTPUT FROM the DTI/1xx. They should be wired to the pins on the CSU connector that correspond to the INPUT TO the CSU

Grounding your CSU. CSU-to-equipment cables are usually grounded at one end only. The DTI/1xx end of the cable does not require grounding. If you choose to ground the DTI/1xx end of the cable, you can attach the common shield to the metal casing of the connector you are connecting to the DTI/1xx.

This will provide adequate grounding to the PC chassis when the cable is attached to a properly installed DTI/1xx board. attach the common shield to a wire and attach the wire securely to the PC chassis Be sure to test your cable for continuity after you have built it.

Crossed Wires. Your particular CSU may use the terms transmit and receive differently than your T-1 network interface board manufacturer does. Be sure you understand which direction the signals are traveling. Mistakes in wiring transmit and receive pairs is one of the most common problems encountered during T-1 installations.

(Dialogic Corporation)

European Tip & Ring

Tip is blue and Ring is green. This information is useful if you want to use your modem to send stuff back to the states over phone lines. For strange European phone systems you'll need a cable with an American RJ-11 male plug on one end (to plug into your modem) and two spade lugs connected to two small alligator clips on the other. Most Radio Shack stores will sell you all that equipment for less than $3. You can now buy conversion cables from some mail order companies.

(Teleconnect Magazine)

Fax Archive Consideration

When you install a fax server, consider the installation of tape backup or some other back-up scheme. Although not widely practiced or available today, fax archiving may be an important fax server aspect. With fax archiving, tape or optical drives record the full images of all inbound and outbound faxes so archival records exist for purposes of simple retrieval (i.e., for re-faxing) or for litigation purposes. With fax machines there is always a paper copy to archive, but that is not so with fax server systems (some fax servers let users auto print all outbound faxes, mimicking fax machines' paper archiving capability).

(Davidson Consulting)

FAX On Demand Installation Warnings

1. **Do not underestimate the time needed** to get a fax-on-demand system up and running. We had one C programmer fiddle with one system -- one of the most popular ones -- for 23 hours before giving up.

2. **Do not underestimate the time to keep it running** -- testing it continuously, updating new documents, improving the voice prompts, etc. You need a fax sysop. It's a full-time job if your system is substantial.

3. **Do not try and get your fax-on-demand system working on a cheap PC.** You need power, plenty of RAM and an industrial grade computer to handle the heat. Do not use the manufacturer's minimum hardware specs -- unless you have all the time in the world. Every document you create on a PC -- in any program -- has to be converted to fax language. Each page takes time. Several pages take oodles of time.

4. **Do not underestimate the popularity of your system.** Once you start getting calls, they'll escalate. Then you'll need to expand your system. You'd better have plenty of software and hardware room to grow.

5. **Do not start with one phone line** for all the reasons above.

6. **Protect your system with surge arrestors, UPSs, etc.** Back up regularly, etc. There's too much here at stake to lose.

7. **Do not underestimate the charm and convenience of a fax service bureau like MCI Mail.**

8. **Do not be suckered into doing all this** -- programming your own fax server, programming and running your own fax-on-demand system, etc. -- to save money. You are doing this to give better service to your customers, or to make money with some unique hot application. This stuff is not trivial.

(Computer Telephony Magazine)

Fax Server IRQ Problems

Installation of a fax server may be laborious or highly automated. Some systems allow network users to be registered automatically via automatic downloading from a network operating system directory (e.g., NetWare bindery emulation), while others require manual input of all user names. With some fax servers, clients and remote fax server nodes can be installed automatically and managed from one remote workstation (or any remote workstation). Most allow phone numbers to be imported into phone books.

Installation of fax boards into fax server PCs can become complicated if there are **IRQ conflicts** and the like. As plug-and-play becomes a reality, fax devices should become easier to install. Some fax boards already are much easier to install than others. With self-contained turnkey fax servers, fax boards are built-in, which also can simplify installation. Make sure you check and re-check the IRQs on all the boards in the server.

In general, a certain amount of effort has to be put into software configuration. Decisions must be made about redialing, configuring phone lines for sending and/or receiving, who will be forced to delay transmissions (or not), whether lines will be send and/or receive (and whether to schedule different phone line configurations by time of day), who will be alerted when problems arise, how to set thresholds that define when something becomes a problem, etc.

(Davidson Consulting)

Fax Server Compatibility Issues

All LAN fax systems are G/3-compatible, which means they can send and receive faxes to and from about 99% of the fax universe. But many intra-system compatibility issues exist. Make sure you test for the following:

1. **Fax servers must be compatible** with customer LANs and/or host computer networks

2. **Server and client software must be compatible** with installed hardware and operating systems

3. **Fax server software must work with suitable fax boards**, machines or other fax devices (and printers too)

4. **Fax servers may have to support the particular file formats which users exploit.**

5. **Fax servers also must support available serial ports and IRQs**.

Plus, where TSRs are used (i.e., in DOS, to provide access to the fax capability without having to exit primary applications like word processors), a slight possibility exists that fax program TSR interrupts will conflict with (be the same as) those of some other TSRs in use with other programs. For fax servers which support multiple fax devices, some features may be fax-device-dependent (e.g., BFT is supported only if the fax device used supports BFT).

(Davidson Consulting)

Fax Testing For Computer-Based Systems

What's cool about regular old fax machines is that they are easy to operate – a five year old can do it. Regular fax machines are also inexpensive and everywhere. For the most part, they work. Computer-based fax (CBF) is more complicated. Sometimes CBF is centralized, as with a fax server. Sometimes it's not, as with mobile computing. It's really convenient (as long as it works) and fairly "invisible" to the user.

You Should Care About Testing. Fax server companies are integrators, because they manage hardware and software resources from many suppliers. There's fax boards, motherboards, operating systems, APIs and application software to integrate. Although SOHO software and low-end modems are easy to buy, the heavy-duty fax feature for your IVR system is not a "check off" feature.

Fax modems are not typically well-tested. Guess who gets to do the testing? You do. The person who is installing a system. There are testing tools and software available to verify protocol, type approval, transmission speed, etc. Perhaps one of the most important things you can do before making such a purchase is to put some basic testing infrastructure in place. Find a way to:

1. **Monitor transmissions**. Make sure you are using the available call logging and report software that comes with most fax server and fax-on-demand packages. Match the records with your PBXs Call Detail Reporting System.

Look for connect times to match-up. Look for repeat dials to the same number, etc. You can even rig a passive listening device on the lines to so you can hear the fax transmission without crashing the line. Time the transmission with your watch and match it up to the reports. You'll be surprised at the results.

2. **Establish traffic patterns and call success rates.** Keep a log of the actual number of successes vs. Failures. Do you notice a pattern? Does time of day, weather or the phase of the moon effect your success rate. If your system is on a LAN, do file server hiccups and other LAN-related incidents trash your fax queue?

3. **Call failures** - is it your fax server or remote machine? Confirm that a remote fax machine is working by faxing to it from a regular fax machine. If your regular fax machine can fax to it, but your server can't -- that says a lot. In many cases, you may need to flick a dip switch on a fax board or get an updated driver. If you are using older or cheaper fax boards, you may have problems with compatibility.

(Genoa Technology)

Grounding - A Common Problem

Over 50% of all the power problems are **not** related to the quality of power at all. They are the result of inadequate or faulty grounding.

(Teleconnect Magazine)

Headset And Wireless Considerations

Consider the following when you're in the market for wireless headsets, LAN connections or other peripherals that use radio frequencies:

1. **Do you need to rewire the building?** Take a look at your cabling requirements.

If your plenum, conduits or cable runs are full – wireless may be the way to go, but even some wireless base stations need wires.

2. **Does everyone need to convert?** Consider you motivation for using wireless in the first place. If it's because you can't use wire – maybe it's not for the whole building.

3. **Don't allow radio channels to overlap.** This goes not only for equipment of the same type, but also different equipment you either already have or are planning to buy. Check with your vendors to chart out the frequency ranges first.

4. **Use the installed base of computers and telephone switches** you already have. In many cases, there are instructions on how to hook-up wireless gear and some computer and switch vendors manufacture or private label recommended wireless gear. Go with what is supported.

(ACS Wireless)

Key System Alert

If you have an electronic key system, don't plug an answering machine, fax machine or modem into a telephone jack. You might blow a fuse on the phone system, damage a port on the card, destroy the device plugged.

(New Pueblo Communications, Tucson, AZ)

LAN Segmentation Strategy

Your one-wire, all-in-one LAN is dying. Reason: Too much traffic. Solution: Segment it. Good news: It's much cheaper than it was. This is a true-life story. It happened in the big city to Computer Telephony Magazine. What brought our big LAN down: We combined all our separate networks into one big network, covering three office floors.

The network has five servers, 10 Macs, 18 Windows for Workgroups editorial PCs, 11 accounting PCs, 10 administrative and management PCs, and about 9 printers, depending on which were working on that day.

We wired our gigantic LAN with six twelve-port Ethernet hubs, lots of Unshielded Twisted Pair (UTP) cable and one run of RG 58 coax.

The result was great connectivity, but all problems at times of heavy traffic: Lost connections, slowdowns, reboots in the middle of programs.

We put a network analyzer (a laptop running Novell's LANalyzer software) to spot causes. We were cluttering people's connections with traffic not for them. We bad connections in the coax, which itself was too long. And one of our file servers was old and decrepit, not unlike its MIS manager. (Do servers get like their managers, as owners get like their dogs?)

What we did to solve it: We installed four six-port Ethernet switches to isolate users from traffic not for them. We distributed the servers and users over the hubs and switches so as to give the shortest possible routes between them. Here's what we learned:

1. **Unshielded Twisted Pair cabling is more reliable than coax.** Two reasons: the physical connections in a UTP setup are more robust. Second, if one UTP link dies, it doesn't crash the network. It does with coax.

2. **Coax cabling is flexible and easy to reconfigure.** You can break up a long run into separate shorter lengths in seconds. But you have to watch the selection of physical wiring and connections, and the overall lengths of your runs. Don't let them get longer than IEEE specs. A coax run should not exceed 600 feet.

3. **Your server and its software can easily become out of date.** While you run the same applications year after year, as you add users and upgrade the power of your workstations, your servers will become overloaded. You need to keep an eye out for server software upgrades.

Upgrade it hardware-wise (faster CPU, more memory, bigger hard disks) on about the same lifecycle as your workstations. Three years tops. Give it to your children.

4. **New Ethernet switches are more cost effective** than older bridging and routing technology in segmenting your network. And prices are coming down. These switches let you expand your networks to new computer telephony apps.

5. **The weakest link in any server is its tape backup system.** The mechanical and software problems with tape give the word "nightmare" a whole new meaning. Always use RAID 5 disk arrays in your server so you never have to depend on tape backup. Always check your server once a day to see one of your disks hasn't failed. If it failed, replace it fast. If you lose two disks, you're up a creek without a paddle. Oops, you do have the tape. And maybe it will work. Maybe.

The Gruesome Story: Last year, we had a peer-to-peer DOS network for our editors to share files. We had a Novell LAN just for our Accounting and Book Sales. We had an Ethertalk network of Macintoshes for the Art Department. Three separate networks. We had no company e-mail, and we moved editorial to the Art Department's Macs with floppies.

By upgrading our cabling to all-Ethernet Category 5 UTP and RG 58 coax, we were able to have a single, totally-connected medium throughout all three floors of our New York offices, on which we could run several networking protocols.

We moved the editors to Novell 3.12 and Windows for Workgroups. We installed a couple of Windows NT Server machines with large RAID 5 disk array for Macintosh file services.

This let us save large Mac Quark, Illustrator and Photoshop files faster. It also enabled Windows-Mac file-sharing through Windows NT's Services for Macintosh, which allows access to the same files from both Windows for Workgroups and Macintosh workstations.

Finally, we installed e-mail by using ON Technology's Noteworks, a wonderful NetWare messaging system that runs in both DOS and Windows. And we extended it company-wide by linking our New York and Southampton, PA offices together with dial-up Novell MHS servers at each end (one dials the other and they exchange messages, then distribute them over the LAN to the respective addressees).

The Problems: At first everything ran well. Traffic on our network was really quite low. Only the Accounting and Books departments were running truly network-based software, that is, programs which ran entirely from the server. All other users were running programs from their local hard disks, and using the network only to retrieve and save files from and to various servers.

However, at certain times of the day, the network began to slow down. We found it happened artists were saving large files to their server. We began using a server-based database program to register attendees for Computer Telephony Expo. A big plus of the common medium network throughout the office was we could sit people down at any PC and get them to enter names into the server database.

The added traffic slowed us down. We had chosen to wire our tenth floor (our first floor) in coax not unshielded twisted pair because we didn't have the space for a rack with hubs, and we liked the idea of the flexibility of coax.

You can add machines to coax by simply unscrewing connectors and adding in a loop of wire. You can split a coax run into two by putting simple 50-ohm terminators on T connectors and feeding the second section from the other end.

There are limitations to coax. The connectors must be the correct size for the cable. You have to lay the cable so it doesn't kink. You must watch the overall length of the coax run. It's really easy with all the twists and turns to end up too long. We ended up with a 932 foot run -- from one floor to the next! We used a Time Domain Reflectometer to measure its length electronically. Neat piece of gear.

The Analysis: We attached a laptop running Novell's LANalyzer software to the LAN to monitor traffic for a couple of weeks. The software tracks packet traffic over the network, counts packets originating and received by all workstations and servers, and tracks errors of all kinds.

We found our network was experiencing lousy bandwidth utilization (approaching 30 percent). We found our Accounting and Books server couldn't keep up -- not enough RAM, out of date LAN and disk drivers, and a relatively slow system processor. The 486/33 it came with in 1992 was a speed demon then -- it's among the slowest machines on our network today.

We found that while our users need access to all our servers, each spends the majority of their time only working with one server. Thus individual and overall performance could be improved by logically segmenting our network into domains.

Accounting and Books users didn't need to see the Art Department's traffic when Art was saving a 30 Megabyte image file.

The Solution: Traditionally, segmentation was done with routers and bridges. However, routers are protocol-dependent, and with our setup, we had too many protocols to make this feasible (IPX, TCP/IP, NetBios, Netbeui, Appletalk, Ethertalk, etc, etc).

Bridges work with addresses at lower layers in the packets, and so are protocol-independent. This would be a better solution.

However, we knew that beyond solving today's problems, we wanted to prepare our network for the more demanding computer telephony applications we're putting in. We will install a LAN-based voice mail system that will let us retrieve telephone voice mail from the screens of our PCs. Lots more network traffic. More bandwidth demands.

We invested in the newer technology of Ethernet Switches, in the form of the six-port Klever Switch from Klever Computers (San Jose, CA, 1-800-745-4660).

An Ethernet Switch takes advantage of today's higher concentration of processing power to create a super-fast, programmable, and externally manageable bridge.

Klever does this by building a box that contains eight 25 MHz RISC processors, one each for each of the ports, one for programmable internal management of the network traffic, and one to allow external management of each individual bridge from a desktop PC running Windows. This uses special management software which they also sell. You can also use standard SNMP management software.

Each port of the switch is able to deliver full wire speed on the network, with very low "latency," or wait time, within the switch while it decides which segment it has to send each individual packet to. It is much faster than a conventional "smart" bridge.

To install it, what you do is to lay out your network topology into the segments you think you need, and then turn on your Klever Switch and connect each segment to one port.

If you have a server accessed by users in many different segments as we do with our Accounting and Books server, you put that on its own segment. If it's primarily accessed by all local workstations within a segment, you put it right in that segment along with them.

We put in four Klever Switches in our network. No user is more than three segments away from any server they need. We have a completely separate segment for our Computer Telephony Lab server (the mega-powerful Windows NT Server 3.5), yet it's still accessible from anywhere in the network. We have broken up our excessively long and overcrowded coax segment into three segments each of which is lightly loaded.

Klever's excellent six-port switches are $3,500, about $1,000 more than you might pay for a conventional bridge. But they offer vastly superior performance and have worked flawlessly since we put them in.

(Flatiron Publishing IS Staff)

Load Testing IVR Apps

Adequately testing IVR apps can be very cumbersome and expensive. If you don't have the budget for a Hammer system and you just don't have enough willing people or phones to do the testing manually... here's what you can do. Design your application such that the main readDigit routine reads from a file containing DTMF digits rather than from the phone directly. Put the DTMF entries in files named xxx.001, xxx.002, etc., where the prefix is a predefined string which defines a set of script files and the extension is the port. The files are simply text files, where each line represents the DTMFs which would be pressed at a particular prompt. To activate this test mode, allow the user to set a "testing" option. When the system sees this flag is set, it will start the scripts for all lines simultaneously and read from these files rather than wait for incoming digits.

(Rhetorex, Inc.)

Meridian 1 PBX T-1 Conditions

Even though the settings all appear correct, applications using T-1 trunks from Meridian 1 PBXs experience inconsistent events. They seem to work but get occasional frame slips, bipolar violations, spurious winks, etc. These can manifest themselves as unexpected conditions during specific protocols, like fax communication. The Meridian 1 T-1 interface cards can be run in a number of different framing patterns.

Depending on the framing pattern chosen, different bit polarity compensation modes are used by default. If D4 framing is selected, the default bit protocol is AMI with Bit 7 stuffing. If ESF framing is selected, B8ZS is the default bit protocol. This information is not immediately apparent. Note also that the Meridian 1 documentation indicates AMI as a selection and default, but it is really AMI with Bit 7 stuffing.

(Dialogic Corporation)

Mistakes To Avoid With Large Systems

The biggest mistake people make implementing high-end computer telephony voice- and call-processing systems is they don't get what they've paid for.

We know. Almost every time someone comes to us and asks us to test their system, we consistently find only 65% of what the user was expecting as far as system capacity.

Capacity Means Several Things. The calls per hour (CPH) the system is supposed to handle. Systems are usually rated at a CPH ability. If you are expecting, say, your IVR system to handle 1,000 incoming calls an hour, you better make sure it can and still provide reasonable response times to callers. A good rule of thumb is to make sure your system can process at least 120% of its designed capacity.

The number of simultaneous calls it can process. If you're purchasing a system with 48 ports (two T-1s), you'll probably expect that all 48 ports can be busy handling calls. Poor naive devil. This is not always the case.

Ability to handle burst traffic. Burst traffic are calls that arrive at your system all at once.

If you are advertising a new promotion during the Super Bowl, you might have hundreds of callers dial instantly into your automated CT call center. Usually the toughest thing a system does, especially if you're using ANI or DNIS to intelligently process the calls, is set up to answer a new caller.

Burst traffic is usually made up of a lot of new callers. This kills many systems that are improperly checked before the real onslaught.

How To Avoid Death. There are plenty of mistakes we see people making. They only find out that features on their system don't work until after they cut it over. Their user interfaces give problem when live people get on the line and nobody can figure out how to touch-tone their way through the applications. Etc.

But by far the most pervasive mis-step people take is not doing "dynamic" load testing up front and addressing the bottom-line capacity issues. It's a shame because misjudging your system's capacity is the most expensive error you can make. And it's also the most easily avoidable.

Basically, "dynamic" testing actually recreates the way a system will get used and abused once you go live with it. This is very much different from old-fashioned load tests like those pumped out by bulk call generators. Those are good for hitting your system with a lot of calls, but they don't exercise it adequately.

Modern CT voice / call processors have too many subsystems -- voice, database, LAN, switching, call-center agents, etc. A true dynamic load test of the system requires exercising all these subsystems together to make sure they not only work individually, but interact correctly as a whole.

Under the real-world conditions of dynamic testing, this is always accomplished because real users will muck with every nook and cranny of your overall app. A true dynamic tester inherently creates incoming calls that hammer the system with variety and diversity. Some calls will act as experts, some as novices. Real IDs and passwords will also be used.

Again the wicked big mistake of straight load testing is it only tells you how many generic calls your system can take -- not how many real calls it can handle. Remember, you're buying your system to treat your customers the way you want them treated, not just to answer and drop them into automated computer telephony hell.

What Goes Wrong. System faults uncovered by dynamic load testing frequently fit into the following areas:

Poor software design. We tested one system that spent so much time paging programs in and out of memory it ran out of gas at 50% of its target capacity. When the code was restructured to run more efficiently, performance and capacity went way up.

I/O problems. Getting stored voice prompts quickly on or off disk in high capacity systems can be a bear. How callers use a system can make it worse.

For example, typing ahead slowly in the middle of prompts can force the system to swap prompts in and out rapidly, decreasing perceived performance. But typing ahead quickly before prompts are played, so the system can figure out what the user is trying to do without playing many of the prompts, can actually increase system performance.

Database access. Many high-traffic automated CT platforms (like IVR systems, groupware servers, call-center servers, etc.) can create a lot of transactions to database servers. Sharing the traffic among multiple servers can increase system throughput and increase survivability.

Switch response time. Heavily loaded switches (PBXs, Central Offices, ACDs, dumb switches) can behave erratically under load. This can create havoc with applications.

Inter-system timing issues. Whenever one CT system interfaces to another, there are opportunities to get out of synch. For example, a switch may be ready to present a new call to an IVR system that has not had time to reset the port. What will happen? Some switches may busy out the port, screwing up your hunting sequence and actually reducing capacity.

What To Do. To make sure you are getting $100 of capacity for every $100 you spend, you need to:

1. **Figure out exactly how your system is "supposed" to function**. This is called setting up the "Service Objectives. "

Service objectives include: the number of calls per hour; the traffic patterns; the types of calls and the call mix (how the callers will use the system when they reach it, e. g. the percentage of expert users, novices, what types of transactions they will perform, what app subsystems they will access); expected response times; and anything else that defines how your system is expected to work.

2. **Dynamically load test** and make sure your objectives are met. Create the tests with the measurements taken at key areas that will illustrate system bottlenecks.

For example, measure the response times associated with complex database access (which is a key ingredient in so many cutting-edge CT voice / call processors today). These measurements then become "instrumentation points" that simplify your analysis.

If a Service Objective is not met, identify the bottleneck, fix it, then retest it to make sure it has been fixed and that nothing new was broken by the fix.

3. **Run the dynamic load test over a period of time.** Many systems work great for the first few hours but degrade slowly over time.

We had one customer that found a problem with a counter overflow only after running several hundreds of thousands of calls over several days. These are the problems that are almost impossible to track down in the field once a system has been released.

(Hammer Technologies, Inc.)

PBX Disconnect Supervision Tones

You can use these tips so disconnect supervision tone can be learned by using Dialogic's CPLEARN.EXE. This procedure was successfully used to learn the disconnect supervision tone supplied by a Panasonic KX-T30810 PBX. This procedure may not work for all switches.

1. **Establish the proper connections to obtain the tone to be learned.** The voice board must be connected to a PBX or Central Office (CO) telephone line that provides the tones to be learned. The ability to take a dialed extension/phone line offhook and onhook is also required. The CPLEARN program requires an analog loop start interface; if your system uses another network interface protocol, you must modify the CPLEARN.C source code as required for your environment.

2. **Before running CPLEARN.EXE**, D40DRV must be executed with the -G option; for example: D40DRV -G150.

3. **Edit the CPLEARN.CFG.** Specify TONEID for the TONEID to learn, MODE 1 for manual mode, CHAN_A for the voice channel to run CPLEARN on, and DIAL_A for the extension/phone number to dial on CHAN_A.

4. **Execute CPLEARN with the -P option** so that CPLEARN will pause between classifying the tone and learning the tone.

5. **Let the extension/phone line ring** until classification is complete. CPLEARN will prompt you to press any key to learn the tone. At this point, take the extension/phone line offhook then place back onhook. This should present the disconnect supervision tone to the Dialogic voice channel.

6. **Press any key to learn the disconnect supervision tone.**

7. **If no errors occur during execution, CPLEARN will learn** the disconnect supervision tone and create a file (named CPTONE.OUT by default) containing a tone description for the tone that was learned.

8. **This procedure first classifies to the ringback tone** provided by the switch and determines if it is single or dual tone. This information is then used to learn the disconnect supervision tone.

If the ringback tone and the disconnect supervision tone are not of the same classification (single or dual tone), then errors may occur during the learn mode phase of CPLEARN. The CPLEARN.C code could be modified to pause between dialing the extension/phone number and beginning classification. This would allow the program to both classify and learn the disconnect supervision tone.

(Dialogic Corporation)

Power Protection On Phone Lines

You need to protect your phone lines from the possibility of AC coming up them and blowing out all or part of your phone system. It happened to us. Someone touched 240 volts to two of our incoming 800 lines. That blew one of our PBX's trunk cards. It cost us $500 to repair the damage. We could have protected ourselves with $40 worth of surge protection inserted into our 66 block.

(Teleconnect Magazine)

Power Protection On AC

Every phone system needs AC power protection. TELECONNECT cannot say this more strongly. Lightning strikes can blow most phone systems. But high voltage spikes don't crash phone systems; they just cause phone systems to act funny. They may knock calls off the air. They may ring extensions when there's no one calling. They may send screwy error messages to the console. Make sure your system is backed by AC surge arrestors.

(Teleconnect Magazine)

Power Failure Transfer

Many PBXs have power failure transfer -- it's cheap disaster recovery. This means if your PBX crashes, you still have some single line telephones to make and receive calls. In most companies, no one knows if power transfer failure works or where the single line phones are. Put a regular "PFT Drill" together and make sure you test it once a month or so. Do the test over the weekend or during slack times.

(Teleconnect Magazine)

Power Outlets

Don't neglect to install plenty of power outlets in the offices you wire up. You don't need heavy wattage. You **do** need plenty of outlets. You need outlets for your cellular recharger, for your laptop recharger, for your modem (if it's external), for your computer, for your parallel printer, for your serial printer (for labels), for your fax machine. The list grows endless as the amount of stuff going into a typical office escalates daily.

(Teleconnect Magazine)

Power Problem Solvers

Power problems crash, upset and mangle phones, computers and LAN far more than anyone ever dreamed. Item: Our voice mail regularly dumped calls because our PBX's power supply was pumping out 9.8 volts DC, instead of 10.2 volts. We boosted it. The problem went away. One Monday, many of our UPS-backed computers and servers died because of a weekend power surge. Turned out many UPSs don't have surge protection! We've had more computer and phone disasters because of power problems than any other single problem (including lousy software).

Power problems come up phone lines. We once got 240 volts AC up our phone lines. Killed a thousand dollars of phone equipment.

Power problems come up power lines. We've had lightning surges. We've had low voltages. We've had brownouts. We've had blackouts. Everyone has caused horrible problems we could live without.

Both phone lines and power lines should be protected. Whatever money you spend will be more than repaid in saved time, aggravation and lost sales.

Do not skimp on power protection! The most important types of protection are: phone and AC line surge suppressors, power conditioners and uninterruptible and stand-by power systems.

Surge suppressors work by catching excess power (sometimes called spikes) rushing through either power or phone lines and shunting them away from sensitive equipment to a ground. Surges happen more often than you think and are caused by such diverse occurrences as lightning strikes, the return of utility power and cars slamming into telephone poles. It's that arbitrary. Our advice: Leaving for the weekend? Unplug equipment that won't be needed. Don't just turn it off.

A good surge suppressor should have a very low clamp voltage and a high surge capacity. Clamping level is the amount of voltage the protector will let through before cutting off anything higher. The lower the clamping level, the better the surge suppressor. Look for suppressors with a UL 1449 rating. They'll have a clamping level of 330V.

Another thing to look for here is some notification mechanism. If suppressors are damaged (happens all the time), they'll continue to pass bad power. A warning light is a good way to recognize a damaged surge suppressor, but best is an audible alarm. You also might want to incorporate a fast acting fuse or auto shut-off feature that will stop all current flow to protected equipment in case of a component failure.

Power Conditioners also handle spikes and surges, while filtering out most electrical noise. This is important because once noise gets into a network, it can cause "hidden power problems," disrupting logic circuitry and causing equipment to behave strangely. PBXs will do strange, intermittent things. (don't hesitate to reboot them.)

PC computer telephony equipment -- voice mail/auto attendants/interactive voice response devices, etc. -- may drop calls. You can get errors in transmission and network synchronization, garbled transmissions and corrupt data. In general, bad stuff.

Conditioners with built-in isolation transformers provide protection against both high frequency normal mode and common mode noise and will work automatically and repeatedly, so no indicator light is necessary.

Line regulators, often grouped into the category of power conditioners, control the voltage pumping to your telecom equipment, but do not necessarily protect against noise and spikes. Before buying any "power conditioner," read the label carefully.

Uninterruptible Power Supply (UPS) systems are the most complete and the most expensive type of power protection. True UPS protection handles uninterruptible backup power supply, line regulation, line conditioning and surge suppression. They provide battery backup to equipment during a full (a blackout) or severe (a brownout) power outage, allowing you to operate your computer telephony system until the power outage ceases or everything is safely shutdown. They're mandatory (for us) on all file servers and critical machines.

There are two types of UPSs -- off-line and on-line. Off line, or standby power systems (SPSs), are ready to turn on and supply battery power to phone systems, PCs, servers, nodes or workstations during power loss. A transfer switch senses the loss of power and turns on an AC source that allows the system to keep running. Off-line UPSs are cheaper because they don't have to be rated for continuous operation and they don't have to be heavy duty. Some don't provide power conditioning.

Some feature a regulated charger. Its job is to make sure that the system's batteries are always fully charged. That way, when an outage happens, you get the longest backup possible. Still other systems are "line interactive." They come with voltage regulators that monitor incoming power. When power drops a little low, the UPS gives the system a boost. The advantage is conservation of battery power. It doesn't always resort to full battery backup, saving the batteries for serious power losses.

An on-line UPS system is always on. It converts incoming utility power from AC to DC by means of a rectifier. DC power is used to charge the UPS batteries so they are always ready. When you use an on-line UPS, your computer is isolated from the utility power source and always has "clean" conditioned power.

Most of the newer UPSs have microprocessors to give you more access and control of your power protection plans. Think of them as computers protecting computers. They have displays that tell you when your batteries are running low or when they sense an impending power failure. They figure that out by monitoring trends. Big problems are usually preceded by fluctuation patterns.

Many have audible alarms and software that monitor the system. UPSs that monitor file servers are capable of sending warning messages across the network so workers can log off and a graceful shutdown can be performed. Automatic shutdown software is also available to protect critical data on the LAN. The software gently closes all the files and prepares the network for a safe discontinuance of power. Look for SNMP (Simple Network Management Protocol) support here.

Overall, LANs are more susceptible to power and noise problems than any high-tech platform. Often times their wiring actually forms ground loops, passing power glitches among network nodes with punishing accuracy. And network operating systems are constantly working in the volatile world of RAM memory and exposed time-critical operations.

Still, the Local Area Network is the most important platform in today's business world. That's where we keep our data. That's where we launch our applications. That's where we're adding intelligence to our phone systems.

The good news is that UPS systems have grown with the technology they protect. They're no long just rigid black boxes. They add their own intelligence to the computer telephony arena. Says Marc Vernon of Tripp Lite: "A smart UPS has now become a management and communication tool as much as protective device. For example, some UPS software packages can even provide paging features to remotely notify network administrators or users of power snafus on the network as they happen. It's cool. They'll use the new LAN CT platforms they're protecting to deliver critical protection information. Get it?"

(Computer Telephony Magazine)

Protection For ISDN Applications

There are plenty of hazards that await you when you hook-up to ISDN. Just as with analog circuits, there are spikes and transients on the phone line, spikes and transients on the AC line, sustained overvoltage on the AC line and brownout/undervoltage on the AC line. Without protection, you will get data corruption, system lockup, damage to components and possible loss of your entire system. It's easy to get protection. You can get surge protection on telecom connections and surge protection on all AC connections with combined protection units. These units provide common grounding between telecom and AC protectors. You can also configure your system with a UPS on critical components to avoid brownout/outage problems. See figure 7.1 for an example.

Remember that when you combining surge protectors with UPS systems, plug the UPS into the surge protector! Batteries are susceptible to transients and voltage spike just like any other equipment. This will ensure protection against from AC surge. You can also protect laser printers or other heavy equipment using a common ground.

(Panamax)

Figure 7.1 - Typical Protection / Backup Topology

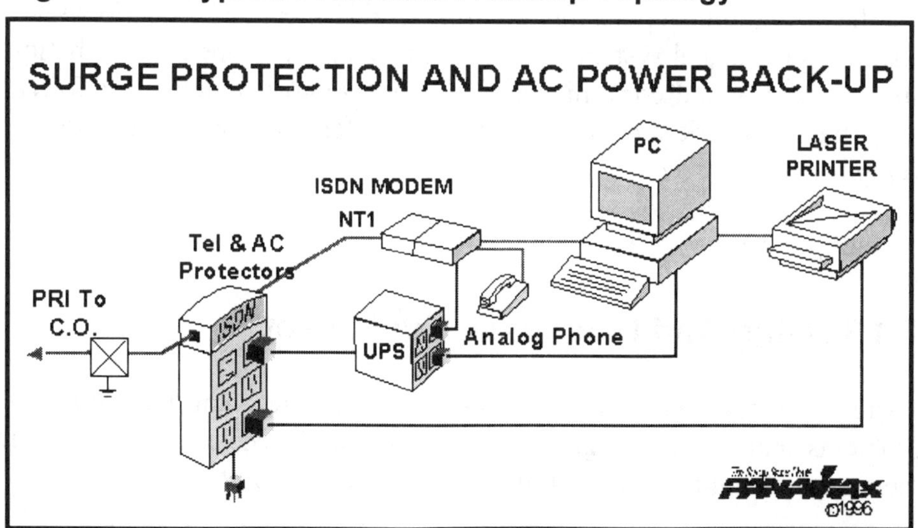

Reboot Your Phone System

The lights on one of TELECONNECT's Mitel consoles stopped. Yet the lamp test showed all the lights were working and the various test were working. And the fuse was intact. **Solution:** Turn the whole PBX off, count to 10 and turn it on. It's the same "fix" we do to our PCs when they act strange. If your phone system starts to act strange, turn it off, count to ten and turn it on again.

(Teleconnect Magazine)

Routing Tables On Your PBX

Check your routing tables for area code routing. Area codes are being added and changed all the time. Many ARS tables are set up to force non-defined area codes to the most expensive route -- DDD.

(Teleconnect Magazine)

Simulators For Testing

Invest in a network simulator for ASR Apps. This allows developers to test the recognition application in a "real world" telephony scenario without disrupting the actual system for which it is being developed. The functionality, reliability and convenience in addition to cost savings on CO/PBX lines make network simulators indispensable for telephony application development.

(PureSpeech)

T-1 Testing And Diagnostic Equipment

After your equipment is finally installed and tested, you may still have problems that relate to signaling protocol and parameters. Carriers limit their engineering responsibility to the demarcation point.

So, unless you have contracted for installation services, you should be prepared to verify carrier line parameters at your installation site. Without evidence indicating a problem with line performance, you may have a difficult time getting your carrier to respond.

You can reduce delays and avoid frustration by owning or renting diagnostic equipment that permits you to examine protocol timing and signal parameters such as signal and tone levels, tone duration, tone distortion, and interdigit delays. To verify T-1 signaling, you can use a T-1 test set such as the Tekelec, T-BERD, or Sage. If you need equipment to verify inband analog signals, the Sage includes an analog option.

T-1 Troubleshooting Equipment Companies

Tau-Tron
Westford, MA
1-800-TAU-TRON

Verilink Corporation
San Jose, CA
408-945-1199

Sage Instruments
Teaneck, NJ
201-836-1004

Telecommunications Techniques Corporation
(TTC)
Germantown, MD
1-800-638-2049

(Dialogic Corporation)

T-1 Trouble - Lock-Out

Check your T-1s (T-ones), which are high speed digital lines. Individual channels on T1s often get "locked out" by long distance companies. One company found that 22 of 48 voice channels on its two T-1s were locked out. Check at least once a week. Better yet, check once a day.

(Teleconnect Magazine)

Test Everywhere

Be testy! Be sure to test the working script out before turning it on live for the customers. You have to test. Things can look good on paper, but be very awkward in use, or not work at all! The design may be in error, or the programming may be faulty, but testing will pinpoint the problems and allow resolution before the customer is affected.

Test everything. The test plan has to include every option - even those that are wrong. Callers are human and they can be depended upon the make mistakes. During testing, you need to ensure that when a mistake is made, the proper response is given and the caller is given the opportunity to correct the mistake. Try every combination of entries, with all types of data.

Host up and host down. If the voice response system will be accessing a host system that is not always available, your test plan will need to include both the host up and host down condition. Even during the host down condition, some generic information (such as directions and hours) may still be available and should be accessible. Attempts to access information supplied by the host when the host is down should be greeted with a polite explanation which includes the expected up time.

Check the output. Look closely at any documents that will be faxed. Proofread them word for word. Check spelling. Make sure your company name is on them. Verify any information that is read out. Phone numbers, directions, addresses, everything that will be given to the customer needs to be verified. Make sure all of the information is current, too. Set up the mechanisms you will need to ensure the information is updated regularly.

Test fax on demand. To thoroughly test fax on demand, faxes will need to sent to a variety of machines and locations. Plain paper fax machines may respond differently to your documents than thermal paper machines do. (End of page signals need to be sent to thermal paper machines to signal when to cut the paper.) Local faxing may work, while faxes to long distance numbers may be inhibited. Find a way to check it all.

Test the recording. Listen carefully to the recorded voice. Is the recording clear, or is there background noise or static? Check the tone and pace of the speech. They should be consistent throughout the menus, not hurried and abrupt in one area while leisurely and friendly in another. Any noticeable problems should be re-recorded.

(Enterprise Integration Group)

Test Platform Economics

We needed a multi-port test platform to test a new fax product we had developed. We needed an 8-line test system that would simulate a central office by generating dial tone, answer the incoming fax calls, receive the faxes that were sent, and log the receive information (fax number dialed, time and duration of call, filename of received fax, etc.) into a local database for later review. One option was to buy a test system off the shelf for $50k. The other option was to build one ourselves. We chose the latter because we didn't need the bells and whistles, and because we figured we could do it cheaper. We succeeded and saved a bunch. We used two PCs connected by Skutch Electronics (Roseville, CA) AS-66 phone line simulator boxes, slightly modified. The second PC is the "Simulator" that we developed using Visual ProVIDE, two 4-line fax cards, and two 4-line voice cards (has a 486-120 motherboard, 16 MB ram, 1.2 GB hard disk).

Here is how it works: A fax phone line on PC #1 goes off-hook, expecting to hear dial tone before it starts dialing. The under-$200 Skutch Box is modified by clipping four resistors so it doesn't generate any dial tone itself, so it detects off-hook on PC #1's side and generates ring voltage toward the Simulator.

The Simulator answers the call, then immediately **simulates dial tone** by playing a high-quality (PCM format) prompt. When PC #1 sees the dial tone, it dials the fax phone number, which gets captured by the Simulator and logged to the database (dial tone prompt gets interrupted as soon as it sees the first touch tone digit). Then PC #1 starts sending the fax, and the Simulator starts receiving it.

We wrote the initial software and tested the first version in about a day, but it took a week to get most of the bugs out. Turns out the biggest problem was having the Simulator PC answer fast enough to keep the fax cards from complaining about not getting dial tone (for awhile we had to insert some delays on the PC #1 side). Remember, we are generating dial tone by playing a high-quality (PCM format) speech prompt recorded with dial tone. You can easily make bigger test systems by using PC's with more slots. Many cool enhancements can be added, like playing busy, fast-busy, and other tones instead of always playing dial tone--this will allow you to simulate less-than-perfect real-world conditions, essential to good testing procedures for outdialing (or faxing) systems. If you use a good CT application development tool, pick up a few modified Skutch AS-66 boxes and put a similar system together for yourself.

(Telephone Response Technologies, Inc.)

Testing The Entire Lifecycle

Testing your applications is a "lieftime" undertaking. Don't just test your application when it's put to work, but at every step of development as pictured in figure 7.2. When should you put "checkpoints" in place?

1. **When you're defining the problem** and developing the requirements specification with the customer. Now is the time to put testing measurement indices in place. Discuss each capability of the system, including L:AN connectivity, scripting and reporting. Write down how you intend to test each item as you are drawing-up the specification.

2. **Inside your functional specification.** Define a test point for every functional specification paragraph. If you don't you'll miss something and end-up testing it by trial and error later.

3. **In your design specification.** Consider all of the intricacies of your design. This includes the type of boards you will use, the platform and the operating system. Try to use suggestions form each manufacturer about how they test their own components. Work these observations into your testing.

Figure 7.2 - Testing Is Part Of "Life"

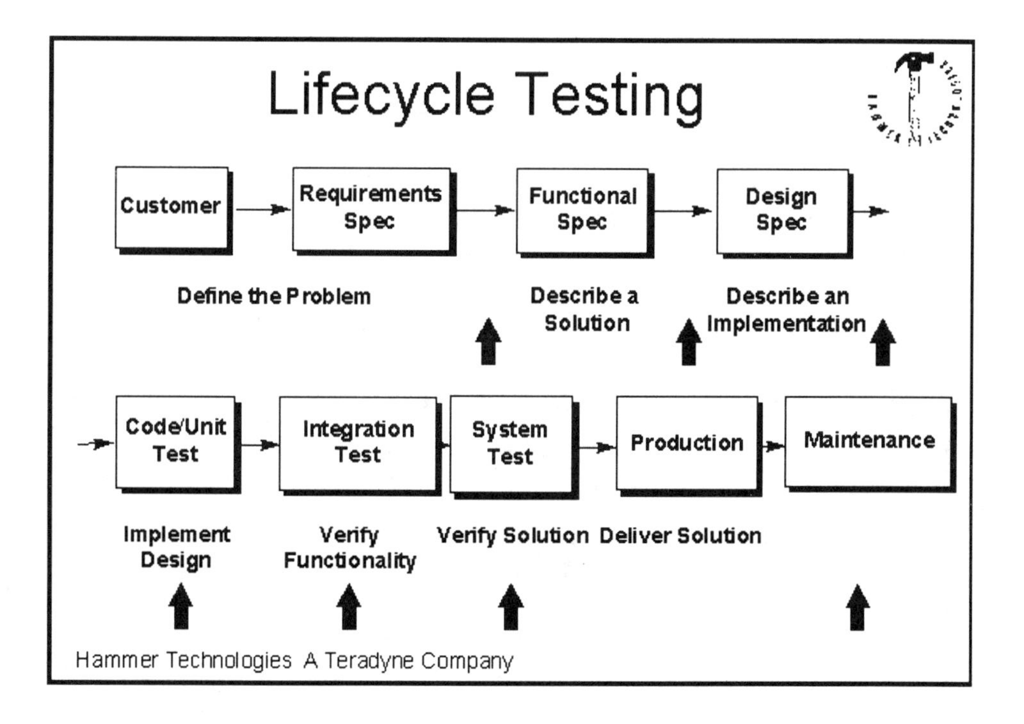

4. **As you verify functionality.** Don't just rip through your "dry run." Many of us have the tendency to be so happy with a project, that we forget to test every inch of each function. Just because it looks like it's working doesn't mean you've tested it.

5. **During the system integration and overall system test stage.** Hook everything up to your system. This includes power protection units, battery back-up, the file server, data circuits and PBX extensions. Your system is part of a bigger whole, so you have to test every link and interaction.

6. **At production.** Get input from installers, users and other vendors when you do your testing. Their observations can be very valuable. Ask everyone who even remotely comes in contact with the system.

7. **All the time.** Yes, there will be changes, updates, and upgrades not only with your own platform – but with those things attached to it. You need to make testing a part of your system's life.

Testing Is Expensive. Over time, the cost to develop a system goes down below the cost to test the system and fix it as shown in figure 7.3. Testing costs go on forever. The best hint here is to put lots of effort up-front in developing a product that infinitely testable. Make sure a lot of work goes into building-in test parameters. Ask yourself these questions:

Figure 7.3 - Cost Of Testing Over Time

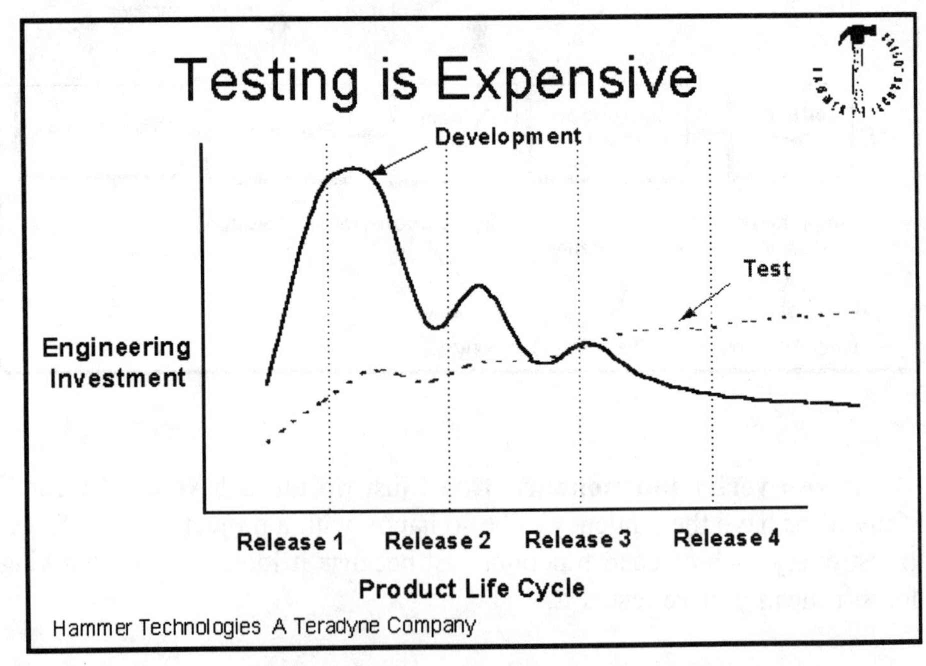

1. **How will we test this feature?**

2. **How will this feature impact other things connected to my system?**

3. **How will this (new) feature affect the performance of my existing code?** How about the old test plan (regression testing)?

The Cost Of Not Being Test-Minded. You can't spend too much time working on a robust test plan. The more you invest now, the less costly testing will be in the future. Just look at the costs associated with removing defects in your product. In order to prepare, execute on and finally remedy one defect, you will have put in an average of 45 hours. Do the math as suggested in figure 7.4. How much is it really costing you **not** to plan for defects with good specifications and "test-minded" development?

(Hammer Technologies, Inc.)

Figure 7.4 - Testing Payment Plan

Effort in Test Creation

Defect Removal Time Utilization

	Preparation	Execution	Repair
Unit	25 hrs/kloc	2 hrs/kloc	10 hrs/Defect
Functional	35 hrs/kloc	3 hrs/kloc	15 hrs/Defect
Integration	45 hrs/kloc	4 hrs/kloc	20 hrs/Defect

Hammer Technologies A Teradyne Company

Thunderstorms

Be safe: When thunderstorms flash and crackle, unplug **every** piece of electronic gear, including and especially your phone system, from its ac outlet – whether they're protected by a surge arrestor or not.

(Teleconnect Magazine)

Troubleshooting for T-1

Segment the T-1 installation process into four steps. In this way you can easily isolate problems to specific equipment.

STEP 1 - Check the Clock Setting. Incorrect clock setting is one of the most commonly encountered problems during T-1 installations. Timing slips will occur if the clock is incorrectly set. The majority of T-1 network interface applications will receive synchronization clocking from the network. You should set your gear so that it will take clock (derive its transmit clock) from the network.

STEP 2 - Test your network interface card. To test whether your device is set to transmit and receive correctly by using a loopback plug. In the case of a Dialogic DTI/1xx, follow these steps:

On a 15-pin subminiature D male connector, solder a wire from pin 1 to pin 3 and a wire from pin 9 to pin 11.

Plug the loopback connector into the female DB-15 connector on the rear bracket of the DTI/1xx board.

Power on the PC in which the DTI/1xx is installed. The green LED on the rear bracket of the DTI/1xx should light. Be sure that the remote loopback test switch on the rear bracket is OFF (switched to the right). The remote loopback test indicator (bottom red LED) should NOT be lit.

Test T-1-to-CSU Connection if the light is green. If this is not the case, recheck the above steps, then call for technical support.

STEP 3 - Test T-1-to-CSU Connection per the CSU documentation.

STEP 4 - Connect DTI/1xx to CSU. As a last step in the process, connect the DTI/1xx to the CSU. Make sure that the transmit and receive pairs from the DTI/1xx are routed to the correct terminations at the CSU. In most cases, once you are getting a signal from your carrier, the green LED on the DTI/1xx rear bracket should be lit if everything is working correctly.

GREEN LED. You can still get a green signal if you have mis-wired the tip and ring connections in your cable. If you are showing a green LED, but you are still getting errors from the network, check to make sure that the tip and ring connections on your cable are wired correctly.

GREEN and YELLOW LEDs. You possibly have a short in the transmit wire pair (the pair connected to the transmit pins on the DTI/1xx). Remove the CSU-to-DTI/1xx cable and test it for shorts. RED and YELLOW LEDs (case 1).

The transmit and receive pairs between the CSU and the DTI/1xx are possibly reversed. Check your CSU-to-DTI/1xx cable for incorrect connections. Consult your CSU documentation for proper pinouts and signal direction.

RED and YELLOW LEDs (case 2). You possibly have a short in the receive wire pair (the pair connected to the receive pins on the DTI/1xx). Remove the CSUtoDTI/1xx cable and test it for shorts.

(Dialogic Corporation)

Troubleshooting PC Boards

If you're having problems installing a complex set of computer telephony boards in a PC for the first time, simplify! Take out all boards, drivers and features except the minimum you need for the call processing cards. For example, re-name or delete CONFIG.SYS and AUTOEXEC.BAT and substitute new versions that just load the drivers for the C-T cards.

Under Windows NT disable the other devices using Control Panel. Remove cards such as LAN cards, I/O boards, sound cards, etc. Disable features such as BIOS shadowing, disable devices such as COM and printer ports which may be on your mother-board (you'll need to get into the BIOS setup program to do this). Chances are that with this "minimal" setup, your boards will work OK. Then you can gradually re-install all the stuff you took out - when things stop working, you know that the last thing you added is causing a conflict with your C-T boards.

(Parity Software)

Trunks - Check For Bad Ones

Always run a trunk summary report from your CDR at the end of each month. You will be able to spot any outgoing trunk malfunctions (no traffic). We had a WATS line accidentally disconnected for two and a half months before someone noticed the lack of activity on the trunk summary report.

(Karen Osuch)

Water Alert

There are devices which sit on your phone or computer room floor. If it feels any water, it squeals for 24 hours. Operates on a 9 volt battery. Costs under $100. Highly recommended. One maker is Dorlen Products (414-282-4840).

(Teleconnect Magazine)

Chapter 8

Buying and Selling Tips

Some of the finest ideas for computer telephony systems came out of the desires we have as customers. Why can't I just call a number and get through instantly? Why can't I just get that product alert faxed to me automatically? When is my order going to arrive? All you need to do is listen to yourself and your fellow employees – and you'll have a computer telephony wish list within minutes.

If you're reading this book, you're either in the market to buy a new system or upgrade, or you want to know how to sell with more panache. You'll get both by reading this chapter. This is perhaps the most comprehensive collection of buying and selling tips ever put together for this industry. You'll get checklists, questions to ask and learn what to watch out for.

Pay particular attention to the sections on "what questions to ask." Feel free to copy them and use them for your own RFP (request for proposal) if you are now budgeting for software, hardware or services. You can also use these lists to perform a sanity check on your own offerings if you plan on developing any of these products.

Alarm Subsystems

Behold the everyday PC configuration as pictured in figure 8.1. There's one processor, one power supply and one hard drive. This is fine for one person, because if it breaks - it only effect one person. However, most service bureaus, telcos and mission-critical call centers need guaranteed up-time beyond what is available from a regular desktop PC. This is due to the fact that thousand of users (over the phone) will be affected if the system fails. Standard configurations on fault tolerant PCs include rack-mounted chassis and dual power supplies. Many come with a pair of mirrored RAID hard drives. All of this is ruggedly packaged in rack-mounted gear that's right at home in a central office environment.

Figure 8.1 - Standard PC Layout

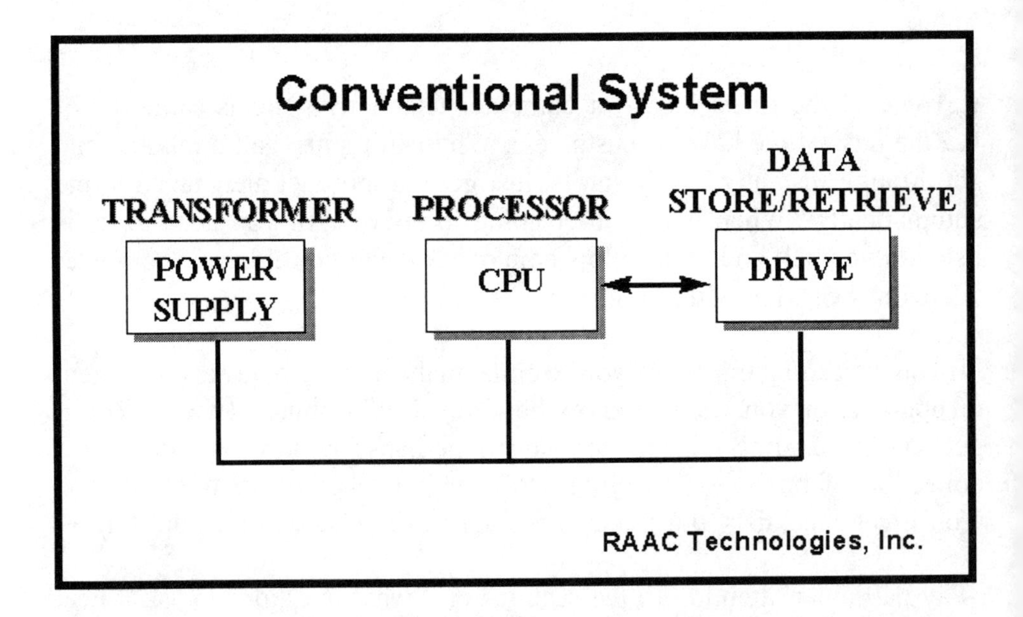

Reasons To Select A Passive Backplane System. Scalability is perhaps one of the most compelling reasons to use passive backplane computers in a fault tolerant environment. Having multiple slots in passive backplanes make it easy to grow the number of lines you need.

You can get ISA backplanes configured for up to 20 Total Slots; EISA - Up to 15 Total Slots; ISA/EISA - 15 EISA / 5 ISA; and ISA/PCI - Up to 19 Total Slots on some systems. Passive backplanes are much easier to service, because of the way you populate the system (there's no bolted-down motherboard). You can easily populate a passive backplane system with one or more single board computers (SBC), peripheral cards, video, SCSI, Etc. Of course, all those slots make it easier to pack the system with telephony cards.

Why Choose Fault Tolerance? A conventional PC uses one transformer, one processor and one hard drive. This is fine for most commercial or personal use applications. but when you design a system that will be used by thousand of people each day, it's a good idea to make a bigger investment. The whole idea is to achieve less down-time, because down-time costs money due to service interruption. A fault tolerant system gets around this problem by replicating critical points of failure. Figure 8.2 shows hardware failure rates by component in typical PCs.

Figure 8.2 - Typical PC Component Failure Rates

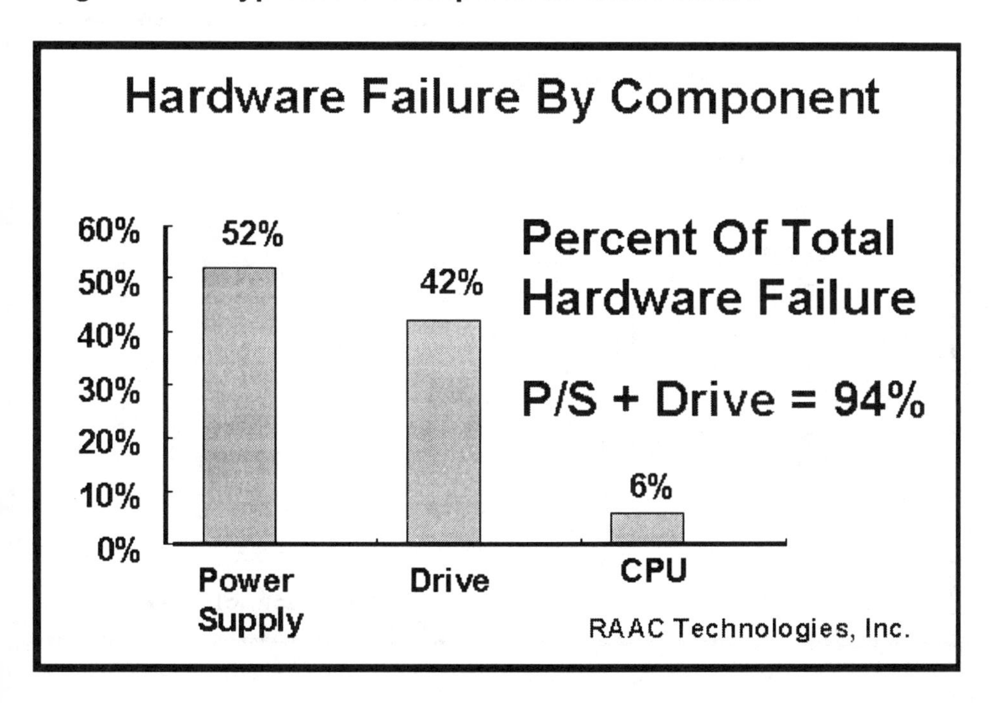

As you can see, the biggest offender (besides fans) is the power supply with over 50% of total hardware failure compared to the other components. The biggest runner-up is hard drives. So it makes sense to concentrate on these components and make them redundant. Notice that CPUs don't fail as much as these other parts. With an SBC card, it's easy to swap-out the CPU anyway. That's the idea in building a fault tolerant system – make the [parts easy to swap-out.

Figure 8.3 shows a typical fault tolerant topology. Notice that there are multiple power supplies, CPUs and disk drives. It's no wonder why these components are duplicated – they are the ones that fail most often. But what if there are failures? Just how are you notified about them and what actions are taken? This is what you have to keep your eye on in selecting a good fault tolerant platform. Processor switchover, alarming and system monitoring functions are what separate a decent fault tolerant platform from a so-so one.

Figure 8.3 - Fault Tolerant PC Layout

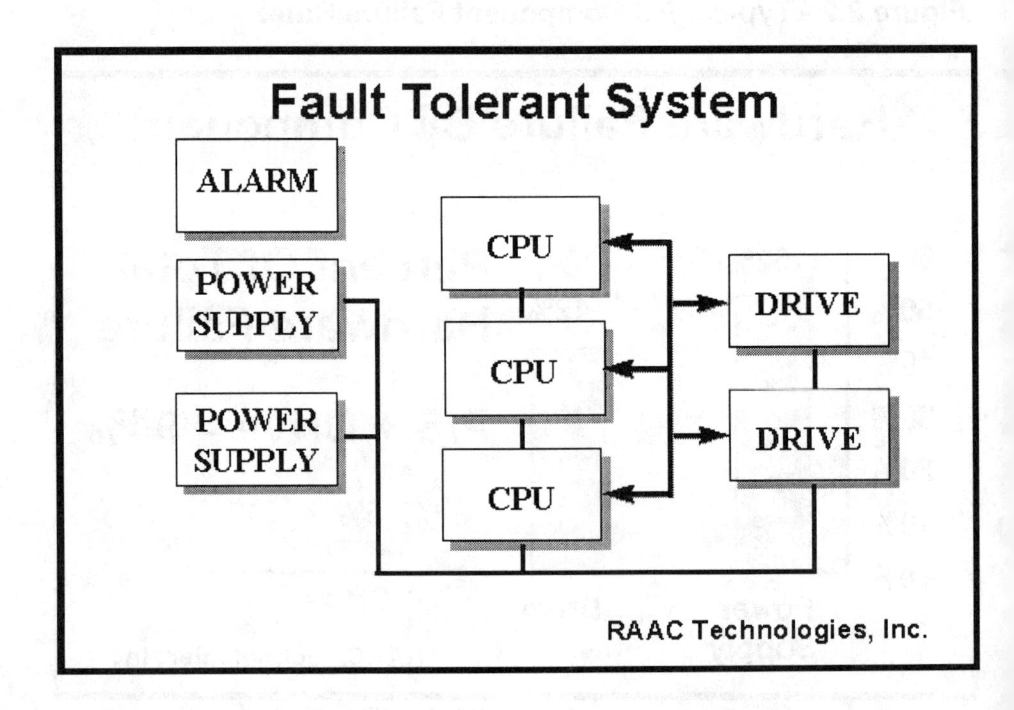

Processor Switchover allows your computer platform to "switch-over" to a stand-by CPU in the event of a primary CPU failure. It lets you recover from a CPU crash no matter what operating system is in use.

This kind of recovery system can be a real lifesaver -- especially for service bureaus and large call centers. If you need an application that handles thousands of phone calls every hour -- you don't want one minute of time to be wasted on fumbling with card-swapping PCs.

System crashes resulting from CPU failure can be triggered by:

- **A software application error that "hangs" the computer**

- **The failure of a LAN, Fax, or voice card**

- **The failure of the CPU itself**

What To Look For In an Alarm Subsystem. First and foremost, look for alarm capabilities that are physically separate from the system's microprocessor and not on the system bus. A well-designed alarm/monitoring system should operate on its own microprocessor in its own chassis. The alarm subsystem should be battery powered, so it will still operate for several hours even during primary power failure or application processor crashes.

The alarm subsystem should be battery-backed and provide a variety of I/O including LED, LCD, Tone, and RS232 output.

The sub-system should have a local display (LCD, etc.) for detecting, reporting and recording critical, major and minor failures. Both audible and visual alarms are a good feature. The system should monitor different points including fans, enclosure temperatures, power supply voltages, UPS, modem and CPU. Look for a system that provides a continuous time-stamp log of events. The alarm should also communicate with the CPU for application error reporting.

In addition to switchover functions, the alarm should be programmable enough to reset the primary system automatically (potentially resolving a software "hang"). You should be able to automatically call an emergency phone or paging terminal. This is especially helpful for unattended, remotely located computer systems. These means that if your customer-pleasing, revenue-producing application fails all of the sudden -- the alarm subsystem will sense this and issue local or national pages to your service staff. You can also make calls via modem to remote terminals. This means better response time to system failure.

Really creative developers can use "user-definable features" like this as an alert of unusual traffic conditions, "winners" on the line and other special events. That's because you can manipulate the paging with instructions from the application. RAAC calls this a. I call it smart.

The cost of adding processor switchover can just over double the original system cost. Plan on an original $13,500 to cost as much as $27,500 to get you switchover technology. This would include duplicate CPU, video and RAID cards. You'll also have to duplicate your computer telephony cards in order to be back on line as quickly as the standby system boots. That's the whole idea -- minimum downtime. When you count your daily transactions in the thousands or millions of dollars, this is a small price to pay.

(RAAC Technologies, Inc.)

Buying Computer Telephony Application Tools

Open Architecture Questions To Ask (From Dr. CT -- a.k.a. Chris Bajorek). Does the software vendor let you add ready-to-use modules to applications with a minimum of hassle? Your application should be able to communicate with host computers, access on-line databases, accept credit-card charges, provide fax-on-demand services, handle line switching and resource sharing, and so on. Virtually all voice software vendors charge extra for add-on features like this; you may not need them now, but explore their depth of add-on module coverage. Find out how easy it is to use the functions these modules offer - if it was done right, it will simply extend the command set of the base package.

Is a C language development tool kit available? Without it, you can only do things that your vendor has already thought about and supported. Even if you don't write the C code, you should be able to support virtually any new hardware device or software process. Ask how much effort is involved in passing parameters (e.g., touch-tone digits) between user-written modules and your application. Some vendors give you a lean interface that is difficult to use.

Application Generator Questions To Ask. How easy is it to use the program's features, day after day? Application generators should make these things easy and intuitive. The user interface should use pull-down menus, choice lists, and command buttons to make the system easier to navigate.

Does the vendor offer on-line help for all commands and application functions? Context-sensitive hypertext help is best. This minimizes the need to look in the technical manuals. When a company says they have on-line help, ask how it works or, better still, get a demo. Some products have a poor concept of what help really means. If the vendor says the entire technical guide is on-line, consider it a plus.

Can you record and edit speech messages and their scripts without leaving the main AG program? Some do, some don't. If the product forces you to run totally different programs to deal with messages, you're wasting valuable time.

Speech Message Questions To Ask. Speech messages, those disk-based *sound bites* that callers hear (and record when they leave messages), are a very fundamental resource to PC-based voice applications.

Can the vendor handle applications that use thousands of on-line speech messages? It is most desirable if the vendor can provide this capability. Today's voice/call processing applications must be able to offer callers a multitude of services in a single system, from voice mail to fax-on-demand. The more caller features you offer, the more speech messages you will need. A minimum should be 50,000 to allow for growth and complex applications.

Is it easy to access messages from separate physical drives, sub-directories, and ram disks? The answer should be yes if you are concerned at all about performance. Operating systems like MS-DOS cannot efficiently search for more than several hundred speech messages in a single subdirectory, and ram-disks are a great place for your most-often-played messages.

Is the location (i.e., drive, directory, and filename) of each speech message *hard-wired* into the application? Hard-wired speech messages are a thing of the past. You can be faced with a significant editing task every time you want to make significant changes to your application if this is the case..

Does the vendor support both discrete and *packed* speech message files? Discrete message files contain speech information for a single message; packed message files (sometimes referred to as Macro Files or VAP files) contain speech message and script information for hundreds to thousands of speech messages. The latter provide improved access-time and thus better performance - if your application needs thousands of messages on-line, these are the only way to go. If your vendor can handle both types of speech messages, you've got a winner.

Can you easily create and edit scripts for all speech messages in your application? Some software tools limit the number of messages that can have scripts, and some don't support script maintenance at all. If you are using an AG product, choose a vendor that lets you maintain scripts directly in the speech message editor.

Can you reference messages symbolically in applications? Script languages always allow you to reference messages using symbolic names like GREETING, ENTRY_ERROR, and the like. Some AG products reference messages by number and others allow the use of names. If this is important to you, be sure to ask.

Script Language Questions To Ask. What do the script language statements look like? Do they look like the BASIC, C, or PASCAL languages you may be familiar with? Ask to see a sample application listing. The more arcane the language syntax is, the harder it may be for you to learn.

The best choice is to have your script language emulate the C language - this should make it relatively easy for your application developers to master it.

Is it possible to easily move an application between the script and AG environments? If the vendor supports both script and AG development, you will want the flexibility of taking script-based applications and loading them into the AG (and back to script again). This is especially important if you will have more than one application developer.

Is an integrated script language development environment available? This saves even more time - you can edit script files in multiple windows, compile with a single keystroke, and have your cursor directed to each compiler error with a minimum of keystrokes. These environments (like Borland's Turbo C++ product) even allow you to position your edit cursor over a keyword and get instant hypertext help.

Run-Time Engine Questions To Ask. After you have developed your application, most voice software products require you to load one or more application files into the companion *run-time* program - that is, the program that executes the step-by-step instructions you created using the development tools.

How many voice lines can be supported in a single PC? Most vendors support at least 24 lines (a single T-1 span) in a single PC-based system. If you need support for more than 24 lines, find out how many PC's are required since each additional PC increases your total system costs. If support is claimed for more than 48 lines, ask how they achieve acceptable performance.

How does your vendor treat system errors? Most applications must run 24-hours a day, 7 days a week. Callers have little patience for equipment or application failures. Be absolutely sure that run-time failures generate a clear, concise critical error message that alerts the system operator immediately. Not all software vendors consider this important. Some vendors even omit documentation on errors altogether.

Ask for a complete listing and description of all possible run-time errors - the longer the list, the more they consider this a key issue.

Does the run-time system accumulate and save call statistics? At some point you are going to want call statistics from your run-time system. Without them, you cannot really know how successful your application is. Figure 8.4 below details a minimum number of CT activity statistics for any system.

Figure 8.4 - Sample Call Statistics

Item	Call Statistic Type
1	Total daily calls for all lines.
2	Total call-minutes (total calls times the length of each call - this gives you a measure of how much on-the-air time has been consumed by callers)
3	Total time all phone lines were busy (this will let you know if your system is not able to respond to all incoming calls).
4	Total calls for each voice line (this will alert you to hardware or phone line problems if one or a group of lines shows low counts).
5	Number of times specific application areas are executed by callers - this tells you which parts of the application are being used (or not) by callers. It lets you quickly implement changes to help callers through areas that maybe weren't so clear.
6	Call detail statistics: time, date, and duration of each call along with credit card, PIN numbers, etc. Some applications are billing callers based on this kind of information.

Is it possible to remotely control your run-time system? Say your customer calls you and needs an urgent change made to his application. How will you respond if you don't have someone at the customer site to help? UNIX application environments have remote console access as a built-in feature of the operating system; MS-DOS platforms do not. Many voice VARs use programs like PC-Anywhere or Carbon Copy to give them dial-in access to customer voice systems. This makes it possible to do everything from uploading new applications to printing statistics reports - all without leaving your cozy office.

Be sure your run-time platform peacefully co-exists with remote control programs. Ask what programs have been validated by the vendor.

Performance Questions To Ask. Run-time engines support multiple voice lines at the same time by dividing up the PC's single processor (e.g., an 80386) into a number of *logical* processors: each handles the application for a single line. Whether your vendor uses preemptive multitasking (e.g., UNIX, OS/2, or a special kernel written under DOS) or a multi-line state machine, it should not matter to you. What does matter is how your vendor defines acceptable performance and what tools you have to measure and improve it for each application you develop.

How does the vendor define acceptable performance ? A general *rule of thumb*: if it takes longer than 1 second for a voice system to act on a caller's touch-tone response, it is not acceptable performance. Ask your vendor if he agrees and, if so, what guidelines you should use to achieve this level of performance or better. Expect guidance on what processor your PC should have, minimum access time for your hard disks, and how much PC memory you need - each can have a significant effect on performance. If performance issues are glossed-over by your vendor, not only may touch-tone responses be seriously slowed but speech messages can actually be broken up on a poorly-designed system.

Are you given tools to measure performance? If a vendor side-steps the issue, side-step the vendor. At least one vendor provides a special run-time screen that shows a performance index display. If there are no built-in performance tools available, you will have to add your own low-level timers and counters and a method of saving and printing this information periodically.

Are tools provided to improve performance of applications that need it? A number of system parameters can usually be changed that will yield a net improvement. While it is true that spending lots of money can help (e.g., buy the fastest PC and hard disk you can afford or buy an expensive memory-intensive caching disk controller), there are more sophisticated ways to approach the problem. One trick: move commonly-accessed speech messages to ram-disks.

Does the vendor understand the effects of file fragmentation? If your system is running on an MS-DOS based PC, your vendor should be aware of the effects of file fragmentation on your hard disks - especially if you record speech messages from more than one caller at a time - this is virtually guaranteed to cause fragmentation. Performance is adversely affected since the disk drive head has to move longer distances (and thus consume more time) while recording and playing the message. If recordings are made to LAN-based file server disks, fragmentation is managed by the LAN software itself (e.g. Novell 3.11). Your vendor should give you the ability to automatically run a defragmentation program as an off-line process once a day.

Debugging Questions To Ask. No matter how sophisticated your tools are, you still should verify every application step to be absolutely sure your logic and speech messages match perfectly. This process is called *debugging*.

Are special tools available for debugging your applications? If not, you will be spending much longer than necessary getting your system up. You will find the depth of application debugging tools varying considerably from vendor to vendor. Some offer virtually nothing in the way of debugging support while a few have fairly sophisticated run-time analyzers to speed the process.

Is *single-step* debugging of applications supported? This feature, sometimes called trace mode, makes it possible to manually execute applications a step at a time, giving you time to compare what is actually happening with what you think should be happening. Typically, a screen displays *source* statements about to be executed. With script-language-only products, you should see several language statements before and after the highlighted one to be executed. If you are using an AG, you should see a representation of the screen step about to be executed and be able to temporarily view application steps before or after this point.

Does the vendor support *breakpoints*? You should be able to set a number of breakpoints - symbolic names for application segments where you want to stop if they should be executed. For example, say you wanted to halt execution if particular choices were made at a specific menu.

This feature lets you run most of the time at real-time speed but manually single-step after a breakpoint has happened. Be sure you can easily add or delete breakpoints in the middle of a debugging session.

Are *watch commands* supported? Watch commands allow you to *watch* the value of selected application variables while the application is executing (e.g., see the last-entered touch-tone digits from a caller). A portion of the single-step screen should display these variables. As well, you want the ability to change any application variable and resume execution - this allows you to override application errors and continue testing without having to edit the application and start your test call over again.

Documentation Questions To Ask. Nothing is more frustrating than trying to plow your way through a poorly-written technical guide looking for an answer to a simple question. Here are a few things to look for.

Can you order the technical guide before you buy the product? Order the actual technical guide first. Expect to pay $50 or less for each guide. Some vendors give you credit toward the product when you buy. Browse through the guide and get a feel for how it is arranged.

Is the table of contents clearly organized? Can you quickly tell what is covered in each chapter by the section titles? Does the guide have a large enough index? Does it make good use of screen captures, illustrations, and tables?

Is a separate Run-time Reference Available? If you plan on selling turn-key systems to end-users, you will have to give a basic operation guide to your users. If your vendor has thought about this and provided you with one, it will save you valuable time.

Is there a separate chapter on error handling? Be sure system error messages are clearly listed and defined in the run-time reference. Some vendors don't document any errors -- if an obscure error happens after-hours (they usually do), you're in big trouble!

Technical Support Questions To Ask. Until you have to call support, you may not appreciate how valuable a resource they can be in your time of need. Especially if the technical guide can't help, technical support folks should be equipped to give you quick suggestions or resolutions.

Does the vendor charge for support? If it's free, great, but many charge for support. Some allow you to purchase support on an hourly basis while others offer two, six, or twelve month plans. You may get a free introductory period after your initial purchase. When checking the value of the support plan, ask about discounts for software upgrade releases - some companies offer this. In general, purchasing support in larger blocks should be more cost effective.

What response time should you expect? This is closely tied to the above question. If you are paying for support, you should receive faster responses to your calls. If a support person can't take your call immediately, expect a return call in one to two hours at the most.

What support services are available? Support services vary from one vendor to another. It may only mean you can call their bulletin board for help. What you really want is a range of options, including access to support specialists who are trained to do one thing quickly and professionally - answer your product questions.

What is the size and technical depth of the support staff? Ask how many full-time support staff are taking calls - if the answer is one or less, be very cautious: your callback may take longer than you want. Ask your vendor how long support staff members are trained - the longer it is, the better the chances they have handled your particular question. Ask how long support staff members have been on the job - the longer, the better.

How can you get the latest bug and new product release information? It is essential for you to get this information periodically. Most vendors print quarterly newsletters.

Some vendors give you unlimited access to this information via an on-line fax or Bulletin Board System (BBS). BBS's are also useful for downloading new software versions if you need it yesterday.

Vendor Questions To Ask. An important part of your pre-purchase research is qualifying the vendor itself. If you want the relationship with your vendor to be good for the long term, here are a few areas to explore before deciding to go with its products.

How long has the vendor been in the voice software business? The voice industry is not actually that *new*, the more perspective a potential vendor has, the better they can anticipate your problems and get you on the right path. Three years should be considered an absolute minimum.

Are price policies clearly-defined? Ask to see a standard price list. If it takes more than a few minutes to determine how much it will cost to put your application on-line, the vendor may have more hidden costs. Most voice software vendors charge a fee for run-time licenses - that is, every time you install a new system, you purchase a license (usually accompanied with a hardware protection device) specifically for the number of voice lines you want to run. See if you qualify for special OEM pricing.

How are product returns handled? Look for a 30-day trial period on all voice software products. If you have to ask for this privilege, be sure you tell the salesman to clearly state the return offer details on your invoice and indicate when the trial period starts (hopefully, it won't start until you receive the package). Keep in mind that hardware purchased from most vendors is not returnable unless special considerations are made beforehand.

Does the vendor practice full disclosure of all known product problems? A particularly annoying practice of some software (and hardware) companies is they rarely acknowledge that problems or bugs exist in their products. The hard reality of complex software products is that it is rare when no known bugs exist. Any company that claims otherwise is clearly more worried about public relations issues than they are about your ability to develop reliable applications.

Ask to see a copy of the release notes for the last major product release. Were past problems and their solutions openly discussed? If not, you will have no opportunity to intelligently work around them.

If you discover a software problem, how long will it take the vendor to fix it and get you a new version? Pose this question to 10 vendors and chances are you will 10 very different answers. What you are looking for here is a general understanding of how a potential vendor treats the subject. A reasonable answer might be anywhere from one to two weeks, but it is common for larger companies to take much longer releasing new software versions.

More Questions To Ask. Look for a company that offers a range of tools. Non-programmers may want to start using forms-based tools, then switch to a high-level script environment as the application expands. The best way to make this transition is to use an integrated environment script product - where the text editor, compiler, and hypertext help are all supported from the same environment.

How long will it take for me to develop an application? Describe your particular application, then ask this. Since it depends on skill level, the amount of time it will take varies: don't expect an exact number but try to get an idea of what it will take. Ask this question two ways: How long will it take me, and how long will it take an experienced VAR?

What roadblocks will I run into developing applications? This is an important question because no software environment is perfect. Look for a truthful answer that acknowledges where users typically run into problems. If the vendor denies that his customers run into any problcms, look for another vendor.

If I am not happy with your software product, what are my options? Look for an answer that won't leave you high and dry if things don't work out. Expect either a 30-day trial period or request a low-cost demonstration version of the product to help you get familiar with it before you commit the big dollars.

If I don't have time to do the application myself, can you find someone to help me right away? Many vendors can refer you to system integrators who will work on your application on a fixed-price or hourly basis. Find out how long they have used the software product you are interested in - one year should be considered a minimum, with two or more even better. Remember, before beginning any such project with a consultant, always state the project requirements in writing and execute a consulting agreement that clearly specifies the project billing rates and terms.

If I purchase run-time support for a specific number of voice lines, how do I later support more lines? Since the vendor's development tools presumably make it easy to develop voice applications without regard to how many lines are being handled, find out what must be done to expand to more lines. At a minimum, you need to add more speech cards. Hopefully, only minimal actions need to be taken with the application itself: you may need to edit a few startup parameters to signify the new number of active lines. Expect to pay for the difference between the new number of lines and what you have running currently. Usually, the vendor sends you a new set of disks and a hardware protection device and you return the old device to them.

Do your software tools allow me to protect my application development efforts from software piracy? Yes, in this day and age there are still people who would love to take your most feature-rich voice application and use it, royalty free. There are two basic ways protection is offered:

1. **Encrypting and password protecting** the distributed application files so unauthorized persons cannot edit them.

2. **Using hardware protection** (dongles) to prevent unauthorized users from running it. Be sure you ask your software vendor if and how he provides this protection.

(Telephone Response Technologies, Inc.)

Application Generator Buying Tips

1. The software should have specific modules and/or script-level routines to access the APIs of the latest in PC switching resources, including cards from Amtelco, Dianatel, Dialogic, Excel and NewVoice.

2. And, because you'll want these types of "interactive" switching/information systems to handle callers based on automatic inputs like ANI and DNIS (besides traditional voice response where callers are prompted for touch-tones), they need to work with digital PC network interfaces as well.

3. This is a new frontier. Before buying, get real-world application examples from the application generator vendor and their developers. Also inquire about special developer seminars. The better vendors hold them for these advanced applications.

4. Pay attention to whether the product offers Caller ID/ANI support, what databases it can connect to and whether it supports specialized switching functions like international callback.

5. Can you monitor and control your running app? Can your clients? Can it be done remotely? Or do you have a Frankenstein's monster on your hands, running amok at the first opportunity, treating callers to a text-to-speech rendition of Henry Miller's **Tropic of Capricorn**?

6. If you want to develop your own system, look for an application generator with good debugging tools. That will simplify the job of generating and collecting usage statistics and maintaining an error log. It should allow non-technical supervisors to monitor system activity and shut the system down in case of trouble.

(Computer Telephony Magazine)

Application Generators For UNIX

Begin by selecting the Operating System platform that your application will run on. Use UNIX (more expensive, harder to use) for hosting multiple applications on the same platform or if your application is deemed to be mission critical. otherwise use a DOS/Windows-based application generator (less expensive, easier to use) if the application is not expected to grow significantly. Some advanced application generators give you the best of both worlds, allowing you to work with easy-to-use Windows-based tools to create and manage applications that an subsequently be run on robust UNIX platforms.

Check that your application generator supports beginner, intermediate and advanced user mode. GUI-based application generators are great for first time users but if you plan on developing many applications, dragging and dropping is not the most efficient way of going about it. Entering script commands can be 2 or 3 times faster.

Select an application generator that comes with simulation and stress testing tools. You don't want to use live callers as guinea pigs or risk having your system crash or slow down under sudden heavy loads.

If selecting a "stick the boxes together" or "tree building" type of GUI application generator, make sure you have the ability to drop down to a script language at any time or anywhere in the application creation process so you won't be limited to only filling-in forms the application generator maker gives you for each open box. Forms-based systems cannot always anticipate all of your future requirements and you may not have the luxury of waiting for your application generator maker to modify its software to meet your unique requirements.

Select an application generator that supports Open Standards. you don't want to be locked into a proprietary (read expensive), single vendor solution. Also check that it supports advanced telephony features like text-to-speech, automated speech recognition, fax, ISDN, etc. and that it gives you tools to interface with wide variety of PBXs and host computers.

The application generator should allow you to easily create and manage multi-lingual voice prompts. You never know when you will need it to build applications for the global market or to deploy applications into bilingual (i.e. Spanish/English) markets.

(MediaSoft Telecom)

Application Tool Software Choices

One of the most important aspects of successfully implementing an IVR solution is choosing an application development tool. As we all know, after all, it's in the design and development of the actual application that, metaphorically speaking, your IVR rubber hits the road. So, given the sometimes confusing array of application tools available in the market today, how do you choose a package that will enable you to develop your application as quickly and easily as possible? Following is a list of key features and benefits to consider.

First, make sure the application package is designed specifically for the creation of interactive voice response application solutions. Otherwise, you'll have to customize an off-the-shelf development package with specific code, which will cost you money and time. While you're at it, be sure to look for a tool set that is portable to a wide range of runtime application server platforms.

Second, give some thought to your developers. How much experience and knowledge in IVR application development do they have? Then choose a development tool that addresses a broad spectrum of user skills and preferences. Generally speaking, this means a tool with graphical user and menu-driven interfaces that also accommodates the incorporation of subroutines and programs written in other languages such as "C."

Third, look for a tool that's rich in functionality. A lot of icon-loaded graphical user interface (GUI) tools are available, but you'll either end up with a palette that's too cluttered to be understandable or be forced to repeatedly select icon after icon to complete a simple function.

Look for a tool that is designed to invoke both simple and complex functions with ease, with both an icon tool bar and menus, and one that features some standard skeletal templates common to expedite development. Do not become so fixated by "ease of use" that you select a tool set that places restraints on your more experienced developers.

There are plenty of terrific IVR application tools on the market and, with a little up-front investigation, you can find one that will really work for you.

(Brite Voice Systems)

ASR Application Opportunities

Here are the top seven vertical markets for ASR (Automatic Speech Recognition) solutions:

1. **Financial**
2. **Travel/Transportation**
3. **Health**
4. **Retail/Catalog Sales**
5. **Insurance/Benefits**
6. **Service/Utility**
7. **Government**

(DSP Group, Inc.)

ASR Benefits

Once confined to laboratories with specially controlled environments, speech recognition in the 1990's has emerged as a commercially-viable technology with many real-life applications. William Meisel, editor of the newsletter "Speech Technology Update" projects a market for speech recognition software and hardware of $9 billion by 1997. Other industry watchers also predict rapid and explosive growth over the next few years.

This growth will come in large part because of the tremendous progress made in speech recognition technology, especially in the development of true continuous, speaker independent speech recognition systems -- what some industry watchers have dubbed as the technology's "holy grail".

If the progress of the past few years continues its fast pace, the next several years will witness the widespread use of such real-life products as voice-activated/voice-dialing cellular and "land-line" phones, voice-driven VCR remotes, and fully-functional voice-controlled computers.

And while existing voice processing applications allow us to interact with a computer by using the touch-tone keypad of a telephone, speech recognition is already providing a more natural method of interaction -- the "simple" spoken word.

One of the leading suppliers of speaker-independent speech recognition technology to voice processing system vendors and integrators is Voice Control Systems (VCS), based in Dallas, TX. The company licenses its speech recognition algorithms and DVM-4s and 2c PC-based board components to such companies as Dialogic, InterVoice, Periphonics, Perception Technology, Brite Voice Systems, IBM and others.

Another speaker-independent speech recognition technology leader, Voice Processing Corporation (VPC), based in Cambridge, MA, also claims a large number of OEMs, VARs and systems integrators for their VPro family of PC-based software and hardware components, including Centigram, Digital Sound, Enhanced Systems, Microlog and Octel. VPC's products are designed to work in both digital and analog environments with Dialogic, Natural MicroSystems and
Rhetorex telephony boards, and can also be ported to proprietary platforms if needed.

Since speech recognition provides for voice control of computer-based machines and devices, the benefits of such control can enhance a number existing man-to-machine-related tasks.

In addition, the technology also has a proper place in many telephony and voice processing-related applications, where voice is the most natural and most easily implemented mode of communication and control.

A third area where speech recognition has been used for many years is for speaker verification in a number of security-related applications, particularly for the military.

The benefits of this technology are indeed numerous. In terms of man-to-machine communications, these benefits include a new freedom for the handicapped. Speech recognition-enabled devices can provide a level of independence for people with physical and visual handicaps that would otherwise be unattainable.

Speech recognition can also provide for eye and handsfree operation of equipment and instruments. Many office and manufacturing tasks can be made more efficient and productive when one's eyes or hands are freed from entering data or pushing buttons and instead devoting one's eyesight and hands for the task at hand.

For people entering data into a computer off of hard copy documents, speech recognition might serve to substantially speed up the process. For technicians whose hands and eyes are frequently tied up working with tools or in making adjustments to machinery, and/or who must enter information into computerized monitoring equipment, speech recognition allows them to do both tasks at the same time.

There are also a number of important telco and voice processing-related benefits. For example, rotary phone access to CPE-based interactive voice processing applications is possible. By eliminating dependency on the touch-tone keypad to control access to voice processing functions, speech recognition allows the rotary phone user to enjoy all the benefits of this fabulous interactive technology. For callers who do have touch-tone phones, speech recognition offers those callers who aren't comfortable "keypadding-in" commands an alternative -- using their voice.

Once large-vocabulary, speaker-independent speech recognition systems become widespread, the touch-tone keypad just might become obsolete for practically every telephony-based application.

The RBOCs and LD carriers are currently employing speech recognition as an enhancement to many custom-calling and automated operator services. Several RBOCs have trials up and running in selected regions of the country.

Information providers can substantially boost their revenue stream (and the profitability) of their information services by serving all callers, rotary and touch-tone. Speech recognition gives them that opportunity.

"Spelling" names or words using the keypad of a touch-tone phone is a lesson in frustration (especially when the word has a "Q" or "Z" in it). The phone's keypad was simply not designed to operate as a typewriter. Speech recognition makes it easy to "input" names and words, solving the touch-tone alphabet dilemma.

Some voice processing applications require knowledge of a large number of numeric codes which must be entered via the keypad in order to access features. Some people, especially occasional users, have a hard time remembering all the necessary codes. Speech recognition makes these codes unnecessary, as a simple voice command will do the trick.

(Excerpts from The Speech Recognition Reference Manual And Buyer's Guide -- Robins Press)

ASR Challenges

Consistent, perfectly reliable speech recognition is difficult for a number of reasons. For one, speech differs from person to person. Most people speak with noises or disturbances (coughs, sneezes, pauses) in their speech, and not everyone pronounces the same word the same way every time. For another, there is often background noise which "pollutes" the voice signal.

Most offices are noise polluted to some degree which can effect the "reading" of a voice signal -- but manufacturing environments such as factories or outside payphone enclosures are even noisier.

To deal with these situations, speech recognition vendors have enabled their systems to recognize speech clearly and effectively despite all the background noise which surrounds us all the time. And for products designed to provide for speech recognition over telephone lines, vendors have also found ways to compensate for various degrees of line noise.

Another challenge is the way most of us speak. When we talk, we commonly make errors (slightly mispronounce words) and insert extraneous words, such as "um", "er" and "ah". Words can also run together when a fast talker speaks. In addition, many words sound alike, and unfortunately, computers can't read our thoughts (at least for the time being).

Speaker Dependent vs. Speaker Independent. Commercial speech recognition systems can be basically divided into two main types: speaker-dependent and speaker-independent (sometimes referred to as SIR systems).

The speaker-dependent side of the technology -- where the system is "trained" through repetition to recognize a certain vocabulary of words and accept no substitute -- is fairly well-established. This technology is based generally on a template, or acoustical, representation of speech.

Users "train" the system in their particular voice patterns by speaking the words, or voice samples, that will need to be recognized. These "voice prints", or templates, are then stored on the system. When the system is working, these voice prints are compared with the spoken command of the user.

If the voice print and the spoken word match, the speech recognition system "recognizes" the word and executes the command. Template-based dependent recognition is used for relatively small to medium-sized vocabularies (generally up to a few thousand words).

Other speaker-dependent recognition systems operate by matching "phonemes" (syllables), multiple words and triphones. The phoneme/multi-word approach is typically used for systems with larger vocabularies, up to tens of thousands of words. Typically, speaker-dependent systems work well with medium to large vocabularies, and with either isolated or connected word recognition.

Speaker-independent recognition systems, on the other hand, have the ability to recognize any words no matter who speaks them. True continuous (normal sentence or phrase-long speech) speaker-independent systems with unlimited vocabularies, no required training, and high accuracy is an extremely difficult process. The system must recognize anyone's speech -- and vocabulary -- without prior training. Until recently, this side of the technology has been limited to isolated words and digits (with distinct pauses in-between) or a small vocabulary. In speaker-independent systems, generic templates are stored for comparison with voice input from any speaker.

Isolated vs. Connected vs. Continuous Speech Recognition. Isolated word recognition is the most basic form of vocabulary matching, and requires the user to speak a single word when prompted by the system.

Connected word speech recognition is a special recognition technique that allows users to speak numbers or words in a connected fashion without pausing between each successive number. The recognition system generates a beep as each digit is captured to provide feedback to the user.

Continuous speech recognition, on the other hand, allows a user to speak words or numbers in one continuous stream. The user does not receive a beep for feedback from this type of system.

Word-Spotting. Developers of speech recognition systems have started to deal with the fact that human conversation is filled with a variety of sounds that aren't really words -- such as "uh" , "ah" and "um". During a usual conversation, we automatically recognize these words as irrelevant, and focus instead on the real words that carry the message.

Computers, on the other hand, aren't so smart and can become confused by these additional sounds. Researchers have begun to teach computers what humans do by nature -- to recognize key words. This technique -- called word spotting -- allows a computer to recognize one word in a phrase, such as "Chicago" in "I want to fly to the Chicago area."

Neural Networks. A bright area of speech recognition research today is in the study of neural networks. While traditional speech recognition explored the vagaries of the speech signal, which are like finger prints with every person having a different signal, neural networks are sort of an analog of the human brain. Just like there are neural connectors between different layers of the brain, the idea is to create similar connections between computer circuitry.

By designing a computer mechanism similar to the way the brain is organized, researchers are hoping to develop a system that can learn vocabulary the way a child does when he is growing up.

Lernout & Hauspie Speech Products (Woburn, MA), a major developer of multi-lingual speech technology, employs artificial neural networks together with Hidden Markhoff modeling technology (a "phoneme" modeling method) in their Automatic Speech Recognition software product (the company has in fact won a patent by the World Intellectual Property Organization (WIPO) for its use of neural networks in its speech products). The company credits the use of neural networks to solving the memory bottleneck common with providing speaker-independent, continuous recognition. A pilot test on a database provided by the National Institute of Standards and Technology (NIST) showed an accuracy rate as high as 96.7%, with memory requirements reduced to about 7% that of equivalent products.

At present, the biggest challenge is training a system to be speaker-independent and training it to recognize words in a string of continuous speech. Although systems can be trained to recognize thousands of words if it's trained to the voice of one or two people, the accuracy of such a system drops if you set it up for speaker-independent, continuous speech.

At that point, the biggest, practical vocabulary is several dozen words or so. As systems' vocabularies continue to increase, so will the technology's possibilities and applications.

(Excerpts from The Speech Recognition Reference Manual And Buyer's Guide -- Robins Press)

ASR Considerations

Shorten Menus. You can shorten long touch tone menus by using speech recognition. Today's recognizers are excellent for lists of up to 1000 words and phrases. This shortens calls and increases capacity.

Low Voice-to-ASR Port Ratio. Speech recognition only occurs during part of the call. Plan on adding one port of speech recognition for every three ports of IVR.

Crawl Before Running. You don't need to speech-enable every call. By asking a few simple questions, an up-front speech application can identify who is calling and determine whether the call can be answered with automated speech recognition. Even automating 10% of your calls can pay for itself quickly.

ASR Needs AGC. When implementing speech recognition, seek hardware that offers automatic gain control. Otherwise recognition accuracy will suffer every time you get an extra loud (or soft) phone connection.

(PureSpeech)

ASR Integrator Guidelines

As speech recognition begins to fulfill its promise of becoming a mass market-ready technology, many companies are rushing to integrate this capability into their product lines.

Companies jumping on the SR bandwagon include PC vendors, independent software vendors, telecommunications equipment providers (including voice/fax processing system vendors), automotive electronics manufacturers, consumer electronics providers and industrial electronics equipment manufacturers.

Some of the important questions a prospective OEM should be asking speech technology providers include:

1. **Will your speech technology run on our existing processors?** In short, OEMs would prefer not to change DSPs or other hardware components if they don't have to. Still others want to run enabling speech algorithms on their existing host processors. The prospect of changing existing hardware architectures can be risky, costly and extremely time-consuming. SR providers should offer algorithms that are capable of running on multiple DSP and host processors.

2. **What is the range of speech technology solutions that you offer?** Ideally, a speech technology provider will have an expertise in speech recognition, text-to-speech and speech coding (compression/decompression) supported by a range of capabilities.

3. **What languages do you offer products in?** OEMs considering international markets can jeopardize revenue potential when offering products not tailored to specific countries. At a minimum, companies should be able to provide their products in major European and Far East language versions, in addition to American English.

4. **Who will help us integrate your speech capabilities with our products?** Providing platform compatibility is only one piece of the puzzle. Other issues include the provision of integration support for host-to-processor communications and API/application links. SR providers should not only provide algorithm functionality, but also the technical prowess to adapt their products for use on an OEM's products.

(Excerpts from The Speech Recognition Reference Manual And Buyer's Guide -- Robins Press)

ASR System Evaluation

In evaluating automatic speech recognition (ASR) systems for accuracy two important caveats: do your own accuracy tests but remember that accuracy isn't everything.

Don't believe the vendors' claims of accuracy. Vendor recognition tests are always done under conditions and with vocabulary sets which will make the results come out in the high 90 percentages for individual words. The only way to determine the real accuracy of the system is to do an apples-to-apples comparison of two ASR engines in the same application with the same vocabulary under the same field conditions. This is generally not a trivial task. But take heart. The results from a fair test of two or three good ASR systems are likely to differ only by a percent or two anyway. At this point, the developer should realize that the really important considerations with respect to the usability and throughput of the system will be determined by the applications design.

This includes choosing easy-to-understand vocabulary sets, dealing with out-of-vocabulary words, creating an intuitive menu structure, and providing for efficient use by both novice and experienced users. A well-designed application is far more important than two or three percentage points of accuracy in virtually any application. The vendor that provides the best applications design support rather than the very best recognition accuracy may be the best choice.

(Voice Information Associates, Inc.)

ASR Technology Evaluation

First and foremost, you should experience using the system proposed yourself. Don't take anyone's word for how good it is. Test it yourself by building database of words for testing.

Work with your colleagues to create a benchmark system that will measure accuracy, noise immunity, rejection rate, and overall performance. This benchmark should equate to the types of words, letters, phrases, etc. you want the ultimate system to understand – and also the types of users who will be accessing the system.

The critical components to evaluate are:

Accuracy. The system needs to be accurate enough to work reliably for any user at any time. It should quickly learn and adapt to any user's voice and yield an accuracy rate higher than 95%. Consider that user dependent systems are more reliable, because they have a closed community of (trained) users.

Hardware requirements. Look for PC-based applications because the tend to be less proprietary and well-supported. A good ASR system will perform well within the hardware and memory constraints of PCs because much of the "power" required to run the algorithms can be "on-board" expansion cards.

Noise immunity. The ASR system should work well with environmental noises like those found in your office, car, airport, exhibition hall, etc. This includes white noises, cocktail party noises, etc. In short, it should work well regardless as to where you are calling from. Make sure you test the system from different (noisy) locations and make note of its performance in each one.

Ease of use. A good ASR system adapts quickly and easily to the user's voice. The "learning procedure" should be quick and painless. If it takes more than a few minutes, then you should look elsewhere.

Language support. An good ASR system needs to be flexible enough to deal with any accent and deal with a variety of languages. This allows you to use your application worldwide.

(ART, Inc.)

Caller ID Buying Tips

Caller ID expert Gordon Hamm of Voice Data Systems (Yacolt, WA -- 206-686-3238) gave us this checklist for buying Caller ID technology:

1. **Does the device need a TSR?** If so, how much memory does it use?

2. **How many Caller ID decoders** does the device have?

3. **What is the maximum number of lines** the device can handle? The more the better.

4. **Is the entire protocol available** if you wish to develop your own integration?

5. **Does the device monitor any outbound events** in addition to monitoring Caller ID?

6. **Is the device an internal or external device?** We prefer external as it uses no slots, works with any platform that has a serial port and requires no TSRs.

7. **How flexible is the software?**

Caller ID Device Things To Look For

1. **Serial Daisy chaining.** To support more than one device per serial port. Needed if a single device won't support the number of lines required.

2. **Inbound DTMF Detectors.** Needed if you want to collect digits dialed by the caller after he's connected. For tracking calls through an auto attendant or other voice-processing application.

3. **First Ring Suppression.** Needed if you can't prevent a phone device from answering on the first ring. CLASS Caller ID comes for your telco *after the first ring*.

They send a very short burst of data at 1200 baud between the first and second rings. The telephony "thing" must be on hook (i.e. unanswered) for the transfer of Caller ID data to work.

4. **Call Duration Timing.** Needed if you want to track lengths of calls to clients.

5. **LCD Information Displays.** It can be helpful to view the information without a PC being connected to the device.

6. **Call Data Buffers.** This if for storing a certain amount of Caller ID numbers when the computer is off-line; especially handy in remote polling apps.

(Computer Telephony Magazine)

Choosing The Right Voice Card: Robins' 21 Top Tips

Before you select a voice card from the ever-growing array of choices, there are a number of special considerations you must make to determine which one is the best for your needs, beyond simply calculating each card's cost-per-port. These considerations involve the type of PC platform used, the operating system used, the type of telephone network interface required, as well as voice card-specific features.

Voice cards and the PCs they run on differ in some important ways, so let's take a look at these differences and figure out how to make sense of them in terms of making your selection.

1. **What buses are supported?** It's important to know which bus your current PC uses, or what bus you will use if you are purchasing a new PC. Some card vendors have a wide selection of cards for 8-bit, 16-bit and 32-bit buses, while others make cards for only one type of bus. There are also cards made for Sun Microsystems SPARC stations (S-Bus compatible) and VME-bus compatible cards.

2. **What is the PC's processor type and speed?** When it comes to the processing demands of voice applications, the more powerful the PC's processor, the better for overall system performance. With prices for low-end 75 Mhz Pentium systems plummeting to the $1200 range, there's really no reason to go with anything less.

3. **What is the operating system used?** If you want to use the operating system currently installed on an existing PC, then you should make sure that the voice card in question has the appropriate drivers to support it. If you are building a system from the ground up, then you can choose which OS works best for your intended system application(s). OS/2 and UNIX are usually the OS's of choice for high-capacity systems due to their true multi-tasking environments. Windows NT and Windows 95 are also gaining ground due to increasing support from developers of CTI and VP application generator products.

4. **Amount and type of hard disk storage.** Voice and fax processing applications are hard disk intensive, in that the system is constantly reading and writing data from and to the hard disk to deliver and store voice information. When choosing a hard disk, look for a drive that has an access time of 12 milliseconds or less. To determine how much storage you need, you must take several things into account: how many minutes of voice storage per megabyte of disk space is supported (this depends on the efficiency of the algorithm used by a voice card to digitize voice signals), how many ports will be supported by the drive, how many applications you are providing, and how many users will be supported.

5. **Number of available slots.** The number of available (free) slots in a PC will determine how many voicc cards can be installed into the system and ultimately, what the system port capacity will be. Remember to take into account other add-on cards you might need, such as a video card, network interface cards, fax cards, speech recognition cards, etc. These will occupy empty slots and reduce the overall system capacity.

6. **On-board processors.** Most cards have an on-board processor (in addition to DSPs, EPROM chips and firmware) which serve a variety of functions.

On single port cards, these processors allow the cards to run in the background. On multi-line cards, these processors help speed up and improve system performance. Like their PC-based brethren, the more powerful and faster, the better.

7. **Amount of on-board memory provided.** Most voice cards have a certain amount of RAM (and sometimes EPROM) memory on board to support various firmware and software algorithms. Generally, the more memory available on-board, the more features can be provided or added on to the system. Memory Chips are also used as temporary storage areas for digitized information coming from the PC and from the Codec on the card.

8. **Digital signal processor (DSP)-based.** Digital signal processors are specialized chips that perform continuous digital signal analysis, dramatically boosting signal quality and overall system flexibility. Voice cards that sport on-board DSPs are designed to support high-quality voice compression and playback, as well as support other applications such as fax, speech recognition and text-to-speech. Some voice cards have "shared" DSPs, where a number of channels or ports uses a single DSP, while others have DSPs dedicated to each port. The latter is the more sophisticated design.

9. **Port capacity.** The number of ports on a voice card determines the number of telephone lines which can be connected to the system and the number of simultaneous calls which can be supported. Voice cards are available in a number of capacities, from 1 to 60 ports and more. It is important to keep the issue of port contention in mind when planning for capacity: some applications, such as automated attendant and outbound telemarketing, generally require dedicated ports.

10. **Dynamic or dedicated processing capabilities.** Some voice cards have dedicated allocation of functions per port, meaning they are voice-only. If you plan on running multiple applications on your system, look for the newer multi-application cards that dynamically allocate functions. This means that, depending on caller demand, each port can provide voice, fax (and even modem) functionality.

11. **Operating systems supported (available drivers).** Some cards are designed to run exclusively under the MS-DOS operating system (mostly single and 2-port cards). Other cards can be run under MS-DOS, UNIX and OS/2, and others OSs. Check whether drivers are available for the operating system you desire.

12. **Available network interfaces.** There are a number of different network, or telephony, interfaces available. These interfaces determine how the voice system connects to the public (or private) network. Available analog network interfaces include E&M, Ground Start, Loop Start, and DID. Digital network interfaces include T-1, ISDN and CEPT. Some cards have the necessary interfaces built-in, while others require a special interface card or module to supply the connection.

13. **International compliance.** For overseas operation, voice processing cards need to be approved by the PTT or telecommunications authority of each country of intended use. If you are interested in developing systems for the international market, make sure the voice cards you are looking at are approved for the countries you intend to market to.

14. **MF and DTMF detection and generation.** MF signaling is the common signaling used in ANI (Automatic Number Identification) circuits for providing calling party number identification, and Feature Group D circuits, which provide called party number identification. This signaling is used in Central Office intercept circuits, payphone billing transactions, integration with paging and cellular telephony systems, and many other inter-switch communications.

DTMF signaling is the Dual-Tone Multi-Frequency signal format used by touch tone telephones, and DTMF detection & generation is a standard feature in all but the most basic single-port voice cards.

If you have a need for MF signaling in your system, check whether MF is available on-board as a standard or optional feature, or whether an additional MF card is needed.

15. **Rotary (pulse) dial support.** There are still many old-style rotary dial phones being used by callers. The numbers will surprise you. By most accounts, more than 35% of all American households still use the P.O.R.T. (plain old rotary telephone). This is the national average. Pulse-to-tone conversion capability allows a voice processing system to detect the rotary pulses, or clicks, from rotary phones and convert them to DTMF tones, allowing rotary callers to use interactive DTMF applications.

16. **Automatic gain control.** Automatic gain control, or AGC, compensates for voice signals (and DTMF signals) that are too low due to long distance "dropout" or just a plain bad connection. AGC helps to reduce unintelligible voice messages, errors in detecting DTMF tones, and improves the overall sound quality of a system. Check to see if the card you want offers AGC.

17. **Noise filtering.** Digital noise filtering on DSP'd cards helps to reduce latent phone line noise (static, for example), and complements AGC. It reduces talk-off and provides more reliable keythrough (DTMF detection) over noisy long distance connections. A nice extra feature available on only a few cards.

18. **Variable compression rates.** Most multi-line voice cards today allow the developer to set a voice compression rate from an available selection via software. The rate can range from 64 Kb per second uncompressed PCM speech to 12 Kb per second. 64 Kbps uncompressed speech produces the highest quality sounding voice, but uses up the most hard disk storage. Some vendors have been able to supply proprietary algorithms that produce respectable voice quality at low rates on DSP cards.

19. **Full call progress monitoring.** Full call progress monitoring (CPM) is essential for intelligently controlling and appropriately responding to the various call states experienced in telephone communications, including dial tone, busy, fast busy, ringback, voice detection, ring/no answer, hang-up, wrong number and other call status conditions. Look for a vendor and card product that supports the full range of conditions. This will insure that whatever applications you build, it will run efficiently and error-free.

20. **Networking capabilities.** If you want to build high capacity systems and want to expand beyond the limitations of a single PC platform, make sure that networking is supported. Check into how many cards can be installed in a single system, and how many systems can be linked in a network.

21. **"Open" standards support.** A present, there are three "open" PC platform standards being used in the voice processing arena, Natural MicroSystems' MVIP and Dialogic Corporation's PEB and SCSA. A number of voice card, fax, speech recognition and text-to-speech board manufacturers have built one or both of these specifications into their own products to allow them to integrate with associated voice card products and software resources.

(Robins Press)

Consultants and Honesty

Many telephone "consultants" are crooks. They get paid under the table by the vendors whose equipment and services they recommend. There are some ways to check you won't get ripped off:

1. Ask your consultant which company won his last 10 recommendations. Not any ten, his **last** ten recommendations. If one vendor keeps popping up, be careful you may have a "consultant" who's on the take from the vendor. (Typically the take is 10%.)

2. Ask the three leading telephone equipment vendors in your community if they will bid on a proposal your proposed telephone consultant will design. If **any** of the three says he won't, be very careful.

3. Don't let your consultant get all the proposals that come in for your phone system **without** you seeing them simultaneously.

4. Make sure the potential vendors know it's your company they are selling the equipment to. **Not** the consultant.

5. Before accepting your consultant's recommendation check out the approximate cost of a telephone system of the size and complexity you're thinking about. You should know roughly what your system is going to cost you. That way you can see if your consultant's recommended bid is way out of line.

6. Be wary of small consulting fees. There are no bargains in this business. You **are** paying, though it may not be obvious **how** you're paying.

7. Check out your consultants' qualifications with his last five clients. Call the clients and ask them very specific questions, such as equipment recommended, cost of services, services contracted for, and services performed, etc. Don't ask them one dumb question, like *"Were you happy with the consultant?"*

8. Bid out your consulting firm's services as exhaustively as they will bid out a phone system for you. Make them give you all the information above.

(Teleconnect Magazine)

Debit Calling Service Buying Tips

Things to look for in the platform:

Capacity and performance. Remember, users expect the system to always be available. Check for the number of lines supported, call throughput, redundancy and back-up options and growth potential. When a big customer asks for a million PIN numbers, you don't want to have to turn them down because your system is already maxed-out.

Standard platforms with intelligent open interfaces. Make sure the platform is heavily standards-based. The key is to be able to work with as much of what already exists at the customer's sight. If you can tap into their existing CT plant in any way, you're golden. But the real important thing is working with their embedded databases, networks and billing systems. They won't forklift any of those out to get into the telephone biz.

Flexible call processing and media processing feature set. Can you customize call flows for different accounts? For different app services? And can the switching backbone perform complex routing. You'll need that to cash in on least cost routing, multiple carrier support, etc.

Business features. Look for things like card-generation invoicing, custom reporting, printing file and flexible rating so that you can assign different charges or billing schemes for different accounts.

Stability. Make sure the technology supports data backups and high-availability. You can't afford to lose a month's worth of billing data because you skimped on the system and someone forgot to back up the data manually. And you don't want to screw up the existing company's core business (like a bank's, a utility's, etc.). Check suppliers for their training programs, technical support and overall responsiveness.

Card Printer. To get the service going, you'll need to find someone to print the cards. The vendor should be able to offer high-quality color printing, since the packaging of the product, the card, is extremely important.

Check on their ability to do rush jobs and the range of cards they can provide based on your clients' needs: relatively cheap cards for mass promotions; premium cards that will be more durable if clients are recharging them. They should also have equipment for bar coding and serial numbers for card tracking.

Network provider. You'll need to strike a deal with AT&T, MCI, Sprint or other long-distance carriers to get the long-distance access. Don't necessarily limit yourself to just one. As long as your system has intelligent optimized call routing (least cost routing), you can use multiple carriers, depending on the rates for various destinations or times of day. And some carriers can offer better rates for a specific coverage area.

(Harris Digital Telephone Systems)

Fax Application Generator Questions To Ask

Fax expert Maury Kaufmann, author of the brand-new book *Computer-based Fax Processing*, says you should use these tips when buying a fax application generator or tool:

1. **Does the tool support fax products from multiple vendors?** Some fax cards are best suited for one-call fax applications. Others are better suited for two-call.

2. **Does the tool have text-to-fax conversion** *without* making calls to external utilities or C libraries?

3. **Does the tool have document processing features** (e.g. the ability to merge data from an IVR application into a free-form template to create customized documents)?

4. **Does the tool support multiple bitmap formats** or just raw T.4 fax coding? Support for PCX, DCX and TIFF/F insures maximum compatibility with document preparation and desktop publishing utilities.

5. **Does the tool support the use of fax boards as a shared resource?** Many fax-on-demand apps need only four ports of fax from a 24 port system. The rest may be used as an outbound fax server. Make sure you choose a tool that lets you use fax devices as a system shared resource.

6. **Does the tool support digital network interfaces to fax products?** On larger systems, you will want to locate fax devices behind a digital trunk (e.g. T-1, E-1, ISDN). Many application generators support only analog front-ends to fax peripherals.

7. **Does the tool provide you with source code for modules which control fax transactions?** Without this capability, a simple feature such as stripping (combining text and bit mapped graphics on a single page) will be impossible to implement.

8. **Does the tool come with example programs that illustrate how to integrate fax into telephony apps?** Without examples, you will spend a lot of time experimenting.

9. **Does the tool come with bundled technical support or is that extra?** Some fax products are difficult to configure, especially in high-densities.

10. **Does the application generator or tool kit come in an evaluation version that allows you to build and test applications?** Some tools are good for some projects, but wrong for others. An evaluation system lets you figure out whether a product will work for you -- before you spend money on development tools.

Regarding price: Don't be penny-wise! Shop features and performance first, price second. A good development tool will save you money by enabling you to get your product to market quickly. A cheap tool will be more expensive in the long run.

(The Kaufmann Group)

Figure 8.5 - Fax-On-Demand User Profile

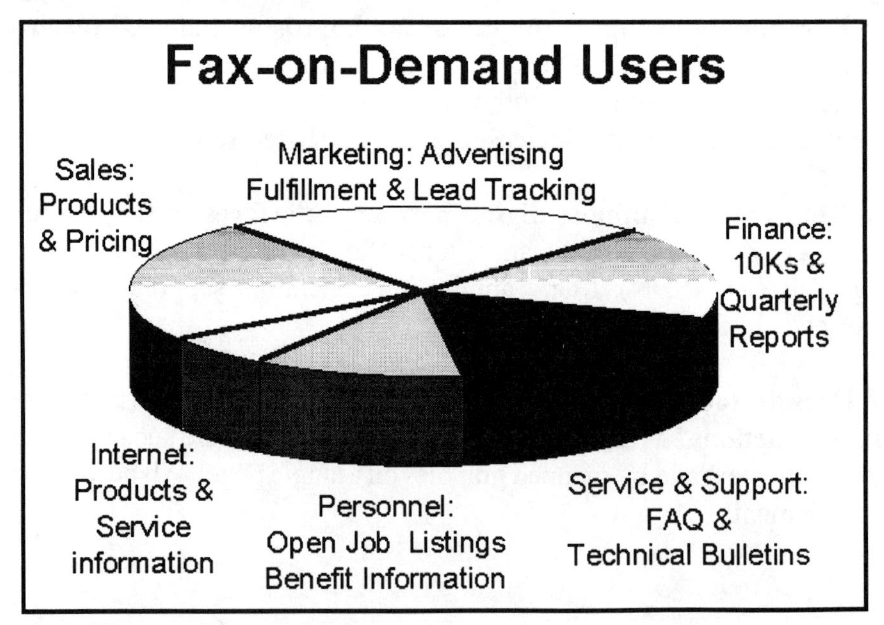

Fax-on-Demand And Who Needs It

Fax-on-demand provides instant fulfillment of a wide range of information. The users of fax-on-demand include customers/prospects, investors and job applicants as illustrated in figure 8.5. Fax-on-demand enables the instant and unattended ability to perform solution prospecting, target marketing and effective analysis.

(Brooktrout Technology, Inc.)

Fax Cards And What To Look For

Drivers. Make sure you can get drivers for the hardware on each platform you intend to develop on. Boards with no drivers are totally useless. Check how new/tested the drivers are. Brand new often means buggy.

Speed. Fax transmission speed is important. A 14.4 Kbps fax session (meaning both sides are 14.4 Kbps) will transmit approx. 50% faster than a 9600 Kbps. This saves on telephone costs. This saves on fax resources. And being faster, it's more reliable.

Compatibility. Not all Group III Fax machines are compatible with each other. (Don't ask why.) Any fax card you use should be able to send and receive faxes from any fax machine. Your fax card maker should be obsessed with compatibility, which means constantly fine tuning his card to new domestic and foreign fax machines. DSP-based fax products can be upgraded via software, as compared to chip-based fax products where you've usually got to replace chips or ROMs.

Translation. Any fax card worth its salt should be able to accept a text file and output a fax image. Some of the better fax products offer cover letter generation tools with replaceable fields.

Compression. Since the fax process uses any feature common between the two devices in a fax session, having many features on the board (i.e. various forms of compression) will cut down connect time.

If you fax a three-page document to 1,000 users, you will probably notice that while some transmit in about 45 seconds, still others take as long as 2 minutes. The two-minute transfers are most likely to old fax machines and the quick transfers are to newer equipment (like some of the high end plain paper fax machines) having compression features.

Three compression schemes are called Modified Huffman (MH), Modified Read (MR) and Modified MR (MMR). They can reduce transmission time from 20% to 40%. Around 90% of the fax machines in the field today support MR and up to 30% of them also support MMR compression. There's no visible difference between a compressed fax image and an uncompressed fax image.

Error Correction (ECM). Any static or line noise during a fax transmission often comes out as a garbled line or a duplicated previous scan line. Error Correcting Mode protocol asks for a repeat of the previous scan line if an error is detected. This can extend the connect time but, it gets it right.

Binary File Transfer (BFT). BFT allows a fax machine to transfer a binary file (e.g. Word 6.0 document, Excel spreadsheet) from one machine to another. Windows 95's Microsoft Fax supports a Microsoft-special version of BFT, which will probably become the standard, since no one else is pushing BFT.

As a way to send binary files around the world, BFT offers major advantages over standard asynchronous data communications protocols -- e.g. xMODEM, ZMODEM, etc. You can easily get a flavor for its advantages by trying this: Go to any country in continental Europe. Send a fax to your fax machine in the U.S. It will go through fast and perfectly. Now try using your modem to call your e-mail server.

Try sending a binary file by modem back to the states. Good luck. We've spent hours trying to direct dial modems in the U.S. from Switzerland, Germany, and France with zero luck. (England, Australia and New Zealand are better.) We were trying to send an article back. When we faxed it back, it went through immediately.

Distinctive Ring Detection. On some lower volume applications, a dedicated phone line is not required. By adding a feature to your phone service, you can get a second (and often a third) phone number that rings the same line with a pattern ring.

If the voice hardware can detect the ring pattern and route the call appropriately, the caller will never know that you don't have a dedicated fax line. Most fax modems cannot detect distinctive ringing patterns and can only answer on a predefined number of rings.

(Computer Telephony Magazine)

Fax Modem Co-Processors

When selecting a fax modem for your network fax services make sure to use high-end co-processed modems. The extra cost will save phone bill $$ since these modems support higher compression specifications, error correction mode, and are compatible with more fax machines than the low-end modems. It's the old saying - "pay me now or pay me later".

(Optus Software)

Fax Server Pricing

Vendors employ a broad range of pricing structures. Here are some typical options:

1. Unlimited-user prices for fax server software designed to run on a single server.

2. Unlimited-user prices for fax server software designed to run on a single server with a single phone port -- but extra charges are levied to add more phone ports.

3. Graduated site licenses for which pricing is determined by the number of users per license.

4. Graduated site licenses for which pricing is determined both by the number of users and number of ports per license.

5. Specific prices for specific bundles of fax server software, fax boards, and sometimes the platform as well.

Prices vary tremendously, from as low as $100 for just the software to as high as tens of thousands of dollars for multi-line large-computer fax servers. Many buyers want fax servers to cost almost nothing and to perform very well. Somebody told them fax is a commodity. Well, in fax servers, the performance of software and hardware and the combination of the two can make huge differences, even if it all seems alike on the surface. The old adage applies: you get what you pay for...

(Davidson Consulting)

Fax Server Retries

When selecting fax server software, be sure to check items like auto redial on busy. Some products only allow you to set the number of retries. More robust products will allow you to also set the interval between retries. You should select an interval and retry combination that will allow for at least a 20 to 30 minute overall time. That way if someone is receiving a long document, you will not exhaust the retry time span before the fax machine on the receiving end clears the document ahead of yours.

(Optus Software)

Fax Server Vendor Questions

Ask your vendor These questions when looking for a fax server.

1. **How complex is the installation and configuration?** Few users know computer and fax technology well enough to make this a simple exercise. Make sure your vendor does.

2. **How well does the server do "cross platform"?** Most fax servers support DOS & Windows users; but many networks also have Mac & UNIX & other workstations to support as well.

3. **How do my users handle Print-to-fax technology with this system?** Often, users have to switch back and forth between having their default printer drivers be the laser printer driver and the fax-device driver. This drives some people to distraction and they cease using computer fax.

4. **What kind of idiosyncrasies have your other customers experienced?** These are usually due to problems with fax software & hardware combinations or software bugs (in the fax software or operating system, etc.), which can suddenly mean that critical faxes can't be sent or received.

5. **What fax devices do you recommend we use?** If your vendor recommends the use of low-end modems, watch out. Many of them can't fax to many other fax devices and which transmit slower than most fax machines, causing fax phone bills to rise.

6. **How problematic is the receipt of faxes on your fax server?** Some manual or automatic method is required to route faxes past the fax server and onto end user mailboxes. The automatic methods either have some catch to them or involve extra monthly phone bill charges from the telephone company. Even then, receiving faxes electronically on PCs isn't everyone's cup of tea: they can be hard to read (impossible for some people) and time-consuming to access and print.

7. **How do I get support?** Getting support can be a problem. If there isn't a local VAR, support is usually remote. Among other things, this often means the customer has to do its own training of end users.

8. **How flexible are the fax phone books?** How many entries can they hold? Are there both system and personal phone books? Can you use your existing e-mail address directory? Can phone numbers from other databases be imported so they don't have to be entered manually? Can they be exported in case you change fax servers?

9. **How does the system handle fax broadcasting?** Can you broadcast over multiple phone lines? How easy is it to create and save broadcast groups? Can it be done "on the fly" by clicking names in a list? Can names be merged to personalize each recipient's cover sheet? Does the system create one file and merge names, etc., onto it, or does it create a separate file for each recipient (if it is the latter, the fax server must do a computer-file to fax-file conversion for each broadcast recipient, greatly slowing down the speed with which the fax broadcast can be completed, regardless of actual transmission times).

10. **How does the fax server handle delayed transmissions?** User interfaces vary; some automatically prompt the time and date of a transmission while others require extra steps to get to sub menus in order to delay transmissions. If delayed-transmission phone discounts are important to you, make sure you use software which either auto prompts users or via which some users can be forced to delay all transmissions.

Many users instinctively send all faxes immediately because it provides a sense of closure. Thus, users may have to be trained or motivated to use the delayed transmission feature.

11. **Is redialing programmable by the user?** Is it programmable by the reason for needing to redial (e.g., redial in 1-minute intervals on no-answer, redial in 5-minute intervals on busy signals, and no redial at all when non-fax devices answer). By the by, it is our understanding that the FCC only allows automatic redialing five times in succession (but many fax servers allow 99 or more times).

Perhaps most importantly, how efficiently does the fax server redial? In other words, if multiple fax jobs are queued and one job doesn't get answered, does the fax server re-queue that fax and transmit the next one in the queue or does it sit idle for the programmed redial interval (e.g., 1 minute or 5 minutes) and then redial the original fax while all the others wait Idly in the queue. In the latter case, if the fax server is programmed to redial five times at five minute intervals, all faxes could be delayed 25 minutes while the fax server redials the original fax.

Meanwhile, redialing only refers to unanswered calls. There is a second issue about re-sending faxes when fax phone calls fail for some reason during mid-call. Some fax servers may not re-send the fax at all (but hopefully list it as a failed transmission in an error report). Some send the whole fax all over again (even if 9 out of 10 pages were sent during the first call). Some send only the unsent pages (like pages 9 and 10 when the phone call failed during the transmission of page 9).

12. **Does the fax software support cover sheets?** Does it have a library of cover sheet templates which users can easily employ? Does the library include suitable cover sheet templates or are they mostly humorous templates? Can the templates be easily modified? Can you create your own cover sheet design in some other draw package then drag-and-drop cover sheet fields on it within the fax software?

13. **How do I attach files?** Is it easy to do and can it be done in one step from the basic SEND dialog box? Can users attach multiple files? Can they see which files are attached?

14. **How valuable are the send-confirmation and error reports?** Can they be activated at end user workstations as audio and/or visual signals? Are they accurate: i.e., they should indicate that the fax server has successfully transmitted the fax or not.

Some servers relay back confirmations of successful sends merely when the computer file has been relayed from the client PC to the fax server (but the fax phone call hasn't actually taken place).

15. **Are error details sufficient?** Error details should indicate why a transmission failed so users know what to do next. Was the fax machine just busy or was the receive device not a G/3-compatible fax device (which might mean a voice number was dialed by mistake)?

(Davidson Consulting)

Fax Server - Vendor Segmentation

Before contacting a list of fax server, fax-on-demand or fax broadcast vendors – save yourself some time by looking at the way these vendors segment the market:

1. **Low-end PC networks** (20-users and under): Delrina, US Robotics (Optus), Global Village, Castelle, Cheyenne/BitFax, Traffic, LANSource, Microsoft

2. **Small Novell NetWare LANs** (50-users and under): Delrina (via acquisition of Intel SatisFAXtion) Castelle, Cheyenne, Optus, RightFAX, Traffic

3. **Bigger Novell NetWare LANs** (50-users and over): Alcom, Biscom, Castelle, Cheyenne, Optus, RightFAX

4. **E-mail-integrated fax servers**: Lotus, Optus, RightFAX, TopCall, TRS, Siren, Resource Partners

5. **Fax machine-based fax servers**: Wordcraft, LA Business, Softline, Canon, Lanier, Pitney Bowes, JetFax, Mita, Ricoh/Biscom, Panasonic, Muratec

6. **NLM-Based fax servers**: Cheyenne, Optus, Tobit, Biscom

7. **NT LANs & servers**: Omtool, Optus, RightFAX

8. **Apple Macintosh**: Global Village, 4-Sight, PSI/Supra

9. **Banyan VINES**: Biscom, Alcom, Traffic

10. **IBM mainframe**: AIFP, Teubner, TopCall, Biscom

11. **IBM AS/400**: TopCall, CMA-Ettworth, Quadrant, Biscom

12. **UNIX-based fax servers**: Devcom, Siren, Softlinx, Faximum, V-Systems, Bristol, Biscom

13. **Production-level fax servers**: AIFP, Biscom, FaxBack, T4, TopCall

14. **Enterprise-wide fax servers**: Biscom, TopCall, Devcom, Bristol, Siren, Open Port, RightFAX, Optus, Omtool

15. **High Performance FAX Boards** for Network Fax Servers PureData, GammaLink, Brooktrout

(Davidson Consulting)

Fax Service Bureau Questions

1. **What enhanced fax services do you offer?** Do you specialize in any one service? Some bureaus (the RBOCs and carriers that offer fax) may specialize in fax broadcasting and actually outsource their fax-on-demand services to others. Make sure your bureau is competent in the fax technology you need, not just a 900 bureau that thinks it understands fax.

2. **How long has your bureau been in business?** Three to five years is the best answer. Less than two is young and over six is really before there were enhanced fax applications.

3. **For fax broadcasting: How many one page documents** can you broadcast in an hour? The only good answer is the one that meets your needs. Will you have others demanding fax broadcasting at times when I want it?

4. **For fax-on-demand: Will I have menu and script approval** before the application is engineered and prompts recorded?

5. **What is your turnaround time?** For Broadcasting, one day to a few hours is normal. For FOD, thirty days to live application is acceptable.

6. **How do you charge: per minute or per page?** Hint: Per minutes, with six second billing is usually less expensive. However per page will be easier for you to budget.

7. **What are your set-up fees**, minimums, storage charges and are there any hidden fees? There are as many ways to price fax services as there are long distance calling plans. Ask plenty of questions and don't be afraid to request a written quote.

8. **What are the phone numbers** of three fax-on-demand applications you currently maintain? If they won't give you three, call another bureau. However once you have them, call them to sample the applications. Do you like them? Do they sound professional? How do the documents look and how long did it take for them to be transmitted?

9. **Will you provide three client references?**

10. **Don't be cheap** -- you will get what you pay for. Like your company, fax service bureaus deserve to make a profit. They know what their costs are and what their competition charges. Try to play one bureau against another to shave a penny per minute and the best bureaus will turn down your business. The top bureaus are busy and smart and refuse to play this game. Look to build a mutually beneficial relationship based on trust, not cost.

11. **Get everything back.** If you decide to no longer use your chosen service bureau (for whatever reason), can you get back all your materials and all the programming done on your behalf.

(The Kauffman Group)

Fax-On-Demand Service Bureau Selection

Two choices are available to the manager who decides to distribute information by fax. One choice is to install a system and select the features that meet the organizations needs. The other is to use a service bureau.

What features do service bureaus offer? Service bureaus offer both broadcast and fax-on-demand services as well as fax mailbox and fax overflow services. Broadcast traffic produces most of the revenue for many service bureaus, but fax-on-demand is a growing part of these businesses. There are a few service bureaus that offer only fax-on-demand services.

Fax-On-Demand. Many service bureaus have an initial fee for beginning operation. This usually includes recording voice instructions for the interactive voice part of the service. Some service bureaus require the customer to arrange for an 800 number voice line for two-call service. Another start-up fee may be charges for entering documents into the database.

Usage fees are for the inbound voice time and the send time for delivering the documents by fax. One company charges by blocks of characters sent while most charge by the page. Different rates may apply during peak hours compared to off-peak hours. The cost per page ranges from $0.30 to $0.50, depending upon the service provider and when the document is delivered.

Documents can be entered into a database by fax or electronic transfer in a word processing or desktop publishing format. The service bureau converts text files into fax format. For high quality results, it is best to convert the text to fax format electronically, rather than by scanning. Word processing and desktop publishing software have fax conversion features that convert the text to fax while retaining the attributes of the document. So, the fax version appears the same as that printed on a laser printer. (The resolution is 200 by 200 pixels per inch for the fax version, rather than 300 by 300 pixels per inch for laser printers.) Some service bureaus offer desktop publishing service to create high quality documents for the customer.

Many features of in-house systems are currently offered by service bureaus. This includes transferring calls to another location, reaching an operator by pressing a key and accepting credit card numbers for customers when there is a charge for information.

Fax Broadcast. Fax broadcasting has been offered by service bureaus for many years. It is demanded by customers who send the same information to many fax numbers in a short time.

For example, mortgage banks broadcast loan rates to 2,000 to 4,000 real estate brokers in only a few hours. This is an impossible task without a computer based broadcast system. Even for 100 fax numbers, a broadcast service is cost effective as it saves the time of the person who monitors the transmissions.

Service bureaus encourage their customers to maintain their own broadcast lists by providing them with software that facilitates the updating of the lists. Many will maintain lists for customers for a fee.

The ability to merge text with a letterhead during the broadcast is offered by some vendors. Other merge options include adding a signature (previously scanned) and a full text merge of variable information into a form. This text could include the name of the person receiving the fax and the company name.

Maintain Broadcast Databases. Service companies prefer to have their customers maintain their own databases of fax numbers. These take a great deal of effort to keep current, and it is in the best interest of the customer to place a high priority on ensuring the accuracy of these lists. To help the customer maintain their databases, service companies offer a microcomputer software package for maintaining lists. Some of these software packages are proprietary, and others are commercially available. Most have an integrated communications feature that simplifies sending a current list to the computer of the service company. Most of these packages are lent to the customer. A few bureaus charge a small license fee.

Two types of lists are used: proprietary lists that usually contain customer's numbers, and a commercial business lists that customers use for prospecting for sales leads. Proprietary lists are compiled by the information providers from sales leads and customer lists. The commercial lists are usually purchased from mailing list brokers or compilers. These are sold for one time use or for a year at a higher fee. Although the number of fax lists is small compared to mailing lists, fax lists are growing because of increased use of fax as a marketing tool.

Fax Forms. Some service bureaus offer the form entry of requests. Name, address, fax and telephone numbers and the numbers of requests documents are entered by hand on a printed form. The form is faxed to the service bureau where the information is read. The documents are then automatically sent to the fax number included as part of the information provided.

Other Enhanced Fax Services. Two other enhanced fax services mentioned above are fax mail box and overflow fax. Fax mailbox service is like the voice mailbox. A fax can be sent to a person's mailbox for later retrieval. Initially this service was provided to simplify receiving a fax when the person is on the move. He can call the fax mailbox and have a fax delivered to the nearest fax machines no matter where he is. The person does not have to wait until returning to the office. Some users send all of their faxes to the mailbox t insure privacy. So, privacy is one added benefit of the fax mailbox.

Overflow fax is intended to overcome the problem of busy fax machines. The person sending the fax would have to repeatedly retry sending the fax until the receiving machine is not busy. With fax overflow service, a busy signal causes the fax to be routed to an alternate number where it is stored. The service retries the fax machine until it is no longer busy and can then receive the fax.

How much do service bureaus cost? Service companies compare favorably with in-house systems. The main appeal of a service company is the low cost of startup and the ease of beginning service. For broadcast service, the fax number list and broadcast documents can be entered into the service company system by modem, and in a few days broadcasting can begin. Similarly, the startup of fax-on-demand services can be quick and low cost. An in-house system requires a significant investment in hardware, software and development of expertise. Also, it often takes several weeks to select, acquire, install and test a system. In the longer term, the cost of using a service bureau may be more than for installing and operating an in-house system. Some service bureaus, however, have developed vertical programs designed to offset all costs associated with enhanced fax services.

System Management. Managing a system takes time and skill. Companies may not want to invest the time and find the personnel to manage such a system. For these companies, a service organization is a better choice as it allows them to concentrate on their primary business and let the service company provide the expertise. Service companies typically invest in redundant systems and back-up procedures so the service is highly reliable. (Today, in-house systems are also reliable, as computer hardware and software are highly developed and work for long periods without problems.) Also, a service bureau can deal with occasional problems more quickly than many in-house ones because of their investment in additional hardware and full-time personnel.

Can I integrate fax on demand into my local area network? Client-server architectures have done much to pave the way for a more seamless integration between computer platforms. This is because client-server programs are written in such a way that each element is modular.

What are network-based fax servers? A network-based fax server is a communications device (typically a PC) that houses multiple fax cards or multi-port fax cards and associated fax communication software. Its purpose is to send, receive, and route facsimile communications on behalf of a given enterprise.

With a fax server installed, each workstation on the network has the ability to create and send documents via the fax server's resources. Fax servers behave much like a shared printer on the network; In fact, most workstation-based fax server software makes faxing a document as easy as printing it. Most network-based fax servers have the ability to process multiple fax transactions simultaneously, up to the number of installed fax ports.

In addition to acting as a shared resource on the LAN for sending faxes, a fax server also has the ability to receive faxes sent to the company. Fax servers can be configured to route incoming faxes to the print queue so they are automatically printed, or electronically route them to administrative workstations, so they can be viewed and then forwarded to the intended recipient.

Another way of routing incoming faxes is the use of DID (direct inward dial) telephone numbers. In this case, the last four digits of the telephone number on which the fax was received are repeated into the fax card by the telephone network. These digits are then used to determine the intended party's extension number.

(ABConsultants and Nuntius Corporation)

Fax-On-Demand Via Fax Server

When selecting a Fax-On-Demand (FOD) system for your network needs, be sure to select a system that allows users to access the fax server portion of the system directly. This will allow you to utilize the same PC, modems, and software for all of your network faxing needs like faxing from word processing, e-mail, or other applications. Also, inbound documents can be directly routed to the intended recipient further increasing productivity.

(Optus Software)

Fax On Demand Software Features

The differences in functionality, performance and quality of the various fax-on-demand systems are found primarily in the software and software support provided by the vendors. Typically fax-on-demand software includes all the basic software functions listed below. Some furnish advanced features and optional software that provides further functionality, networking operation or improved performance.

Most systems include the following software:

1. **Interactive voice response** and voice messaging that allows users to record greetings, instructions and messages. Includes programmable voice tree, password and PIN protection and caller identification. May allow callers to leave voice messages.

2. **Fax distribution** that provides auto-redial on busy, auto-redial on failure, document selection, queuing, merging of cover sheet and documents and on-the-fly conversion from ASCII to fax format. Many provide area code or phone number lockout.

3. **Document storage and retrieval** that accepts the following: documents from a fax machine or from an image scanner; word processing, spreadsheet, or other PC files in ASCII format; and files in other formats. Retrieves a file upon request and allows previewing of documents and images in storage.

4. **Call data recording** that logs calls and maintains records of calls, callers, incoming phone numbers, fax numbers, materials requested and caller messages.

5. **Diagnostics** that provide remote and local troubleshooting of equipment and systems problems.

6. **Configuration and installation options** (required with systems that allow user installation and configuration) that may allow users to select a single-call or two-call system, plus other options.

Optional and Enhancement Software To Look For. Except for the low-end products, fax-on-demand systems include some of the following types of software:

7. **Fax broadcasting** allows a fax document to be sent to multiple locations. Some systems include conversion from E-mail and PC formats to fax formats for delivery of computer messages to a fax. A few provide a mail merge capability to merge documents and a list of names, address and other individual or company specific information.

8. **Fax mailboxes** store fax documents in mailboxes for addressees to retrieve at their convenience. Retrieval is similar to the process used in fax-on-demand. A touch-tone phone is used to key in identification and password security data.

9. **OCR and OMR** are used on specially designed forms that allow fax-on-demand systems to automatically recognize the requester's fax numbers and the numbers of the document being requested.

10. **Multiple language support** (voice tree) provides callers the option of selecting a preferred language for voice prompting to improve understanding (especially useful in international applications).

11. **Image editing** that provides conversion from various file formats to fax formats such as TIFF (scanners), PCX (PC Paintbrush), ESP (PostScript), PCL (HP printers), EPC (Epson printers) and WPG (WordPerfect). Provides image manipulation and editing functions such as image cutting and sizing, rotation, de-skewing, mirroring, de-speckling, adding or deleting pixels, changing contrast, edge smoothing and creation of halftones from scanned gray scale images. Graphics editors for the creation of letterhead and signature files are sometimes provided.

12. **Credit card billing** provides on-line capture of credit card numbers. Some systems also validate the transaction on-line (that is, obtain approval from the bank card system). In most systems the actual charge is accomplished later and off-line. Credit card billing provides more flexibility in the charge structure (such as different prices for each type of document) and overcomes some negative aspects of 900 services.

13. **PIN** (personal identification number) and password protection or other security provisions.

14. **Direct inward dialing** allows for the use of a large telephone numbers on a group of incoming lines. Each assigned number selects a unique voice greeting and voice menu.

15. **The ability to operate on networks**, such as those using Novell's NetWare and TCP/IP (common network protocols), is provided by many higher-end products.

16. **Automated Speech Recognition** (ASR), used with interactive voice, is available for requesting information and documents. (This is important in countries where rotary dialing is used.)

17. **Hooks for application development** (usually in C language).

(ABConsultants and Nuntius Corporation)

Fax On Demand System Pricing

Fax-on-demand systems are marketed in three basic configurations.

1. **Software only**, sometimes shrink-wrapped (the user or system integrator supplies the PC and boards)

2. **Kits** combining software and PC board(s) (the user or integrator supplies the PC). Sometimes integration and testing with the customer's PC are "bundled" with these kits.

3. **Turnkey systems** (software and boards are integrated and tested with a PC provided by the vendor).

Entry level systems. Several entry level systems are currently available. These fax-on-demand systems are defined as:

Low cost (less than $800 for software and board(s)), and no more than one voice port and one fax port or one universal port.

Most products in this category now use DSP (digital signal processor) technology so that one port on one board will provide both voice and fax operation.

Several of these products allow both one-call and call-back operation. The capabilities of these products are increasing, except most do not currently provide an upgrade (for expansion or growth) path.

Personal communicators are products intended for individual PCs or desktop workstations. They frequently provide automatic attendant (voice answering), voice message recording, fax send and receive capability, fax/voice switching, fax operation and data modem functions. They normally incorporate phone (and fax) directories and other office enhancements. These products normally have only one telephone port or line that is used for fax, data and voice. Most personal communicators are sold shrink-wrapped with PC board(s) and software.

Some current personal communicators have fax-on-demand or fax mailbox functions. These units use state-of-the-art chip technology (DSPs) that integrates fax, voice and data in one low cost unit. Although their fax-on-demand functions are limited, they may be effective for very low volume applications. Prices for personal communicators with fax-on-demand functionality are in the $200 to $400 range.

Microcomputer based systems. Vendors of microcomputer based FOD systems have been the major driving force behind development of the fax-on-demand market. Many of their produces are primarily for one function -- fax-on-demand. These systems are normally marketed as departmental systems for one application -- delivery of information for a department or small enterprise.

High end microcomputer based systems provide other enhanced fax functions such as broadcasting, fax mail boxes, fax forms or fax overflow. These systems are frequently marketed as fully integrated systems sold on the corporate level for several applications rather than solely on the departmental level as is generally true with the lower cost microcomputer FOD systems.

Pricing: Because of the quantity, various numbers of available features and the degrees of integration, prices of these systems vary greatly. A system with two voice and two fax ports may cost less than $2,000 for boards and software, while a fully integrated system with four voice and four fax ports may cost $15,000 or more. Larger high end systems with full enhanced fax functionality generally sell for $1,500 to $2,500 per port.

Voice Processing Systems with Fax Enhancements. Most vendors that provide corporate telephone systems with voice processing functions, such as automatic attendant and voice mail boxes, offer fax options with fax-on-demand and other fax enhancements. These options, which typically consist of fax boards plus software, are usually available as factory and field installable kits to existing voice computer systems. Currently the fax-on-demand options on voice-processing systems lack some of the fax-on-demand features available with micro-computer based systems.

Custom systems. Research of the fax-on-demand market shows that there are a rapidly growing number of systems integrators and VARs that provide custom fax-on-demand and enhanced fax systems. Some of these integrators started as resellers of microcomputer based FOD systems (often software only). Finding that the fax portion of the fax-on-demand software did not meet all their customer requirements, a few integrators developed their own software. These integrators tend to concentrate on vertical markets.

(ABConsultants and Nuntius Corporation)

Fax-on-Demand System Sizing and Pricing

As you can imagine, there is and enormous range of applications available for Fortune 50, high-end Telcos and service bureaus to small business & SOHO. Depending on the system you choose, expect to pay as much as $50,000+ to less than $1,000. Of course, this is the range between high-end and SOHO solutions. Here are some examples of systems with differing complexity and prices:

High-End Fax-on-Demand. These applications appeal to the Fortune 50 companies. They are complex, high density, multi-channel installations of custom hardware and software. The number of fax lines they handle can be anywhere from a dozen to hundreds of lines. An entire staff of individuals is on hand to update the document database every day, handle circuit problems, and re-do prompts.

In addition, many of these systems hook-up to mainframe computers and other databases. MIS department programmers and outside consultants are often used to create, change and maintain these systems. They start at $50,000.

Mid-Range Fax-on-Demand. These applications are multi-channel installations ranging anywhere from eight to 24 lines. They can be made-up of either custom and/or off-the-shelf hardware. Most of these systems are PC-based and use off-the-shelf hardware components. It is also typical for fax broadcast and fax routing capabilities to be built-in. $10,000 - $50,000.

Workgroup, Small Business & SOHO. Some utilitarian "no frills" programs are available for smaller companies and the home office crowd. The basic "touch 1 for this data sheet and 2 for that data sheet" stuff is there. These are plug & play office appliances that require little administration. They are fairly simple to use and cost about $1,000 per port.

(Brooktrout Technology, Inc.)

Getting Started Successfully

Before you purchase a computer telephony system or jump into a development project, just look at yourself first. Ask yourself these important questions posed by Jon Shapiro:

1. **Do you have a successful application?** If you intend to automate a certain process, or if you are about to put a new process into place, make sure it is a successful one. Automating something that is already inherently broken doesn't make a lot of sense. Get together with the people in charge of (and the people who execute) the process in question. Are there flaws? Does the order entry or information delivery system you have do the basic job? Explore this first.

2. **Do you understand telecommunications?** Every computer telephony system uses elements of telecommunications. The phone. Your PBX. E-mail. If you don't understand telecom, then find someone in your company that does and put him or her on your discovery team.

If no one at your company has telecommunications experience, then call 1-800-LIBRARY and order some books on the subject or sign-up for a few seminars on telecom. You'll save thousands just understanding the terminology.

3. **Do you understand computers, software development and networks?** If not, it's a good idea to bone up on this subject. More and more, computer telephony systems borrow heavily from this discipline. Today, IVR systems, voice mail and call routing devices interoperate with LANs, mainframes and the Internet. There are also good primer books on this subject. Pick someone on your staff that's comfortable with computers. This will be the one that everyone migrates to when they have a problem. They're easy to find – the ones who seem the most helpful when things go wrong with a computer. Put this person on your discovery team, too.

4. **How much time do you have?** Researching and choosing the right computer telephony system is time consuming. Brace yourself for an exhaustive project. There's plenty of vendors and technology to chose from. Most are good. Many are poor. If you have the time, get involved with every aspect from research to application, installation and upgrades. Your computer telephony system can add a significant productivity boost to your company and make your customers very happy. It's worth your "sweat equity." If you don't have time, then be prepared to pay someone to put time into the research and purchasing process.

Here's Shapiro's three laws of CTI success:

1. **The value on the information must be high.** Consider the type of information that you want to get to your customers or users. If it's really valuable information, people will do just about anything to get to it and call for it any time of day. Information like technical support tips, brochures, and time-sensitive messages can have a high value. Talk to your customers and employees about the kinds of information that is critical to a smooth operation. This is the kind of information that you should target for computer telephony assistance.

2. **Time must be important.** No amount of automation is necessary unless the information can be delivered when it is needed. Inventory inquiry data is only relevant if it is used to make on-demand decisions about ordering. Prospecting information is only important if leads are followed-up before they get "cold." ANI (Automatic Number Identification) data is only good if it helps you to route calls for better service or curb abuse in a real-time way. Consider these things before you automate. If the data is not time-sensitive, then maybe you don't need to automate.

3. **You should have monopoly on the information.** If the information you wish to deliver is easily gotten in a variety of other ways, it may not make sense to put a computer telephony spin on things. You must have a monopoly on the content, or at least package the information in a unique and value-added way. Do not bother packaging information (especially non0time sensitive information) for use in computer telephony if it is commonly available.

Develop a Mission. Before you embark on your computer telephony voyage, make sure you know your goals. Consider the following:

Clarity. Do you and your colleagues have a clear vision of what you are trying to accomplish? Do you want to decrease call abandonment, extend your service hours, or give access to hard-to-reach data? Or do you want to do all of these things and more?

Benefit. Make sure you can clearly articulate a benefit to the proposed users of the system: "By using our new service, you will be able to get your order status quicker;" "our new IVR system will save you time, because you no longer have to wait for your information;" or "Now your sensitive information is secure, and you can retrieve it without having to ask embarrassing questions."

Time. Consider the amount of time it will take you to source, install, debug and improve your new application. There's also a training period and time needed to promote its use before it is accepted by the users. Be realistic about the "ramp-up" required before it is accepted. Sometimes this takes up to a year.

Measurement. Set-up realistic indices of measurement so you can quantify the systems' success over time. For example, you can count the number of service complaints, number of (increased) transactions or proposals. You can also count how many fewer manual calls are taken or made or a decrease in manual fax traffic. Tie these measurement indices to your articulated objectives and stated benefits. Put them on some kind of a chart or graph and plot the data over time. You can also measure hours saved, sales volume, new contracts, etc. and try to link this all together.

Define the Project. Development projects vs. production projects. It is important to decide up-front whether the system in question is part of a development project, or whether it is meant to solve a specific problem today. Development or "skunk work" projects don't have to stand-up to the same demands or measurement indices in order to be successful. Development projects are great for proving concepts and creating ideas, but they rarely deliver value to the users in predictable way. If you want predictability, then go for a production-level system. If you are tolerant of many changes, service disruption, and complaints - by all means go the "development project" route. You either have time to experiment, or your system has to be up and running within days or weeks. Make that decision now, and you won't be disappointed.

Really cool versus utilitarian. Betamax has a better picture, sure, but when was the last time you got your Betamax unit repaired or rented a beta tape? Don't get blinded by the shinny new thing or claims of the best or coolest performance. Not many of us can predict the future, so try to stay away from "state of the art" features and flashy gadgets. These are reserved for those of us who can deal with ongoing "development projects." There are plenty of rock-solid systems out there that will get the job done. Go for the ones with hundreds or thousands of installations. Keep you development projects off to the side – not in your production system.

Crawl, then Walk. Before you buy or develop a system based on tools, take a look at the documentation. Either you can explain this stuff to your mother, or you can't. If you cannot explain it to your mother, or if she can't read it and get the idea - then it is probably too complicated.

Testing and Documentation make all the difference. If your new system is well documented, then it's probably been well-tested. The two go hand-in-hand. If there are no manuals or other complete documentation, then you should run the other way.

User interfaces separate good systems from bad. If you try a system and cannot master its use within fifteen minutes, then it is probably not designed very well. No amount of training is going to make you use something you just don't like. The human factor is paramount in computer telephony system design. If it was left out - walk the other way.

Can I hook this up to a Sun Moon Star key system? A lot of hard work goes into making a computer telephony system work with common telephone systems. Whether you have Centrex, a hybrid key system or a PBX - make sure that the proposed system integrates. It's no fun to "force" a system to work with your phone. If you're looking at a voice mail system and most of your messages are picked-up at work - insist on a fully integrated solution that lights your message waiting lights and automatically routes callers to your phone. If many changes, upgrades and special patches are required to make this happen - keep looking. Most application suppliers have a list of PBXs and Key Systems that their solution integrates with. The good ones have a list of at least 20 integrations to their credit. Any less than this spells an immature product.

(Alliance Systems, Inc.)

Headsets -- The Wireless Kind

First Select a Comfortable Design. The most comfortable are the wireless type, since you don't have to worry about cords getting caught in your clothes or clips, etc. Try one on for size. Ask some call center manager and telephone workers you know. Consider how long you can wear the headset with comfort. Shoot for 4 - 6 hours of comfort. If you can wear a headset that long without it disturbing you, then you've found a good one.

Second, you should look for good sound quality. Listen for the same or better clarity that you would get from your regular telephone headset.

Third, make sure you get some distance out of the battery life of the product (if it uses batteries). Make sure you can re-charge the unit during the stretch of time you will not be using it.

Lastly, go for the aesthetics or cosmetic appearance of the headset.

(ACS Wireless)

IVR Buying Reasons

1. **It's faster to "talk" to a machine.** You don't have to be polite.

2. **It's confidential.** The machine will never snicker when it tells you your paltry bank balance.

3. **It's more accurate**. It won't make a mistake with your balance, though you wish it did (in your favor, of course).

4. **Machines hear better.** Humans can't always understand you. There is no single human language. Everyone speaks in different languages -- mid-west, southern, etc. There's only one universal language -- the language of the machine -- **touch-tone**.

5. **Machines are "repeatable."** You can punch in the same buttons again and again. Humans get bored. Machines relish boredom.

6. **Machines are interruptible.** You can punch in touch-tone while the machine is talking... and it doesn't get annoyed or give you a lecture on politeness.

7. **Machines are stupid.** You know they're stupid. So you treat them stupid. People are stupid, but they think they're not.

(Teleconnect Magazine)

IVR Selling Approach

Interactive voice response (IVR) provides easy access to information and uses the world's most ubiquitous terminal - the telephone. When selling IVR to a prospect, talk to them about how IVR is perfect for casual or infrequent data users. Discuss ways to automate routine data entry and inquiry. Ask if there are ways you can help to free humans for non-routine tasks.

The best reasons for implementing IVR are to:

1. **Achieve some measure of business re-engineering**.

2. **Increase customer satisfaction**.

3. **Enrich employee participation**.

4. **Decrease costs**.

5. **Generate revenue**.

6. **Gain competitive advantage**.

These are the most important hot buttons. If your client does not want any of these things, then you are wasting your time. Discuss this list with your client and find out which of these things they want to achieve.

That's what was discussed with Drexel Heritage. They manufacture fine furniture that is sold through dealer showrooms. Before they used IVR, the in-stock reports were 3-5 days old when received by dealers. This meant that every time a customer was ready top purchase an item from the showroom floor, the sales person had to call the warehouse to check on availability. The 19 customer service representative were always overwhelmed with routine stock availability inquiries.

Since major furniture purchases are typically made by husband & wife together, they tend to shop in the evening & on weekends. Unfortunately, this is when the agents were not available to answer the dealer inquiries. When buyers say: "Do you have the wingback chair in blue leather?" – they are clearly articulating a buying signal. Before IVR was implemented, the typical dealer response was: "I can get you that answer next Monday."

Now (with IVR), the dealers have 24 hour x 7 day access to stock availability by making a simple phone call. The stock information is available at the crucial point in the selling situation. At this point, over 11,000 automated inquiries per week are taking a big load off of the agents, who can now provide important support and non-routine service for dealers. Now, the centralized service agents can go the "extra mile."

Now, let's take a look at those reasons for buying IVR one more time. Drexel Heritage did achieve some measure of business re-engineering, since they linked the stock availability database to the IVR system. This fundamentally changed the entire way their call center handled calls.

They were also able to increase customer satisfaction, but giving superior support to the people who had non-standard questions – and also by enabling their dealers to provide instant information on availability. They were certainly able to enrich employee participation, because everyone involved in the transaction loop has a tangible benefit (less stress, easier work environment, etc.). By freeing-up the agent resources on the routine stock calls, the firm was able to decrease costs, since more agents were not needed to handle the mounting overload of calls.

The dealers are now able to close sales when they get the buying signals (weekday nights and over the weekend), which helps to generate revenue. Lastly, the firm gained a competitive advantage buy supplying on-the-spot answers to prospects.

(Expert Systems, Inc.)

IVR System Cost Justification

In order to be accepted by an organization, and IVR application must meet one or more of the following objectives:

1. **Increase revenue.** Often, an IVR application will speed up process, so more can get done in less time. For example, a company uses IVR for order entry. The system is available around the clock, so customers can place orders whenever they have time. The company saw a 10% increase in the number of orders. Most of the increase came from customers ordering outside of regular business hours.

2. **Avoid costs.** A company experiencing an increase in telephone calls must often add staff to handle the load. An IVR system can often handle the increase, saving the company the expense of hiring extra staff. Secondly, the staff that were handling calls can now be re-deployed to other valuable tasks. By getting more done with the same staff, a company avoids an increase in costs.

3. **Improve service.** Companies can improve service by giving their callers quicker access to information and more control over how their call is handled. Service is also improved when customers can get information 24 hours a day, have no limit on the number or amount of time a transaction takes, and do not have to wait on hold for service.

4. **Save money.** Companies can save money by eliminating personnel and their overhead. Many companies are averse to using technology to replace people , so this approach should only be used when the company has indicated it as an objective. Additional cost savings come from reduced WATS line charges and lower mailing costs. Reduced paperwork results in lower costs as well.

5. **Improve corporate image.** Some companies will implement an application just because they want to be perceived as being on the "leading edge."

Others are similarly motivated because their competitors did it. Still others have to be lead by their customers.

(Voysys Corporation)

OS Buying Tips For Fault Tolerant Telephony Platforms

Here' some questions to ask when selecting the operating system environment for your mission critical platform:

1. **Does the OS have all the "Me Too" stuff?** Everyone talks about it – but just to make sure, your system should be able to handle *multitasking* (ability to do more than one job at the same time). You should be able to write code that allows for *priority-driven scheduling*. Unless the system is completely isolated with no need for user interfacing, a *graphical user interface* (GUI) will make it easy to access by system administrators and users. Make sure it can handle true *32-bit memory addressing* for superior file and memory access. If you are doing a lot of real-time processing of transactions, call routing, etc. – make sure it handles *real-time* processing.

2. **How well does the system perform?** Overall system performance includes fault detection and fault recovery. Ask how this is handled by the OS. Get examples. Ask for a step-by-step explanation on how fault recovery works. Ask for written benchmarks that discuss and quantify, for example:

Does it have a 3.9 millisecond context switch?

Does it have a 4.4 millisecond interrupt latency?

Does it have a 4 Mbytes/second write performance?

Does it have a 5.3 Mbytes/second read capability?

Does it have a 1.6 Mbytes/second network throughput?

3. **How flexible is the OS?** Does it support a plug & play model. How easy is it to install and configure RAID, VME, STD 32, PC/104, PCMCIA, and PCI components, for example? Ask if it can handle multiple simultaneous file systems and shared network links. Can you load drivers for fax, voice, network, and disk technologies on the fly?

4. **How scaleable is the environment?** It should be able to grow with your customer without changing the OS. This means you should be able to go from a single platform with just 12 lines to multiple interconnected platforms with hundreds or thousands of lines (without changing-out your application). You should be able to add new technology quickly and upgrade core components on the fly. Ask how this is achieved and look for a real-world example of how a customer did all of this. Call the customer.

5. **How maintainable is the OS?** You should be able to logically separate "MMU" protected processes, so they can be spawned or shut down without effecting overall performance. It should be able to handle multiple redundant file systems. The processing paradigm should allow for fully distributed load sharing. Ask also how to achieve multiple load-balanced network links and a dynamic "hot-swappable" software environment. All of these factors go into "maintainability."

6. **How committed is the company to standards?** Look for POSIX, TCP/IP, NFS, RPC, X Window System, Motif, CORBA, SNMP, STD 32, VME, PC/104, ANSI, SQL, ODBC, PCMCIA and CDPD in the alphabet soup of standards compliance. This is a huge investment.

(QNX Software Systems Ltd.)

PBX or ACD Buying Tips

Today's leading PBXs and ACDs provide robust and feature-rich capabilities that enhance voice systems for large and small organizations with multi-line telephone systems. However, there are ways to differentiate these components and determine whether or not your PBX or ACD investment is sound.

1. **Defining the system and its components.** Determine what is needed to support your voice system and evaluate whether your organization needs a PBX, ACD or key system. Keep in mind that a PBX and ACD are only components in an overall voice architecture and do not represent the entire voice system.

2. **Consider all options.** Open the evaluation process to multiple vendors and consider the variety of PBX and ACD systems on the market today. Witness first-hand the capabilities of each system and learn how they will fit in to your overall voice architecture. Inform the vendors you are working with that you are considering other systems to help secure more competitive pricing.

3. **Open non-proprietary architecture.** Make sure the PBX or ACD you buy can be integrated with other systems and components within your present and future voice architecture. This is critical when adding emerging technology components to your system and support of telephony-based applications.

4. **Computer-Telephone Integration (CTI).** Ensure that your PBX or ACD vendor provides a full compliment of CTI features to support intelligent data based routing, graphical phones, skills based voice and data transfers, and integrated inbound and outbound operations.

5. **Highly scaleable.** Avoid proprietary linkage of hardware and software and ensure a multitasking operating system exists. Your PBX or ACD should be able to talk directly with your PCs or workstations.

6. **Open transportable software.** Focus on open software that is transport able across multiple, standard hardware and operating system platforms to take advantage of increased power and capacity and reduced pricing.

7. **Universal card slots.** Determine how easy it is to make changes or up grades to the system. Card slots should be universal to facilitate maintenance and upgrades which will protect your investment in the long run.

8. **Support capabilities.** Challenge your vendor on feature support-related issues. Conduct customer site visits as part of a complete product evaluation. Talk procedures and maintenance with technical support people before making your purchase.

9. **Product cycle status.** Determine where the vendor is in its product life cycle. If the product is early in its life cycle, the system is likely to have bugs or start-up problems. If it is late in the cycle, the system will not receive feature enhancements to support emerging voice options. Don't buy at the top limit of a systems capacity to save money since your system will most likely grow.

10. **Real-time reporting capabilities.** Ensure the component provides real time statistics for integration with sales, forecasting, scheduling, and other management applications to enhance call center productivity and profitability.

(Technology Solutions Company)

PC Component Selection

When it comes to computer telephony hardware, not all PCs are created equal (or compatible). And because most CT hardware *and* software vendors don't mention this in their user guides, it's unfortunately a rite of passage for new CT developers.

On the bright side, at least you discovered this before your app went into service. That's when things can really get ugly. It's not uncommon for less-than-ideal PCs to install and apparently run okay until the CT system is loaded up with more call traffic than usual. Then it's kaboom. Then you've got some real problems on your hands.

Here's just a few reasons why different PC motherboards are sensitive to CT hardware and, worse, why they can't handle the "load."

Incompatible BIOS. There are certain brands and versions of BIOS chips that seem to be purposely designed to confound those configuring CT systems.

Look for AMI BIOS dated 1990 or newer whenever possible. Recommended because you can enable / disable memory address options. Other BIOS makers only let you adjust a few (at best) of the memory configuration options for shadow RAM, etc. The big issue is working with memory addresses and setting a stable, non-conflicting page frame. BIOSs other than AMI just won't let you control how the C000, D000 and E000 segments are used. This is very critical in regards to where you can load the voice hardware, page frame, etc. For my money, the AMI BIOS has the best flexibility.

Inability to handle high interrupt loads. CT systems can generate a high rate of interrupts and DMA transfers. Most desktop software doesn't, which explains why some CT-deficient motherboards escape into the marketplace – they aren't tested with CT applications running. If you want to set up a network server, Novell certifies motherboards that have a demonstrated ability to handle the processing load. If you want to get a PC that will work well in high-interrupt environments, get a certified PC (or motherboard).

Sensitivity to bus noise. Lots of high-speed digital signals on a poorly designed PC bus can cause unwanted "noise" that confuses the PC's circuits and corrupts critical signals on the bus. This happens less frequently on normal PC motherboards. But on passive backplane PCs (industrial PC systems that have many slots) this fault can cause all sorts of system instabilities. Typically, moving cards to different positions on the backplane can "improve" things, but I get real nervous when problems are fixed with such black-magic techniques.

Poor PC motherboard design. This doesn't happen too often anymore, especially with the high level of chipset integration that is used on PC motherboards today.

Still, very fundamental digital-circuit timing requirements are sometimes violated, making such PCs more sensitive to temperature and power supply voltage variations. Sometimes they just don't work at all.

Latest is not always greatest. Don't rush out and upgrade to the new high-speed and powerful VLB and PCI motherboards available today. Remember, we're still working with 8-bit and 16-bit voice and fax hardware, so these high-speed slots would go unused except maybe for an HD controller, video or network card.

You'll also find that these great new motherboards use IRQs and memory addresses that normally are not occupied on a simple 16-bit motherboard. This further complicates the already troublesome task of configuring a voice / fax system that needs these few-and-far-between hardware interrupts and memory addresses for the conglomeration of cards in a CT-based system. Recommended base platform: a high-powered 486 on a generic motherboard with 16-bit-only slots and, of course, an AMI BIOS.

Stay away from integrated peripheral junk. Newer PCs are typically coming with built-in video, DSP modems and other peripherals (often proprietary) integrated right on the motherboard. These modules can't be easily removed or disabled if they go haywire or cause memory or interrupt conflicts. And they usually have specialized BIOSs which can add even more to the confusion factor of integration.

Get answers from your vendors. Hopefully your CT hardware or software vendor can recommend PC platforms they are confident will work. Don't expect every vendor to give you a long list, though. It's going to take time to convince many CT companies in the "components" *and* hardware business to keep this list current.

It's not all their fault. Developers on the front lines are the best candidates to discover integration problems. And they don't always let their component makers know what nasty surprises they've found.

Load test. Assuming that everything installs okay on a given brand of PC, testing your CT system under load is the wisest thing you'll ever do.

No matter how many voice lines your system supports, run it for several hours with all of them active. See the system run without any detectable problems (like speech messages that sound broken-up, system crashes, etc.). To really test it, you'll need an intelligent "dynamic" load tester, like Hammer Technologies' (Wilmington, MA -- 508-694-9959) Hammer product. Among other things, it will actually mimic the way "real" callers use the system, not just pump generic bulk call-types through the lines.

Another thing you can do is elevate the temperature of your CT system while operating it under load. Especially with the cheaper PC motherboards, some components may have been inadequately tested and are more likely to fail. Higher temperatures can accelerate the failing of flaky motherboard components already on the verge of collapse.

A simple way to heat up your system is to put it under a cardboard box for a short while. **Caution!** If you run a PC with expensive CT cards in it at temperatures that are too high, you can cause permanent damage to both the PC and your expensive cards.

For best results, monitor the temperature inside the cardboard box (cut a small hole in the side and stick a thermometer inside) and make sure the temperature doesn't get higher than about 85 degrees (F). Then run your CT system for a few hours and see if it survives without crashes or other anomalies. Just be careful. Smoke coming from under the box is not -- we repeat not -- a good sign.

(Telephone Response Technologies, Inc.)

Predictive Dialers And Answering Machine Detection

Just about every predictive dialer offers some kind of answering machine detection, usually based on "Voice Cadence." The theory assumes if there is a short burst of words followed by silence ("Hello!"), it's a person.

If there is a long burst of words ("Hello. This is the Stone residence, thank you for calling.") it's an answering machine. Many manufacturers say their answering machine detection is 90% accurate, but in reality it can be no better than 60% to 70% accurate.

The best systems electronically measure the voltage and line noise that is produced by an answering machine, using a cadence measurement as an auxiliary test. This type of answering machine detection is often referred to as "Frequency Detection," and it can be over 98% accurate within 2/10ths of a second. Only a few out of the 30 major predictive dialing manufacturers (such as Results Technologies) offer "Frequency Detection" on their systems.

(Computer Telephony Magazine)

Predictive Dialer Buying Tips

Predictive Dialers use sophisticated technology. Shopping for the right solution is not trivial. Here are 20 buying tips. Read them carefully and shop (and dial) smart.

1. **Call abandonment may close you down.** Many lower priced predictive dialers hang up on 20% to 40% of prospects, turning your call center into a public nuisance. People with Caller ID will get the phone number of your call center and complain. Phone companies can track the source of abandoned calls. List owners who receive complaints may stop renting you their lists. Moral: If possible, buy a system capable of a zero abandonment rate, a necessary precaution in case future legislation demands this capability.

2. **Slow call transfers can cost you sales.** The prospect shouldn't have to say "Hello" three or four times before the agent hears it. Your system should transfer the call so quickly that the agent hears part of the first "Hello."

3. **Lower conversion equals fewer overall sales.** Having an unlimited supply of leads, with very little wait time between calls, causes the agent not to try as hard as they do when they have limited leads.

Companies using predictive dialing therefore often have a lower sale conversion rate (for example 10%) than with manual dialing (11%) or automated "power" dialing (14%). You need a contact control management capability that finds the optimal contacts per hour and control the contacts per hour to produce the most sales at the lowest cost.

4. **Make sure you have the unlimited right to sell your hardware or software to a third party.** Some predictive dialing manufacturers have a clause in their purchase contract that specifies that you cannot transfer the software license to another party without the manufacturer's permission.

One company who wanted to sell their used system found that the manufacturer wanted $3,000 per seat to transfer the software license, making the sale impractical.

5. **Don't be forced to install extra telephone lines.** Predictive dialers simultaneously dial multiple phone lines to improve connect rates. One of the most expensive systems requires three phone lines for every agent on the phone, most require two phone lines per agent, and only a handful can run okay at 1.5 lines per agent.

If a 16-station system needs three phone lines per agent, that's 48 lines or two T-1 spans. Even at two lines per agent, you need 32 phone lines or almost one and a half T-1 spans. Since T-1 spans come in multiples of 24 lines, you would actually have to install two T-1s. However, if the system only requires 1.5 lines per agent, you would only need 24 phone lines or one T-1 span.

Depending upon mileage and location, the charge for a T-1 can be about $500 a month. If you need two T-1's instead of one, you're wasting $500 a month on T-1 rental, plus all the cost of dialing twice as many lines every time you try to place a call.

6. **Beware of hidden equipment costs.** Besides a T-1 span or two, many predictive dialing systems use analog technology, requiring a channel bank to be connected to the phone lines.

Renting one from the phone company costs $300 to $500 a month. On the other hand, a predictive dialer with a digital network interface can be connected directly to phone lines without a channel bank. If your system requires two or three lines per agent along with a channel bank, that's an extra cost of almost $1,000 a month, a "hidden cost" of $60,000 over a five-year period.

7. **Don't become a captive customer.** Some manufacturers boast that their system is better because it uses all proprietary technology and parts. While that may sound like an advantage during the buying process, it may quickly become a nightmare during the calling process if your manufacturer goes out of business.

You'll lose telemarketing revenue if you have to wait for a special repair person to travel to your location. The manufacturer may even elect to double or triple your maintenance fee after a few years or overcharge for new feature upgrades. Perhaps there never will be an upgrade! You're at their mercy. Solution: Use PCs, open APIs and cards by Dialogic, NMS, Rhetorex, NewVoice, etc.

8. **Does the system allow the agent to enter changed phone numbers and instantaneously redial?** Alternatively, can the system detect a disconnected phone number and put it into a queue and not pass it to the agent? This seems like a minor function, yet people with changed phone numbers typically produce higher sales conversion rates, since they are being called less frequently.

9. **You need inbound/outbound blending.** The system you finally select should have the ability for every agent to handle both outbound and inbound calls from the same terminal.

This leads to higher list penetration and higher sales. The outbound telemarketer can leave a return phone number with a message explaining what the call is about. The person who returns a call is more likely to end in a successful call result. Likewise, if you need inbound call handling, your agent can automatically be passed outbound calls during slow inbound call periods.

10. **Use digitized recordings.** Buy a system that can simultaneously record all agents, for an entire day, and can quickly rewind, fast forward and retrieve calls for playback.

It's now possible to digitally record on a 1" x 2" computer cartridge everything that all your agents say on the phone, during an entire day, along with all the prospects' responses. You can now hear how a specific agent is opening or closing calls, what complaints are called in, and you can let trainees listen to your experienced agents or let your experienced agents critique trainees.

11. **Get call tracking and custom list loading.** Your system should keep a record of call attempts to be certain that each attempt is at a different time during the day and on different days. Some weaker systems have no call tracking abilities at all, and every time you load the dialers the same people are always called first.

Some systems are so poor that on large lists you may have tried part of the list seven or eight times, while leaving as much as 20% or 30% of the list with no call attempts. To get higher list penetration during the day, the system must be able to do schedule callbacks and have list analysis capabilities, so that home rates can be determined for different zip codes or different days of the week and at different time intervals.

12. **Make sure transition times are short.** Predictive dialing should reduce the amount of time between when your agent hangs up the phone and when they are back on the phone talking to another person, at a given low abandonment rate, say, 2%.

Only a few predictive dialers can achieve list penetration level of 50% with daytime calling, while maintaining an average wait time between calls of nine to 20 seconds at a 2% abandonment rate.

13. **Get full branch scripting.** The best scripting systems allow data and calculations to be performed in the script and the screen routing is based on the answers given -- not only on the current call, but on previous calls as well.

Also when an agent hits a key to change screens, it must be instantaneous, no matter how many people are on the system. A multiple-second wait time is unacceptable. The system should also provide function keys for access to questions most likely to come up at each specific place in the script.

14. **Get fulfillment capabilities.** An agent should be able to easily generate a personalized letter, mailing labels, invoices, sales cards or sales sheets and compete an actual order by laser printer.

15. **Get security.** The system should have security at the system level and project level. Individual screens should be protected by User ID.

16. **Get a message bulletin board.** There should be a message bulletin board for individual agents, groups of agents or all agents to inform people of vital news before shifts start and after breaks end.

17. **Get voice mail.** This should be incorporated into the system as well as message on hold while a person is waiting.

18. **Try faxing.** The system should be able to do fax-on-demand.

19. **Ensure campaign flexibility.** There should be no limit to the number of campaigns or number of projects that can be called. Every operator should be able to call a different campaign at the same time.

20. **Make sure it's expandable.** What if your company becomes a big success? Plan for future growth.

(Results Technologies)

Professional Services Fees

When buying software for a computer telephony app, the software company may charge for Professional Services on a daily basis. This may cost $1,000 per day or more.

Make sure your contract clearly states what these services will accomplish and whether or not a "not-to-exceed" amount can be negotiated. Determine whether any other outside Professional Services are needed for the project and what the cost of them will be. For example, the software company may not have staff knowledgeable on telecommunications issues.

(DIgby 4 Group, Inc.)

Selling Application Ideas In An Organization

It's time to sell your application ideas to the company. You may think it's a great application. But what about the rest of the company, especially the person with the money? If you've done the right discovery, this will be a straightforward task. When presenting the application to the decision-makers, be sure that you can clearly state its objective (what we are doing and why) and the benefits to the company and their callers.

The two concerns of most companies considering IVR are: can we justify the cost and how will it impact service? The two issues are intertwined. Be prepared to address them in detail.

In some organizations, you may need to step back and discuss the overall benefits of IVR. Be sure that management supports the general idea of implementing IVR. If not, the remainder of your efforts will be difficult, if not futile.

(Voysys Corporation)

Selling Fax On Demand

In today's information age, knowledge workers are swamped with information. They are finding it increasingly difficult to organize all this information so they can quickly find information they need. They are experiencing an information overload.

Knowledge workers need better ways to quickly obtain information when they need it -- that is, they need information on demand. The burgeoning growth of facsimile has led to new opportunities for fax-on-demand to reach prospective customers with information and advertising. By calling a telephone number listed in newspaper or magazine advertisements, potential customers can request desired information (to be delivered on a fax transceiver) on the product or services advertised. This quick and easy method of providing information to prospective buyers greatly enhances the value of traditional advertising. Fax-on-demand may revolutionize marketing in this decade.

Fax-on-demand systems have proven beneficial as a convenient and cost-effective way of delivering up-to-date information -- especially for information that changes frequently. It is used to distribute technical information; financial information, such as stock market data and charts; government reports and publications; government and tax forms, medical information; restaurant menus and ordering forms; real estate listings; and other time-urgent information.

Most high-tech companies that delivery product literature by fax-on-demand save substantial money over mailing the literature. As a result, many in-house systems pay for themselves in six months to a year while companies using service bureaus find that the financial benefits are almost immediate because of the low startup costs.

Most Offices Have a Fax. Nearly every office has access to information in a fax-on-demand database, for a fax is found in most offices. The low cost of fax machines has increased the population of these units in the office during the last five years. With the increasing number of fax servers, the number of PCs with access to fax sending and receiving is even larger. All of these fax machines and fax/modem devices are compatible. They follow the CCITT group three fax standard that all fax manufacturers meet.

The number of faxes in use in the United States is growing strongly, with more units using plain paper technology. The trend is for continued strong usage growth by both businesses and in the home.

The growth of PCs with fax modem cards or chip sets installed is even more dynamic. Fax modems are being included in many portable and notebook PCs. Fax capability is a convenience for the traveler. With fax capability, the portable user will have access to fax-on-demand databases.

By the end of the decade, there are estimates that 100 million fax machines, PCs, portable PCs and notebook computers will be able to send and receive fax documents. Then fax-on-demand can be a part of every business's facilities and its uses will have multiplied and become as widely accepted as any facility in the office.

Who is using fax-on-demand and why? The first fax-on-demand systems were installed to distribute product literature and customer support information in high-tech industries. These enhanced fax applications grew quickly because the benefits were quite clear:

1. **Product literature could be delivered immediately**.

2. **Literature could be delivered at a lower cost**.

3. **Product support could be handled without adding staff**.

4. **Prospects and customers liked the service**.

Marketing organizations. Fax-on-demand is used to automatically deliver company, product, technical support and related information to help sell products and services and to support current customers' money and allowed the company to expand its support without adding support personnel. The service has been remarkably successful.

As fax-on-demand products and services became available in 1990 and 1991, more computer hardware and software marketing organizations introduced product support and literature fulfillment services to meet their expanding needs. As a result, the market for fax-on-demand developed first in these high-tech marketing applications.

Benefits. Organizations experience many benefits using fax-on-demand for marketing applications. The benefits usually include:

1. **Prospective customers get literature** when their interest is current -- when it is at its highest level.

2. **Delivers literature immediately** to prospective customers 24 hours a day and seven days a week to any place in the world where there is a telephone and a fax.

3. **Delivers literature at substantially lower cost** than the present method.

4. **Increases the effectiveness of space advertising** by giving immediate access to additional product, technical and cost information.

Information providers state that the benefits of using fax-on-demand for product support include:

A. Provides first level support immediately upon request by users with virtually no telephone line "busies" and no waiting in a queue for a service support staff person,

B. Provides support to users and customers located any where in the world where there is a telephone and a fax, 24 hours a day, seven days as week,

C. Helps meet the demand for a rapidly increasing level of support without increasing staff,

D. Satisfies a large percentage of support requests without the use of support staff. For other requests, it improves the user's understanding of their problem so that when support staff talks with the user, the user has more intelligent questions) thus saving time and money, and

E. Substantially reduces cost per support request.

Publications. Fax-on-demand use has also expanded rapidly in the publications industry. Prevalent applications are for literature fulfillment for the advertisers in the publications and for delivery of article reprints. Other applications in the publications industry vary considerably.

Magazine Publications. For years, magazines have had "bingo cards" coded to reference numbers printed on advertisements to allow readers to request more information. This approach has worked well but the information takes from two to eight weeks to reach the reader. Advertisers and prospective buyers want more timely information delivery.

Electronic News Publications. Newspapers accumulate a substantial amount of news and supportive material that is never printed in the newspaper because of cost ands space constraints. These "news libraries" contain information valuable to the business community, the government and consumers. It is possible to categorize much of this information and package it for delivery by fax-on-demand.

For instance, a company involved in the export of scrap metal would be interested in news that affects scrap metal prices. An electronic file could be set up to accumulate information on scrap metals appearing in newspapers or on the news wires. Titles appear on a list that the subscriber could access. The subscriber would select the titles of interest and then information would be sent to the fax transceiver.

A wide range of services can be provided in categories such as financial, medical, exporting, importing and proposed legislation affecting industries of the subscribers. Several methods for paying for "electronic news" are used:

1. **Internal financing by the newspaper to increase circulation** or to save print space in the paper (information is free to user).

2. **Charging the information requester by use of 900 numbers**, on line credit card charging or through subscriptions.

Government. Federal, state and local governments are large providers of information. It has been estimated that 25 percent of government employees are involved in providing information to the public. Most of this information is made available to the public at little or no cost. The challenge is to publicize the availability of information and provide a cheap and efficient means of delivering it.

The federal government has a constant demand for providing information to the American people. Publishing and delivering this information is expensive) especially for rapidly changing information. Some agencies and departments have established electronic bulletin boards to deliver this information. Some make information available on Internet. An increasing number are making information available by fax-on-demand. Delivery through bulletin boards and Internet is limited to owners of PCs with modems that can be difficult to use.

Several federal government organizations have set up fax-on-demand services for distribution of reports, trade information, forms, regulations and standards.

City and local governments have an interest in providing information quickly to the residents in their jurisdictions. Cities are experimenting with the use of fax-on-demand as a way of providing this information. It is too early in the operation of these services to decide whether they will be accepted by businesses and residences in the cities. If accepted, services should expand rapidly because they can be self-funding when operated as a 976-number service.

Another promising application is the distribution of lists of products to be purchased by governments. Some cities have voice-mail services to announce requests for bids on products. The next step is to use fax-on- demand as a way to list scheduled purchases and to invite competitive bids. A specification can be a part of the bid announcement in a fax delivery system. Acceptance of these services in the next few years will set the stage for growth in these applications.

Financial and commerce organizations. Stock, bond and commodity market data is highly time sensitive. Customers want information regarding company performance on domestic or international trade or on their individual or business bank accounts or investments. Fax-on-demand is now being applied to these many financial and commercial needs for instant information.

Health and medical organizations. In our health conscious society there is a desire for information on prevention, treatments and developments in health matters. Also, doctors seek to stay current on treatments and drugs. Leading medical organizations and hospitals are using fax-on-demand to provide information to doctors and the public.

Travel, sports and leisure time. Various applications for fax-on-demand in the home are beginning to appear. These applications are expected to expand rapidly, especially when considering the grow in personal access to fax machines. Within the next two to three years the population of faxes in the home is projected to grow to nearly ten million units. In addition, many people have access to a fax machines in their offices or businesses that they can use for personal use.

(ABConsultants and Nuntius Corporation)

Selling Fax Servers

Consider these applications when you're getting ready to sell a new fax server product. Speak to your customer base and look for companies that do a lot of inbound fax processing and outbound faxing. Look for lines at the printer and fax machine.

1. **Core business application integration.** Medical labs can automatically fax out test results, insurance agents can automatically file image documents with the home office, and real-estate agencies can provide remote searching of residential listings for relocating executives.

2. **Fax publishing.** Publishing company newsletters by fax can save all the hassle of printing them, stuffing them in envelopes, adding postage, etc.

3. **Fax-on-demand.** Basic document fulfillment requests can be automated. Some fax server platforms provide the engine and shared phone lines to support fax-on-demand databases. Such databases allow callers to call in and select documents and direct the system to immediately fax the documents back to the callers' fax machines.

4. **Fax-on-command.** Great for sales departments. Image-document databases (documents stored in fax format for quick transmission without conversion), telemarketers and customer service agents can fax brochures and diagrams without leaving their seats and telephones. And without relying on customers to call a fax-on-demand system.

5. **Business-form faxing**. Overlay forms can be displayed on end users' PC screens and they can just type in the data and then the whole form can be faxed. OCR or OMR-based paper forms can be filled out remotely, faxed from fax machines to specially-equipped fax servers. The servers then read the data on the forms. The data is then tranferred to computer systems in order to update databases or spreadsheets. This is a form of semi-automated data entry.

6. **Production-level broadcasting.** Multi-line fax servers can execute broadcasts rapidly and efficiently.

(Davidson Consulting)

Selling Customers Is Knowing Them

The best way for VARs to make money in CT is to know your customers and, most importantly, how they relate to their customers. You must understand how they make their customers happy. And you must help them make their customers happy.

Computer Telephony is all about improving business communications in creative ways -- so the more you know about your clients' communication bottlenecks, the easier it is to offer them CT-based solutions. Here's an example:

A typical medical clinic. The benefits of voice mail should be obvious, given that doctors, nurses and many other clinic staff are away from their desks much of the time. Voice messages can be retrieved when convenient, usually in between appointments.

Because it's so obvious, I wouldn't really consider it a "staple" CT product that your company should offer. The reason is this: voice mail is a commodity product, which means many firms offer very feature-rich voice mail offerings at very competitive prices. You won't make any money selling voice mail.

Something less obvious? How 'bout when folks make appointments to see their doctors? They usually write the time and date down on their calendars so they don't forget. Unfortunately, many do forget, leaving a non-revenue-generating hole in the doctor's schedule.

Many clinics assign a staff member to the task of calling all appointments the day before to nag them into coming. This greatly increases attendance, but it also consumes a live person for quite awhile, especially when the calls are being made for a clinic full of doctors. Is there a more efficient way?

Let's assume the clinic has a Local Area Network with a patient database somewhere on the network. When appointments are made, the receptionist enters the time and date in the doctors schedule and "points" to the patient's own database record.

You can create a CT application that peeks at the appointment schedule database and automatically calls the patient delivering either a reminder voice message or a reminder fax, whichever they prefer. If the system is developed with the "right stuff," it can be seen as very friendly by those receiving the calls.

Here's the features for this medical computer telephony product idea:

1. **Directly link to the existing patient and appointment databases.** CT systems should seamlessly hook to existing corporate databases. It's always a good idea to make your CT systems fit into your client's existing phone, network or database environments. CT systems that are adaptable in this way will sell more quickly.

2. **Deliver a voice, fax or e-mail reminder message**. Delivering the voice reminder message can be done by asking the answering party to get the person who needs to be notified.

3. **Use text-to-speech.** You could announce the name of the person without having to record a speech prompt for every patient's name in the database. With automatic speech recognition, you could ask the person who answers to say "yes" or "no" if they are the right person. After announcing the date and time of the appointment, you could then ask "would you like me to repeat that again?" and do so if they say "yes."

4. **Use Fax.** Not everyone has a fax machine at home, but almost everyone does at work. Many prefer faxes because it's hard copy – you can take it with you as a constant reminder. Your application could merge the appointment time and date fields into a one-page fax that gets sent to the patient if they have indicated faxes as the preferred method of being reminded.

5. **Use E-Mail.** If you want to be really high-tech about this, you could offer a third message delivery option: send them a reminder via e-mail.

6. **Make calls when patients want them.** You can enter a "notification profile" for each patient, identifying when would be best for making reminder voice calls. Calls could be easily be scheduled after normal clinic business hours, evenings, weekends or early mornings. For my family, often the best time to catch us is between 6:30 AM and 7:20 AM, before everyone leaves for work and school. An automated system could do this very easily.

7. **Built-in reports.** Every morning a report could be generated showing which reminders were successfully transmitted and which ones were not. Reports are a important part of any CT system. Managers, who are the ones who pay the bills, need to see the reports so they can analyze how well the system is performing and whether it's paying for itself (the bottom line).

This system would maintain a special results database that is used to generate these reports, but these results could also be viewed on-screen, allowing the system operator (perhaps the receptionist) to quickly call the few individuals who did not receive their reminder message.

Breaking It Down. The first thing to do is develop a brief cost justification report for your client, showing them all the benefits of the system and how much time (a.k.a. -- money) the system will save them. Specifically suggest what the payback period would be, using the loaded labor cost of the person or persons who normally make the calls. If they do not currently make such calls, get them to tell you what their monthly lost revenue figure is due to appointment "no-shows." The next step is figuring out which program modules need to be developed and tested.

Database connectivity to the patient and appointment databases is crucial, so write a short program and run it on a workstation on their LAN, making sure you can properly access the records you need. Many times, you may find the record-locking methods used by some programs are going to require special techniques so that simultaneous access works (that is, your CT system accessing the same database as their own workstations). The best time to find this out is right at the start, before any real CT design has started.

For a project like this, use the Microsoft Visual Basic language, given its excellent "Jet Engine" database connectivity -- if you can't hook to the target database using VB's native database support. This is also a good path if you can't find an add-on database driver that makes it easy (if that database format is really obscure or, worse, totally proprietary). In this case, your client will have to do daily exports to a more standard format that you can use.

Next, sketch out a block diagram of the different parts of the application, showing how messages will be delivered and the processes (like database look-ups) that will be used to create those messages. Develop a speech prompt script and show exactly what will be heard during each reminder call.

As far as the computer telephony tool kit to use, there are a good number of Visual Basic tool kits that allow you to deploy applications without paying royalties, although you will have to pay more for versions that support more phone and fax lines. Choose products from companies like Stylus, Parity, Pronexus or TRT. Most base packages support one or two lines which should be sufficient to perform your initial tests.

Choosing your CT and fax hardware vendors is not too hard -- companies like Bicom, Dialogic, NewVoice, NMS, Pika and Rhetorex all have credible voice cards suitable for this project. Fax card vendors like Brooktrout, GammaLink and PureData are good bets and all have equally good third-party software support in the tool kit arena.

Collateral technologies like text-to-speech (TTS) and automatic speech recognition (ASR) can really improve the usability of your CT application, but the price you will pay for them can vary greatly.

In both these cases, you can now make a choice between software-only solutions (less expensive) or hardware solutions (more expensive but capable of handling more simultaneous lines). For this particular application, given that the number of lines you will likely be deploying is eight or less per system, go with the software-only approach.

Even with eight lines running at the same time, the odds that more than four lines will need either ASR or TTS resources at exactly the same time is slim. Berkeley Speech Technologies has an excellent software-only TTS solution and Voice Control Systems has "VRSoft," a software-based product (also available through Dialogic) that can provide four shared channels of ASR.

Developing a system like this will take some time. If you do it right, you can resell it many times over, only making small changes to the "database connection" portions for each of your clients.

(Telephone Response Technologies, Inc.)

Selling With Simulators For Demos

If you are developing, testing or demonstrating call processing systems, get a CO simulator! These simple devices, available for under $100 from several vendors, allow you to hook up a telephone directly to your computer. Otherwise you'd need 2 lines: one to dial out and one connected to the PC.

(Parity Software)

SOHO-Perfect PC Phone

Rick Luhmann has the perfect PC – it's a phone *and* a PC. Here's his success checklist and SOHO recipe.

1. **The Basic PC Platform.** Well, maybe not all that basic, but here's a list of stuff you should have:

(A) A decent Windows PC (486DX100-based or so with 12 Megs of RAM, a Gigabyte hard drive, VGA monitor, CD ROM drive, etc.);

(B) A Telex or other cheap PC microphone (which will sound pretty good when you plug it into a DSP-based speakerphone / sound card, but, if sound is super-important to you, you might spring for a noise-canceling headset for better-quality hands-free communication). Check out ACS (Scotts Valley, CA -- 408-438-3883), Andrea Electronics (Long Island City, NY -- 718-729-8500), GN Netcom (Eden Prairie, MN -- 612-932-2992), Hello Direct (San Jose, CA -- 408-972-1990), Jabra (San Diego, CA -- 619-622-0764), Plantronics (Santa Cruz, CA -- 408-426-6060) or Unex (Nashua, NH -- 603-598-1100);

(C) A set of speakers (from Labtec, Altec Lansing, etc. -- only needed in case of board level DSP resource, external models work fine without them);

(D) An HP or Canon ink-jet printer for printing general documents and hard-copy faxes, call-logs, etc. We like HP's new $500 LaserJet 5L.

2. **The DSP Resource:**

(E) Centrepoint's (Ottawa, Ontario -- 613-235-7054) T-Modem is built on the multi-function Sierra chipset. And it will "DSPize" a single PC, adding voice / fax / data / sound tasks. But it's more than just another Caller ID-compliant computer speakerphone and fax station and sound card, etc. It can also act as a "micro PBX" out to your SOHO extensions, which, in the end, makes it a more complete package for smallish offices (up to 10 people) than most other systems.

It comes in both an external box (with status LEDs, $495) and PC board version (the PCT-Modem, $349). The external box is nice for easy install (no opening the PC) and because it works as a mini-switch with the PC on or off. Other than that, both are TAPI-compatible devices bundled with ScanView for Windows for setup and advanced faxing apps. And both support almost identical features. This includes the ability to actually switch calls from up to three single lines (though one or two should usually be enough) out to four extensions.

You could, in theory, run the external box without an attached PC. That still would give you this switching, plus its automated attendant, voice mail and fax-on-demand (which can be programmed in via touch-tones). But then you'd be missing out on the all important computer-phone aspects and the third-party fax, data, voice and multimedia software that has emerged for DSP products.

Still, the prime benefit here is the switching, especially in the home office. Many times a "business" call will be answered by a "personal" line, like when a west-coast home-office worker picks up a call from his bedroom phone at 5 AM and hears a fax tone.

Instead of scrambling to get to the line with the DSP resource (and turn on his PC), he can immediately transfer the call over to that extension, where the DSP will recognize it's a fax call, switch into that mode and take the transmission. You can then go back to bed and view the fax at your leisure.

3. **The CT Software Resources:**

(F) AlgoRhythms' PhoneKits, not all of its modules -- just the PC Phone first-party call-control mechanism, selected because it comes with a "kit" for building it any way you want. You can keep it small and remove the traditional "dial pad" from it entirely. You can put in a bunch of speed-dial buttons and launch other applications with it;

(G) SoftTalk's TAPI-compliant Phonetastic Pro, for setting up sophisticated real-time and rules-based call control apps. It interfaces nicely with existing applications for contact management, database management and even word processing and spreadsheets via DDE (dynamic data exchange), ODBC (open database connectivity), a macro language and OLE (object linking and embedding);

So even a one-person office can have customer histories and preferences immediately pop up from an application. Customer records and notes -- even voice files -- can be updated, stored or have other action taken on them. We also like the way it visually shows you what's going on in your CT big picture as it happens, while giving you point-and-click call control at all times;

(H) Connectware's PhoneWorks software for message management -- looks nice and puts the no-brainer "tape-recorder" voice tool at your fingertips at all times;

(I) Pacific Image Communications' SuperVoice Pro for Windows because it makes setting up the advanced call-handling applications that DSPs give you easy (particularly fax-on-demand) -- visually construct trees and keep track of things as you're building and creating;

(J) Communique Laboratory's 01/FAXCOM for Windows for fast connections to BBSs and the 'Net and its enhanced COM-port juggling and fax scheduling mechanism;

(K) Elan's Goldmine PIM (Personal Information Manager) for super advanced contact management;

(L) Midisoft's AudioWorks software for incorporating the sound card functions into our complete CT platform.

4. **The Anti-Couch-Potato Resource:**

(M) AIMS Lab's TR 200 ISA board and software puts a real-time boob-tube window into your home-office computer-phone screen (CATV cable or antenna plugs directly into the card). At least it will keep you off the couch and at your desk, where you can occasionally glance at ESPN, CNN, Oprah, etc. while still getting some work done. It's also an FM radio.

(Computer Telephony Magazine)

Switching System Selection

When selecting a switch for your enhanced service platform or other wireline or wireless system, make sure to identify the physical requirements of the switch. There are switches out there that fit into a compact footprint, yet support high port capacities and call rates, offering the best of both worlds: the greatest capacities in a neat, deployable package:

1. If your prospective customers range from the small central office to the large interexchange carrier, look for a vendor that support a family of products that span a broad capacity spectrum -- all this with the same application programming interface. There are switch manufacturers who offer several models, each supporting a different range of port capacities, all using the same API. For the developer, this means you can develop one solution and run it on any of those compatible platforms.

As a result, you broaden your market potential to include every opportunity from small CO or CPE environment all the way up to major switching environments supporting hundreds of thousands of ports.

2. If you are deploying switching systems in the international marketplace, such as the wireless local loop or for single-number services in major international metropolitan markets, you will need a switching platform that can address the variations in E1 R2 signaling that occur from country to country (even region-to-region, on a global basis). You could get stuck in the field and not be able to connect with the local public network, even though the software adheres to the local specification. Don't get stuck: Look for a switching environment that allows your programmers to modify the switching software wither in the labor in the field.

3. If you are developing a solution for the Intelligent network, you are going to need a switching platform that gives you both the best performance and the most flexibility for your application. Look for a switch that has an internal solution for SS7, because an SS7 capability is going to perform at its best only if it is truly integrated into the switching platform. Also, look for SS7platforms that give the developer access to the switching software environment, to differentiate your SS7 solutions by providing customizable SS7 signaling services.

4. When selecting a switching system, make sure you identify your port requirements and evaluate the switching platform capacities correctly. Determine how the system uses its DSP resources. Some switches require that you count the number of ports plus additional ports for the DSP resources that are used in a given call. In other architectures, the DSPs are a system-wide resource, accessed by all available ports, and operating on their own timeslot interchange. That way, DSPs do not take up valuable ports in the switching matrix. The total port capacity will change dramatically based on these features.

5. To get the most flexibility out of your common channel signaling environment, pick a switch that separates the common channel signaling from the line signaling.

These switches offer the greatest flexibility in configuring systems. Common channel signaling services, including both ISDN and SS7,interface to the line signaling cards. This enables the system to be configured most cost-effectively and offers the most flexibility for the customers.

(Excel, Inc.)

Text-To-Speech System Evaluation

When choosing a text-to-speech (TTS) system it is important to recognize that there are four relatively independent parameters which determine the value of a particular text-to-speech system for a given application. They are intelligibility, pronunciation, intonation, and naturalness.

Intelligibility is obviously the most critical being the sine qua non of a TTS system. Strange as it may seem, some quite robotic-sounding text-to-speech systems are really fairly intelligible , so naturalness isn't everything.

Acceptable (versus the subjective - correct) pronunciation is important, especially where proper names are involved. If Houston Street in New York City is pronounced "Hewsten" it may not do the driver trying to navigate in Manhattan very much good. Some text-to-speech systems have much more capability in this respect than others.

Intonation or stress on certain words which gives human speech its natural prosody makes a significant difference in text-to-speech applications where full sentences of text must be read out. It can be very tiring and reduce intelligibility if the TTS system lacks any attempt at good prosody.

The last measure of a TTS system is naturalness. That is, how much does the system sound like a real human being or how much does it sound like a drowning robot. If the application is one in which the user may be a customer, naturalness may be an significant consideration. Although this is functionally the least important parameter it also has been the hardest to achieve in most TTS systems.

(Voice Information Associates, Inc.)

Universal Messaging Considerations

What with the increasing magnitude of messages throughout the modern enterprise, it's no wonder that you are sometimes frustrated at the mounting confusion. That's because there are so many sources of messages nowadays. In addition, the many types of media adds to the confusion. There's faxes, e-mail, voice mail and regular mail. And with all of the new messaging devices and their associated access points, it's hard to have a "universal view" of your communications. Many of us also need to access messages from anywhere, so that's where the concept of universal messaging comes into play.

You'll also hear terms from different vendors like mixed-media messaging, and unified messaging. These are more or less synonymous with universal messaging.

Some users are "PC centric" and use their PC for virtually all messaging. Whether they are "PC centric" or "voice centric" you'll find that users have a strong preference for their messaging interface.

Figure 8.6 - Universal Messaging Topology

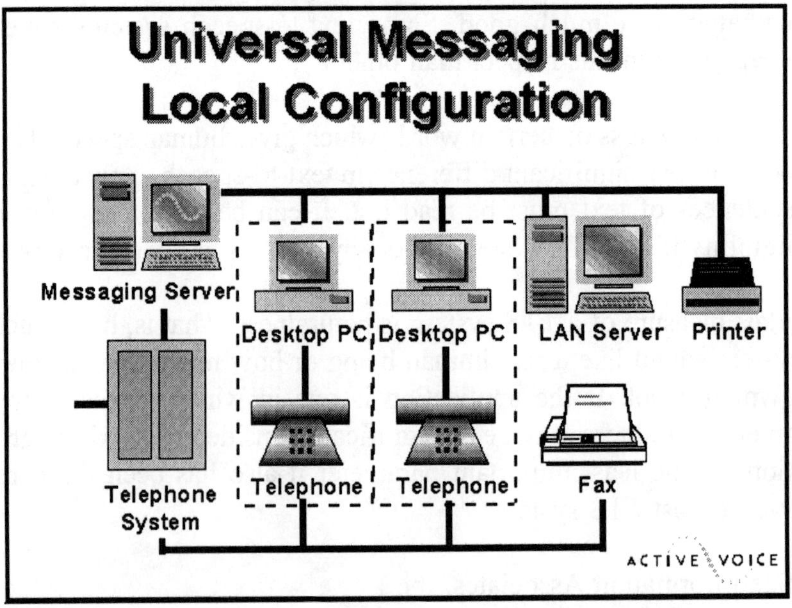

Universal messaging allows for access to all types of messages from a single point via any available communications device. A typical topology of a universal messaging system is pictured in figure 8.6. Here are some technical concerns to consider when evaluating a universal messaging system:

What type of communications device are you using? You should be able to continue to use your most cherished communications device when you make the transition. For example, you should be able to access messages via the phone or your modem.

What type of access software? Make sure the software you choose uses some easy-to-use interface. It should be non-proprietary and work alongside your other programs.

Where's the data? Messaging objects come in all forms and can be stored on your LAN file server, on a Web server, or on your PC's hard disk. Your universal messaging software needs to be able to locate and attach to these files.

Of course, if you consider using the technology, you yourself will be a customer. Most individuals who ultimately consume messaging products have the same concerns:

Does it work? Make sure you talk to other people who use the proposed platform. It either works well or it doesn't. Maybe you can "test drive" someone else's system or get and "audio tour" as a guest user on a working system.

Does it improve productivity as promised? The bottom line to any universal messaging system is its ability to make your life easier. Unless it substantially improves your productivity, then all the features in the world are worthless. Make a list of all of the things you want. Make sure the software actually performs to your expectations. Don't be wowed by gadgetry.

Can I afford it? Look not only at the acquisition price, but also the follow-on services, maintenance and training you will need to get up and running at your company. This includes additional phone line, data line, etc. charges. If you're considering a product that allows you to download messages from the Internet, then also consider those monthly access charges and the requisite software to do so.

Why Do You Need It? Make sure you really need a universal messaging system before you invest. Try matching the message types you manage to the volume of transactions and message content. For example, if you infrequently get faxes and e-mail – you may not need a universal messaging system at all. If, on the other hand, you get lots of mixed message types throughout the work day, this is a key consideration.

If using a modem outside the office is difficult, or you feel uncomfortable using one, then consider a universal messaging system that allows you to access your e-mail and faxes with a regular phone. Audio Messaging for the "voice centric" among us can still be powerful with a universal messaging system. You can still access all message types, and your messages are sorted. You can browse headers with voice prompts and listen to e-mail via text-to-speech. You can forward faxes & e-mail to your location and even use voice recognition if touch tone telephones aren't available. The one big disadvantage is that this type of messaging does not have a visual element. The prompt: "You have 46 New Messages" taxes the very bandwidth to the brain. Consider making friends with you PC, so you can get visual indications of your messages and listen to them, redirect them, etc. with simple mouse clicks.

With a PC, you can instantly browse and prioritize messages. If you're on the road, you can even hear and leave messages with a multimedia device if you're not using a phone. This also leaves your telephone free for inbound and outbound calls. You can also get visual and immediate fax notification. This helps to make your faxes more private (and even password-protected in some cases). Your faxes can also be forwarded to another location and redirected at your command.

If you are traveling (especially abroad) and a data line is easier to use (and less expensive) than a long distance phone call, then some universal messaging systems let you download your voice and fax messages for review on your laptop. Faxes sit quietly in an "in-basket" and are sent to machines, not people with universal messaging systems. You can retrieve faxes at your command in a secure way. Also, when you consider the cost of hotel fax reception (sometimes as much as $5 per page), this type of system may pay for itself in less than a year.

For the more "PC centric" among us, E-mail is the standard for messaging on and off the Internet. Some modern systems allow you to access voice and fax messages with an e-mail viewer. Mixed media messages look just like an attachment with a unique media message. Some messaging viewers can be customized to your liking.

(Active Voice)

Used Equipment Makes Sense

Reasons to buy used:

1. **Dollar savings.**

2. **Response time.**

3. **Inventory in stock.**

4. **Long warranties** (sometimes longer than new).

Some tips on buying used:

1. **Buy from a reputable dealer.** This is self-evident.

2. **Ask about their inventory.** If you buy something, is it in stock? Can it be shipped today? Many secondary dealers advertise stuff they don't have, then go and try to buy it from someone else.

3. **Ask for a warranty.** You should get a minimum of one year. How good warranties are depends on how much the dealer has in stock and how sophisticated your dealer is at repair.

4. **Ask for advance replacement.** If the thing you get sent doesn't work when you get it, your supplier should be prepared to Fedex you another one before they get your broken one back.

But you'd better demonstrate some intelligence about checking the card you got. Secondary dealers tell TELECONNECT that most of the "bad" cards they get back really aren't bad. They're just installed wrongly.

5. **Does your supplier understand** the technical aspects of your phone system so they can match appropriate revision levels.

6. **What about software?** Manufacturers use their software "upgrades" as a bargaining chip to keep their customers away from the secondary market. There are ways to obtain software upgrades while still buying on the secondary market. Can your dealer help you? You might also ask yourself:

7. **Does your supplier offer technical assistance over the phone?** When you get your stuff and it doesn't work, will they "talk you through" your problems until you finally get it working?

8. **Do they refurbish?** And to what level? Some secondary dealers clean their stuff and do nothing else. Some clean it, test it and add small but important things, like new designation strips, instruction manuals, etc.

9. **Be careful with price.** One dealer quipped, "Do you look for bargains on parachutes and brain surgeons?"

(Teleconnect Magazine)

Videoconferencing Buying Tips

1. **Always take the equipment for a test drive.** Always test what you want to do with it. If you want to make remote PowerPoint presentations, make sure you can. Check that the other end can see all your slide (they can't with Intel's ProShare).

2. **Always use PC-based stuff on the fastest PC** you can get your hands on. Don't ever believe in minimum configurations. Minimum, awful performance is what you'll get.

3. **All audio on PC-based ISDN conferencing is delayed and not good quality.** Better to make a separate audio call.

4. **All long distance calls that use two BRI channels are charged** as if they were two long distance calls.

5. **When you install ISDN stuff, always ask your phone company for two SPIDs,** one for each BRI phone line. You'll need two SPIDs to install most equipment.

6. **Watch out for Isochronous Ethernet.** It's a new standard, promises to put 96 ISDN BRI channels above a normal 10 meg Ethernet data LAN and promises great LAN video conferences. But its problem is that National Semiconductor is the only maker of IsoEthernet cards and is presently charging its OEMs $500 a card, which is ridiculous.

7. **We prefer multi-purpose hardware.** For example, you can use Vivo's hardware for tasks other than videoconferencing. You can't with Intel's ProShare.

(Computer Telephony Magazine)

Video Conferencing System On The Cheap

Here's The Recipe for a real live videoconferencing system. Be the first on your block to "call home" with a picture. Actually, this stuff is great for point-to-point video meetings, conferences and other business-related stuff:

One Vodem (Creative PhoneBlaster, Diamond Telecommander, Spectrum Office F/X) ($275)

One Internet Hookup ($20/mo)

One DataBeam's FarSight ($49)

One JABRA-Net (Includes JABRA EarPHONE and VocalTec's Internet Phone $89)

One Connectix QuickCam ($90)

One FreeVue (free) Video Conferencing over the Internet Communicator (www.freevue.com)

Total Cost, Approximately $515

With DataBeam's FarSight, you can hook up via modem or Internet and collaborate with whiteboards or share applications -- even if one of the nodes doesn't have the application. FreeVue let's you VideoConference over the Internet.

The JABRA allows private, hands-free communication, and removes echo induced by desktop speakers.

The Connectix QuickCam provides inexpensive video input without an expensive video capture board--and at the 4-15/fps transmission rates, you don't need a video capture board.

And the Vodem's provide you with Voice Mail, Integrated Fax and Email, Caller ID, Telephony and modem communications. It's a "Welcome to the 90's" SOHO and Video Conferencing Bundle, available today. With Internet Telephony, and TCP/IP, you can talk, collaborate, send video, and file transfer simultaneously over 28.8 on the Internet.

With DSVD, you'll be able to do all the above over land lines, so you'll have better quality audio. And as ISDN proliferates, you'll be able to do all the above with 2 lines, and higher data bandwidth.

(JABRA Corporation)

Voice Mail System – Keep It Simple

When choosing a voice mail system, buy one with features that your business actually needs, particularly from the point of view of customers and prospects who phone the company. Superfluous extras add to the price and often make using it more difficult. Typical cost is $100 to $250 per telephone.

(Harry Newton)

Voice Production Outsourcing

There's application programming and then there's scripting and recording your voice prompts. According to GM Productions, a leading voice talent agency and recording studio, there are many reasons to outsource your "voice programming." Before you try to script, record and program your voice prompts, consider the benefits of using a professional agency:

1. **Reliability.** Professionals are accountable for what they are paid for and take pride in their work. I you ask your secretary or other office worker to be your voice talent, you're asking a favor - not contracting for a job well done. With employee turn-over and other personnel issues at hand, your inside talent may not always be the most reliable source.

2. **Cost effectiveness.** There are hidden costs in scripting, voicing, recording and engineering your own prompts. Your office manager may be a good writer, or even someone in your marketing department. But these people typically do not earn a living by "story telling." And that's what a good script writer does. In addition, your staff will probably have to re-write the script many times, thus cutting into valuable time they could be doing what they do best.

By recruiting an MIS employee or engineer to handle the recording aspect of the job, you are paying way too much in engineering resources. A professional recording engineer is a specialist who can complete the task quicker and more reliably. In the end analysis the cost is minimized with the outsourcing approach.

3. **Higher quality.** You'll get much higher quality by using a professional recording studio. First, the engineering of high-quality sound is a specialty with these people. ^They know how to filter out unwanted noise and hiss. They know what equipment to use. They also know how to save and archive recordings properly. You just can't reproduce studio quality sound with a sound board and several hundred dollars worth of software and cables from Radio Shack.

Figure 8.7 - Face-To-Face Voice Production Meeting

4. **Timely turnaround.** If all you did all day was produce high-quality voice talent recordings and scripts, you'd get pretty quick at this. Most recording sessions take several hours and the follow-up engineering work takes a little bit longer. If you've never done it before, it will easily take five times as long.

5. **Enhanced image.** Nothing beats the sound of a voice talent professional. Their inflection and tone must be consistent. They must have a pleasing voice, and above all their voice must be intelligible. This takes years of practice to master. If a first-time caller hears a professional voice on the voice mail or IVR system, they are more likely to perceive your organization as being a professional one.

(GM Productions)

Voice Talent Agency Selection

Here's what questions to ask when you outsource voice talent, script writing and production of your professional prompts.

The Script Writer. Yes, a professional writer. They know how to "tell a story." That's what they do – probably much better than you can. Make sure you ask these questions:

1. **What Voice processing experience do you have?** Make sure the writer you are interviewing has plenty of experience in scripting for interactive voice response applications. This is a special kind of script writing that must take into account the heuristics of doing business over the phone.

2. **How well do you understands our audience?** Does the writer understand the business you are in? Make sure he or she has some notion of the profile of the callers.

3. **Do you understand what we sell?** Your writer must know what it is you sell, or else the information dispatched on your IVR system could be of little value. This is especially important for Audiotex and "On-Hold" systems.

The Recording Studio. It doesn't have to be real glitzy, but if it doesn't look like a studio (and you should tour it if you can) – then it isn't. Ask these questions:

1. **Do they use broadcast quality gear?** Take a tour of the facility. Either it's a real studio or it's not. Ask for a list of broadcast customers. This will help to ensure that the equipment they are using is top-quality.

2. **What voice processing experience do you have?** If they are at all experienced, they will have catalogued dozens of generic voice prompts for time, dollar, auto attendant, etc. If they do not have any canned material for voice processing, then they probably have very limited experience.

3. **Tell me about your engineers & directors?** The background and resumes for the company's staff should be impressive. Look for plenty of experience in radio, theater and movie production. If there is no director or engineers on staff, you're looking at a very small outfit that's understaffed. They may not be around too long.

Voice Talent. You can get male, female, loud, squeaky or commanding. It's your choice. Just make sure you ask these questions:

1. **How available is your team of voice talent experts?** Predictability is very important with voice talent. Either they do this as a career, or they don't. Ask how long the person in question has provided voice talent. The best talent is always available for re-recordings and updates. Look for proof that the person in question has been available to customers over long periods of time.

2. **How "directable" is the talent?** You don't need a prima donna. You talent should be able to take suggestions from you and the director. If they don't seem flexible, keep looking.

3. **How good is the person's tonal quality and intonation?** This is subjective, but you should get samples. A good studio will have recorded samples of their talent from which to choose. If you don't like what you hear, then your customers probably won't either.

Studio Digitizing Capabilities. If you talk about voice digitization and telephone line transmission of recorded voice – your vendor should not have a glazed look about the face. Look for these things:

1. **What file formats do you offer?** Good studios will be able to handle PCM, U-law, A-law, and ADPCM files. If they don't, then it is not likely they have experience in over-the-phone recordings.

2. **What sampling rates do you support?** Make sure the studio is comfortable with various sampling rates including 24K, 32K and 64K. This is important, since you may use the prompts on systems that use different rates.

3. **What voice processing boards do you support?** Make sure your agency can handle and has experience with Dialogic, Rhetorex, Natural MicroSystems and other boards. An experienced firm should know the ins and outs of all of these boards. They should also have each of these boards in PC that they use for testing. If they don't have any voice boards in house, then they probably don't have a lot of experience in over-the-phone prompts and scripts.

(GM Productions)

Chapter 9

Fraud and Money Saving Tips

There's many simple, almost obvious things you can do to save money and curb fraud. This chapter is dedicated to beating the people who want to beat the system and take you for all you're worth. Computer telephony technology can be abused. So can your plain old, simple telephone system. The more sophisticated technology gets – the more savvy the rip-off artist becomes. It's not too hard to beat them at their own game if you follow these simple suggestions. There's also a handful of money-saving tips borrowed from some of the smartest people in the industry for you to enjoy. After reading this chapter, you should sit down with your employees and discuss ways to save money and stop phone abuse -- immediately. A few simple tweaks and you'll be happier and richer, too. Unless otherwise noted, these tips are courtesy of Harry Newton.

Airplane Phone Warnings

Every telecom device which works through the air -- **cellular, cordless and airplane phones** -- can be listened into by someone else.

There are now devices which will listen to these phones, wait for touch-tone signals -- such as your credit card -- and throw those numbers on a screen or into a database. With those numbers anyone can make a long distance or international phone call.

Auto Attendant Security

Don't allow digits that match your trunk access codes to be used on your auto attendant. This is one sure way of letting hackers use your system as a "leaky PBX" or poor man's DISA (Direct Inward System Access) system.

Block Calls To 900, 976, 970 Etc.

Block calls to "Dial-it" services and area codes where you don't do business. Most businesses routinely block to the area code -- 900. Be careful with the 212 area code which has expensive exchanges -- 394, 540, 550, 970 and 976.

Caller ID Curbs LD Abuse

Zeus Phonestuff (Atlanta, GA -- 404-263-7111) offers a standalone combination Caller ID and paging product called Always-In-Touch.

This little $99 box plugs into your incoming phone line and traps Caller ID. Once the call is complete, it sends the calling number to your digital pager. This is great when you want to know who just left you voice mail or tried to get in touch with you. It does not send an alpha pager message, but instead sends predefined codes for special cases like private number and out-of-area calls. It can also page you with the number from an outbound phone call or just long-distance outbound calls.

If you suspect long distance phone abuse and want to catch it, put the box on a suspected abused phone. When an outbound long-distance call is made from that line, the box will page you and tell you what number was dialed.

It can also be used as a speed pager system by letting anyone on that phone line pick up the phone and press 77**. The box will pick up the line once they hang up and send a code to the pager to tell you to call home.

Dialback Security

This is security feature that more intense hacker-prevention systems can give you. If a person calls in wanting remote access to either LD lines or CT plant, the system asks for a password. Once it receives a correct password, it hangs up on the caller and dials back a pre-defined remote number, only then giving the caller access. Obviously, unless the hacker has you tied up in your living room, it makes things very secure. It's even better when it's combined with specialized password features like voice recognition and extremely secure passkey technology from MicroFrame and others.

DISA And Voice Mail Fraud

Best toll prevention: Kill your DISA (Direct Inward System Access) phone lines and ports altogether. If not possible, at least kill them on weekends. If not possible, check them regularly for abuse or use software which alerts you to unexpected surges in dialing. Meantime, use the lengthiest authorization codes your switch will allow. This will minimize any hacker's chance of success.

Make sure there is no way someone can dial through your voice mail and make long distance calls.

DISA -- And PBX Maintenance Ports

Who can dial into your PBX's maintenance port? Do you know? Have you changed the password? Or is it still the same one your PBX came from the factory with? A crook who gets the number and the password could dial into your maintenance port and turn all your lines into DISA (Direct Inward System Access) ports, thus allowing people to dial in and use your phone lines for dialing long distance and international.

Fax And Its Hidden Cost

Beware of the hidden cost of fax. By far the largest lifetime cost component of a server based fax device is the cost of the long distance charges to send faxes. An intelligent, higher performance fax card will save the average company $2500 or more per year per fax line! Don't be penny wise and pound foolish.

(GammaLink Division of Dialogic Corporation)

Fax Boards And Wasteful Speed

While you have seen 28.8 data/fax cards everywhere, in reality the 28.8 kbps standard for fax transmission was only approved by the International Telecommunications Union in 1996. (Those data/fax cards did 28.8 data but only 14.4 kbps fax). Since 28.8 fax machines will only be introduced this year, and at a premium price, and there are 42 million existing 14.4 or slower fax machines in the world, it will be a long time before 28.8 is a significant factor in the fax market. Consider that 5 years after its introduction, 14.4 fax machines still represent less than 20% of the market.

(GammaLink Division of Dialogic Corporation)

Fraud Buster Top Tips

Malicious people can inflicted hundreds of thousands of dollars of unauthorized calls on your company. Some people even set up "weekend phone companies," re-selling calls to foreign countries dialed through voice mail / auto attendant / IVR systems. It's a serious problem. No one wants to walk in on Monday to a $250,000 weekend phone bill. It's happened. Here are ***The Top 55 Tips*** from Marc Robins on ways to prevent it.

These tips will help you create a program for combating misuse, abuse and fraud on your voice processing system. Like most preventative maintenance, the steps are only effective if you carry them out quickly and diligently. If you wait for disaster to strike, you've waited too long.

1. Don't publish 800 numbers or voice mail access numbers in company brochures, newsletters, or other "public" documents that may fall into the hands of hackers.

2. Many systems boast of an on-line voice mail directory for callers who don't know the extension number of the person they want to leave a message for. While this is a worthwhile service, it can also be a boon to hackers if mailbox numbers are the same as the extension numbers, giving hackers a list of every mailbox. One alternative is to provide a "department" directory on-line. Callers will be answered by a secretary who can offer to forward the caller to the called party's mailbox.

3. Passwords should not be a social security number or someone's birth date. These may seem like safe numbers, but such information can be gotten from various sources.

4. Many systems can enforce the minimum number of digits users select for passwords. This minimum is determined by the system administrator. One vendor suggests a ratio of 10,000 possible passwords for every actual mailbox. This means that if a system has only one mailbox, the password should be four digits long (9,999 possible passwords). If the system has ten mailboxes, passwords should be five digits in length (99,999 possible passwords). One-hundred mailboxes would require six- digit passwords, and so on.

5. If a voice mailbox will remain unused for an extended period of time; e.g., vacation, out-of-office for two or more weeks, disability; consider forwarding calls for that employee to a secretary rather than the voice mail system. If you were a hacker and listened to a voice message that said the user would be not using the mailbox for an extended period of time, wouldn't you be tempted to "borrow" the mailbox?

6. Limit access to PBX ports used for automated attendant service by using toll restriction. Callers can also be denied access to trunks using a class-of-service restriction.

7. Each application on the voice processing system -- e.g., out-dialing, paging or networking -- should have its own dedicated group of ports. This limits hackers from using one of these services as a gateway to the world.

8. System administrator passwords should be as long as the system allows. Twelve or fifteen digits is *not* unreasonably long. Some systems provide several levels of password protection, depending on the activity or function performed. This is a good feature.

9. Change system passwords whenever system managers change.

10. System administrator passwords, if written down, should be kept in a secure place like a locked drawer or safe. (Not written on a Post-it note stuck to the screen of the system-manager's terminal.)

11. Make sure the screen of the system administrator's terminal blanks when not attended. A screen full of information makes a tempting target to passers-by.

12. If your system provides for several "levels" of prompts, opt for a level that reduces feedback to callers logging on to the system. Omitting prompts that can guide hackers through the system may discourage less experienced hackers. There's no reason to "tell" hackers that extensions are four digits long, or passwords are five digits in length.

13. Using a "self-destruct" feature that erases all messages from a mailbox whenever the password is changed is an effective means for providing security. However, this may encounter resistance from some users (particularly those with a habit of forgetting passwords.) Let users know the consequences of forgetting passwords. This should improve their memory.

14. Install a screen-saver program (very inexpensive), that causes the screen of the system administrator's terminal to go blank after 10 or 15 minutes of non-use. Hitting any key will bring back whatever was on the screen, but a hacker doesn't necessarily know that.

15. Review reports periodically that highlight inactive mailboxes, and disable those that remain inactive for longer than one month, unless the user can substantiate the need for a mailbox.

16. Review "bad password attempt" reports on a daily basis. No mailboxes should collect more than three during a 24-hour period. If any do, you should be looking for signs of hacker activity.

17. It may seem obvious, but instruct operators, receptionists and secretaries to use discretion when giving out voice mailbox numbers. If a caller requests an individual's mailbox number, offer to transfer the caller to the mailbox rather than simply giving the number out.

18. Also obvious, pilot numbers (the special number users call to get directly into the voice mail system), should not be given out over the phone, regardless of who's calling. The same goes for access codes. If a caller claims to have forgotten how to access their voice mailbox, the secretary, receptionist or operator should offer to have the voice processing system administrator call them back and explain it. This will discourage all but the most tenacious hackers.

19. Delete mailboxes immediately when no longer needed (i.e., retired, fired, long-term disability and temporary employees).

20. Review lists of authorized mailbox users periodically (every two or three months). Verify that users no longer employed by the firm; e.g., summer interns, former part-timers and retirees; no longer have mailboxes assigned to them.

21. Passwords should be changed as often as possible. A bare minimum is four times each year, but monthly would be preferred. This may not go over well with users, but it is for their own good.

22. Some systems can automatically force users to change passwords after a certain period of time defined by the administrator.

If your system has this capability, use it. If users complain, explain to them that it's the system and there's nothing you can do about it.

23. Restrict the use of outbound trunks as much as is feasible without restricting authorized company business. If there is no need to be calling Europe, don't allow the PBX to complete calls to those international area codes. If there is no need to allow long distance calling at night or on weekends, don't allow them.

24. During regular business hours, route any outbound calls to non-essential destinations (South America, Far East), to the Operator. If there is a valid reason for the call, the Operator can place the call.

25. Don't include executive mailbox numbers in any on-line directory. (This won't make industrial spies happy, but too bad.)

26. Check all authorization codes and passwords when the installation of a new system is complete. Factory default codes and passwords are often used. Experienced hackers know these codes. Don't make it too easy for them.

27. If you have a remote maintenance port, use a secure, two- line dial-back modem.

28. Warn users about SHOULDER SURFERS (hackers who hang out in airports and public areas with binoculars spying on pay phone users as they enter voice mail number, mailbox number and password). Users should attempt to shield the key pad as much as possible to prevent a clear view of the numbers dialed.

29. Beware of DUMPSTER DIVERS (criminals who scout trash containers; e.g., dumpsters; looking for lists of passwords, authorization codes, mailbox numbers, whatever). If these kinds of records must be printed, be sure to shred them before discarding.

30. If the number of "failed attempts before disconnect" is a flexible parameter on the voice mail system, set it to three or less. If an authorized user can't get it right after two attempts, perhaps it's time for remedial training.

31. Program the voice processing system to wait at least five rings before answering the dial-back modem connected to the remote access port. This will discourage hackers and auto-dial programs.

32. Insist that the call-back line on your two-line modem is set up for out-dialing only, and cannot receive incoming calls. (This must be requested from the local telephone company. Note: May not be available in all areas.)

33. Instruct operators and receptionists to give outside lines to people requesting them only if they recognize their voice, or call the party back at their extension.

34. Some users have short memories and need to write down their passwords. Tell them to keep these little scraps of paper in secure places, such as their wallets, purses or the locked laptop drawer of their desk. Explain to them that these little pieces of paper are as valuable as a credit card (especially if a hacker gets a hold of them).

35. The out-dial feature for voice mail should not be used, unless absolutely necessary. Limit and monitor its use carefully.

36. Ask your voice mail, automated attendant system and PBX supplier about any loopholes they may know of that would allow callers to grab an outgoing PBX trunk from the voice processing system. Various features may be ACTIVATED or INACTIVATED when systems are shipped from the factory. Make certain that features you don't want activated are deactivated when the system is installed. After this, test the system yourself.

37. Check the list of authorized mailboxes against the system- generated list of active mailboxes (if one is available). They should agree.

If they don't, investigate immediately. Someone other than the system administrator may be opening up boxes.

38. If you don't already have a call accounting system tied into your PBX, get one. Call accounting packages can provide not only records of what extensions made what calls, but they can also generate lists of calls that fit parameters that you select. These can included all calls over a certain length (60 minutes, for example), all calls to certain area codes (like the Caribbean), and excessive numbers of calls within a definable period of time. These exception reports can be generated on a daily basis and can detect most fraud long before the phone bill arrives.

39. Never allow "default" passwords to be used. Many systems can generate lists of mailboxes using default passwords. Contact each user and threaten them with the loss of their mailbox if they don't change their password immediately.

40. Insist that new users select a password as soon as they log on for the first time. (This should be within hours of when the mailbox is setup by the administrator.)

41. Many systems force new users to select a password, and some even select the password for them from a random number generator to prevent users from selecting "1111" or "1234" or some other obvious number. This is a good feature. Use it if your system has it.

42. Never leave a voice processing system unsecured. Even if it's a PC, lock it up in an equipment room. Restrict access to as few employees as possible (i.e., technicians, system administrators).

43. Secure manuals and lists of mailboxes and other specific system information safely.

44. Encourage users to check their mailboxes on a regular basis, even if they aren't receiving messages. If a hacker has taken over their mailbox, you want someone to know as soon as possible.

45. Change the number of the administrative mailbox. Often, this is a default number assigned at the factory, and used by every new system installed by that manufacturer or distributor. Hackers are well-informed on all default settings. This is one mailbox in particular that you don't want hackers cracking.

46. Monitor usage levels of the system. If available storage message suddenly drops, investigate. There may be hackers leaving extended messages somewhere in the system.

47. Limit the length of messages that callers can leave to a reasonable minimum (three or four minutes is more than adequate). Hackers will find it more difficult to use up system capacity this way.

48. Explain to users the danger of programming passwords into their telephone speed dial (particularly if they have a display phone). Anyone can walk by, press a key, and read the password on the LED display.

49. Limit the number of people given administrative privileges. Four or five is too many. Two is reasonable.

50. Consider additional security measures. There are keyboard locks on the market that disable the keyboard of the system administrator's terminal, making it impossible for anyone to do anything with the system. This may seem like overkill, but it is an inexpensive alternative to changing every password in the system after a hacker commandeered the password file.

51. Many manufacturers are beginning to accept responsibility for alerting users of potential security problems. (Northern Telecom offers a video dramatizing security and toll fraud issues.) Don't be afraid to ask what your supplier is doing to reduce the risk for abuse and fraud. If they respond with, "There is no problem." consider a different supplier.

52. Provide training in securing voice processing systems to all employees when a system is installed, and furnish remedial security training on a regular basis (annually).

53. Distribute fliers to all employees every two or three months covering tips for preventing security problems. Check with the voice processing system manufacturer or distributor. They may be able to furnish fliers.

54. Keep an emergency list of who to call near the system manager's terminal so that, in the event that a hacker makes his presence known, you will have the names and numbers close at hand. Make sure you have off-hours numbers too in case you must contact someone over the weekend or holiday.

55. If your present system doesn't offer an adequate number of safeguards, get a new system.

(Robins Press)

IVR Logic Flow Security

The level of security needed depends on the type of information and functions available. You need to be able to assure your customers that other callers will not be able to access information about their accounts or conduct transactions without their permission. No security is needed for obtaining generic information like directions and hours, but account information may require the entry of some distinctive code. Some transactions may need more than one level of security.

(Enterprise Integration Group)

PBX Fraud Stoppers

A checklist of how to stop PBX fraud:

1. Delete all authorization codes that were programmed into your PBX for testing and service.

2. Treat authorization codes as you would treat credit cards. One code to one person. Treat them like gold.

3. Consider replacing your PBX's remote access feature with a virtual private network or telephone calling cards.

4. Don't use phone numbers, social security numbers or employee ID numbers as authorization codes.

5. Always use the longest PBX authorization codes available.

6. Your PBX's DISA numbers should be unpublished, bear no relation to your other numbers and should wait at least five rings before they answer.

7. Don't use a steady tone as the prompt to input an authorization code. Use a voice prompt or no prompt. Whenever an invalid authorization code is entered, the call should be ended instantly.

8. Restrict international calls to those countries you don't have business with, especially obvious places like Colombia, Pakistan, India, Burma, Thailand and the Caribbean.

Shoulder Surfing Fraud

Shoulder Surfing happens when the guy behind you in the payphone line copies down your credit card number as you dial it and then steals it.

Solution: Dial your credit number and then keep dialing about five or six extra digits. The system doesn't care how many extra numbers you dial. Just don't press the # key. The extra numbers will confuse thieves.

Switchroom Security

Elementary security will stop most casual thieves. Lock your switchroom. Put different locks on your PBX's cabinets. The original installers still have keys. The factory-installed keys are the same as on all cabinets they ship.

Telex, Fax And Directory Scams

One big scam: Sending companies an invoice for an appearance in a fax, telex or businesses directory that **is never published**. No kidding. TELECONNECT itself receives at least six "invoices" a year for directories that are never published. We **still** receive one for a directory listing for our telex machine -- which we got rid off at least seven years ago!

Toll Abuse Responsibility

If your phone system gets attacked by "hackers" who rip you off for thousands of dollars, you're responsible for paying the phone bill – even though you didn't make the calls. AT&T told TELECONNECT: *"AT&T cannot take responsibility for tariffed charges billed when the customer controls the security of the compromised access. AT&T does assume the responsibility when AT&T controls the security of the compromised access. This policy refers solely to regulated services and does not address claims relating to unregulated services or equipment."*

Toll Restrict These Numbers Instantly

They are the country codes and costs for calling ships via Inmarsat. Ouch. These country codes will cost:

011-871....$10 a minute
011-872....$10 a minute
011-873....$10 a minute
011-874....$10 a minute

Voice Authentication Alert

Voice authentication is a technology that uses features of a person's voice to decide if they are the person they claim to be. With telephone networks and cellular telephone companies losing more than $1 million a day, voice authentication is becoming a hot technology. Toll fraud of all types is at $3 billion a year.

Security is all about what you have, what you know and who you are. Examples of what you have include keys, IDs, keycards and tokens. These things can be lost, stolen, damaged or not on your person at the time. Examples of what you know include PINs, passwords and personal information. Someone else can learn what you know. They are things that someone else can discover (and things that you can forget).

To get around these problems, you can use Biometrics. The "who you are" is what Biometrics concentrates on. This includes automatic fingerprint, retina scan, palm print,. handwriting/signature and of course: Voice.

Voice authentication can be used to control access to telephone networks, computer networks and control access to sensitive data. You should consider using long distance service from companies that offer voice authentication. Even if someone steal the long distance calling card – they can't use it. Calls can only be made after the authorized user speaks his or her unique (voice authenticated) password.

(Veritel Corporation of America)

Appendix A

Recommended Resources and Bibliography

The publications cited here offer a comprehensive and detailed understanding of the subjects covered in this book. The publisher would like to acknowledge the authors of copyrighted material for the permissions granted to use portions of their work as solicited by the research staff.

Great Books on Computer Telephony:

236 Killer Voice Processing Applications,
Edwin Margulies, Flatiron Publishing, Inc., 1995.

Basic Book of Information Networking,
Motorola Codex, Motorola University Press, 1992.

Basic Book of ISDN,
Motorola Codex, Motorola University Press, 1992.

Client Server Computer Telephony,
Edwin Margulies, Flatiron Publishing, Inc., 1994.

Complete Traffic Engineering Handbook,
Jerry Harder, Flatiron Publishing, Inc., 1992.

Computer Based Fax Processing,
Maury Kauffman, Flatiron Publishing, Inc., 1994.

Computer Telephony on the Sun Platform,
Patrick Kane, Flatiron Publishing, Inc., 1996.

Customers: Arriving With a History and Leaving With an Experience,
Andrew Waite, Flatiron Publishing, Inc., 1996.

Customer Service Over the Phone,
Stephen Coscia, Flatiron Publishing, Inc., 1995.

Handbook of Telecommunications,
James Harry Green, Flatiron Publishing, Inc., 1993.

International CallBack Book,
Gene Retske, Flatiron Publishing, Inc., 1995.

Local & Long Distance Billing Practices,
M. Brosnan & J. Messina, Flatiron Publishing, Inc., 1993.

Newton's Telecom Dictionary - 10th edition,
Harry Newton, Flatiron Publishing, Inc., 1996.

PC Telephony,
Bob Edgar, Flatiron Publishing, Inc., 1995.

Predictive Dialing Fundamentals,
Aleksander Szlam, Flatiron Publishing, Inc., 1996

Reference Manual for Telecommunications Engineering,
Roger L. Freeman, John Wiley & Sons, publisher.

SCSA - Signal Computing System Architecure - 2nd Edition
Edwin Margulies, Flatiron Publishing, Inc., 1996.

Speech Recognition,
Pete Foster, Flatiron Publishing, Inc., 1993.

Telecommunications Management,
Harry Green, Flatiron Publishing, Inc., 1993.

The MVIP Book,
GO-MVIP, Flatiron Publishing, Inc., 1995.

Telephony For Computer Professionals,
Jane Laino, Flatiron Publishing, Inc., 1994.

Testing Computer Telephony Systems and Networks,
Steve Gladstone, Flatiron Publishing, Inc., 1994.

The Guide to T-1 Networking,
Bill Flanagan, Flatiron Publishing, Inc., 1990.

Understanding Computer Telephony,
Carlton Carden, Flatiron Publishing, Inc., 1996.

Understanding Data Communications,
G. Friend & J. Fike, Flatiron Publishing, Inc., 1993.

Understanding Telephone Electronics,
J. Fike, Flatiron Publishing, Inc., 1993.

Visual Basic Telephony,
Krisztina Holly & Chris Brookins, Flatiron Publishing, Inc., 1995.

Voice Processing,
Walt Tetschner, Artech House, 1991.

Appendix B

Tipster Contact Informaton

Do you want to get more great tips, hire a consultant, team-up with a strategic partner or buy some great products? This is the contact information for all of the companies that contributed tips, secrets and shortcuts to this book. They are in alphabetical order.

4-Sight, L.C.
1801 Industrial Circle
West Des Moines, Iowa 50265-5557
515-221-3000
515-224-0802 (fax)
www.4sight.com

ABConsultants
940 Saratoga Avenue, Suite 220
San Jose, CA 95129
408-243-2234
408-921-2236 (fax)
71732.2776@compuserve.com

Access Graphics, Inc.
1426 Pearl Street
Boulder, CO 80302
303-938-9333
303-546-3259 (fax)
www.access.com/TeamCTI

ACS Wireless
10 Victor Square
Scotts Valley, CA 95066
408-461-3219
408-438-8769 (fax)
www.acs.com

Active Voice Corporation
2901 Third Avenue
Seattle, WA 98121
206-441-4700
206-441-4784 (fax)
www.activevoice.com

Adtran, Inc.
901 Explorer Boulevard
Huntsville, AL 35806
205-971-8000
205-971-8699 (fax)
www.adtran.com

Alliance Systems, Inc.
16801 Addison Road, Suite 120
Dallas, TX 75248
214-250-4141
214-250-0921 (fax)
www.asisys.com

Applied Language Technologies
215 First Street
Cambridge, MA 02142
617-225-0012
617-225-0322 (fax)
www.altech.com

ART, Inc.
43 Brodezky
P.O. Box 61398
Tel Aviv, 69128 Israel
972-3642-7242
972-3-642-5887 (fax)
100274.3223@compuserve.com

Aspect Telecommunications
1730 Fox Drive
San Jose, CA 95131-2312
408-441-2200
408-441-2260 (fax)
www.aspect.com

Brite Voice Systems, Inc.
Perception Technology Division
40 Shawmut Road
Canton, MA 02021
617-821-0320
617-828-7886 (fax)
www.brite.com

Brooktrout Technology, Inc.
144 Gould Street
Needham, MA 02194
617-449-4100
617-449-9009 (fax)
www.brooktrout.com

Caléo Software, Inc.
5255 Triangle Pkwy., Suite 150
Norcross, Georgia 30092
770-453-9680
770-453-9686 (fax)
www.caleo.com

CenterCore, Inc.
1355 W. Front Street
Plainfield, NJ 07063
800-220-5235
908-561-3442 (fax)

Cintech
Tele-Management Systems
2100 Sherman Avenue
Cincinnati, OH 45212
513-731-6000
513-731-6200 (fax)
cintech@one.net

Computer Telephony Magazine
12 West 21 Street
New York, NY 10010
212-691-8215
212-691-1191 (fax)
www.computertelephony.com

Crystal Group, Inc.
1165 Industrial Avenue
Hiawatha, IA 52233-1120
319-378-1636
319-393-2338 (fax)

Computer Telephone Division
Dialogic Corporation
100 Unicorn Park Drive
Woburn, MA 01801
617-933-1111
617-935-4551 (fax)
www.dialogic.com

Cybernetics Systems International
2600 Douglas Road, Suite 700
Coral Gables, FL 33134
305-529-0020
305-443-2335 (fax)

Davidson Consulting
530 North Lamer Street
Burbank, CA 91506
818-842-5117
818-842-5488 (fax)
davidsonco@aol.com

Davox International
6 Technology Park Drive
Westford, MA 01886
508-952-0200
508-952-0202 (fax)
www.davox.com

Dialogic Corporation
1515 Route 10
Parsippany, NJ 07054
201-993-3000
201-631-9631 (fax)
www.dialogic.com

Diamond Head Software, Inc.
1217 Digital Drive, Suite 125
Richardson, TX 75081
214-479-9205
214-479-0219 (fax)

Diamond Multimedia Systems
Supra Communications Division
312 SE Stonemill Drive, # 150
Vancouver, WA 98684
360-604-1400
360-604-1401 (fax)
www.supra.com

DIgby 4 Group, Inc.
370 Lexington Ave., Suite 1510
New York, NY 10017
212-883-1191
212-370-5369 (fax)
janeccc@aol.com

Digital Systems International
6464 185th Avenue NE
Redmond, WA 98052
206-881-7544
206-869-4530 (fax)
www.dgtl.com

ENSONIQ
155 Great Valley Parkway
Malvern, PA 19355-0735
610-647-3930
610-647-8908 (fax)
www.ensoniq.com

Enterprise Integration Group
1320 El Capitan Drive, Suite 20
Danville, CA 94526
510-328-1300
510-328-1313 (fax)
eig@a.crl.com

Excel, Inc.
255 Independence Drive
Hyannis, MA 02601
508-862-3000
508-862-3030 (fax)

Expert Systems, Inc.
1301 Hightower Trail, Suite 201
Atlanta, GA 30350
770-642-7575
770-587-5547 (fax)
www.easey.com

Genoa Technology
5401 Tech Circle
Moorepark, CA 93021
805-531-9030
805-531-9045 (fax)
www.gentech.com

GM Productions
8 Piedmont Center, Suite 101
Atlanta, GA 30305
404-237-9700
404-237-5522 (fax)

Hammer Technologies, Inc.
226 Lowell Street
Wilmington, MA 01887
508-694-9959
508-988-0148 (fax)
www.teradyne.com/hammer/

Harris Digital Telephone Systems
300 Bel Marin Keys Boulevard
Novato, CA 94948
415-382-5000
415-382-5428 (fax)
www.harris.com/harris/comm_sys/

Ibex Technologies, Inc.
550 Main Street, Suite G
Placerville, CA 9566
916-621-4342
916-621-2004 (fax)
www.ibex.com

I-Bus PC Technologies
A Division of Maxwell Laboratories
9596 Chesapeake Drive
San Diego, CA 92123
619-974-8400
619-268-7863 (fax)
www.ibus.com

Intel Corporation
2111 NE 25th
Hillsboro, OR 97124
503-696-8080
503-264-9027 (fax)
www.intel.com

JABRA Corporation
9191 Town Centre Dr., Suite 330
San Diego, CA 92122
619-622-0764
619-622-0353 (fax)
www.jabra.com

MediaSoft Telecom
8600 Decarie Blvd., Suite 215
Mount Royal, Quebec H4P2N2
514-731-3838
514-731-3833 (fax)
www.cam.org/~mst

Micom, Inc.
4100 Los Angeles Avenue
Simi Valley, CA 93063
805-583-8600
805-583-1997 (fax)
www.micom.com

Natural MicroSystems
8 Erie Drive
Natick MA 01760-1339
508-650-1300
508-650-1350 (fax)
www.nmss.com

New Pueblo Communications
660 South Country Club Road
Tucson, AZ 85716
520-322-6556
520-322-0631 (fax)

Novell, Inc.
1555 North Technology Way
Orem, UT 84057
801-222-6000
800-453-1267 (fax)
www.novell.com

Nuntius Corporation
8048 Big Bend Blvd., Suite 110
Street Louis, MO 93119
314-968-1009
314-968-3163 (fax)
www.nuntius.com

Octel Communications Corp.
1001 Murphy Ranch Road
Milpitas, CA 95035
408-321-2000
408-321-2100 (fax)
www.octel.com

Optus Software
100 Davidson Avenue
Somerset, NJ 08873-9931
908-271-9568
908-271-9572 (fax)
www.facsys.com

Panamax
150 Mitchell Boulevard
San Raphael, CA 94903
415-499-3900
415-472-5540 (fax)
www.hooked.net/panamax

Parity Software
1 Harbor Drive
Sausalito, CA 94965
415-332-5656
415-332-5657 (fax)
www.paritysw.com

PIKA Technologies Inc.
155 Terrence Matthews Cres.
Kanata, ON Canada K2M 2A8
613-591-1555
613-591-1488 (fax)
www.virtualmarketplace.com/pikae.html

Plantronics
345 Encinal Street
Santa Cruz, CA 95060
408-426-6060
408-426-1850 (fax)
www.plantronics.com

PureSpeech
100 Cambridge Park Drive
Cambridge, MA 02140
617-441-0000
617-441-0001 (fax)
www.purespeech.com

QNX Software Systems Ltd.
175 Terrence Matthews Crescent
Kanata, Ontario K2M 1W8
613-591-0931
613-591-3579 (fax)
www.qnx.com

RAAC Technologies, Inc.
219 N. Milwaukee Street, 3rd Floor
Milwaukee, WI 53202
414-277-1889
414-277-9876 (fax)

Results Technologies
499 Sheridan Street
Dania, FL 33004
305-921-2400
305-923-8070 (fax)

Rhetorex, Inc.
Subsidary of Octel
200 East Hacienda Avenue
Campbell, CA 95008
408-370-0881
408-370-1171 (fax)
www.rhetorex.com

Robins Press
2675 Henry Hudson Parkway
Riverdale, NY 10463
718-548-7245
718-548-7237 (fax)
robinspr@ix.netcom.com

Rockwell International
Switching Systems Division
1431 Opus Place
Downers Grove, IL 60515
708-960-8000
708-960-8165 (fax)
www.rockwell.com

Siemens ROLM Communications, Inc.
4900 Old Ironsides Drive
Santa Clara, CA 95054
408-492-2000
408-492-2160 (fax)
www.rolm.com

Soft-Com
140 West 22, 7th floor
New York, NY 10011
212-242-9595
212-691-6223 (fax)

SoftLinx, Inc.
234 Littleton Road
Westford, MA 01886
508-392-0001
508-392-9009 (fax)
www.softlinx.com

TALX Corporation
1850 Borman Court
Street Louis, MO 63146
314-434-0046
314-434-9205 (fax)
www.talx.com

Technology Solutions Company
205 North Michigan Avenue, Suite 1500
Chicago, IL 60601
312-819-2250
312-819-2299 (fax)
www.techsol.com

Teknekron Infoswitch Corporation
4425 Cambridge Road
Fort Worth, TX 76155
817-267-3025
817-571-9464 (fax)
www.teknekron.com

TELECONNECT Magazine
12 West 21 Street
New York, NY 10010
212-691-8215
212-691-1191 (fax)
www.teleconnect.com

Telephone Response Technologies, Inc.
1624 Santa Clara Drive, Suite 200
Roseville, CA 95661
916-784-7777
916-784-7781 (fax)
www.trt.com

The Kaufmann Group
Fax Consultants
324 Windsor Drive
Cherry Hill, NJ 08002
609-482-8288
609-482-8940 (fax)

TKM Communications
60 Columbia Way, Suite 300
Markham, Ontario L3ROC9
905-470-5252
905-470-7008 (fax)

Veritel Corporation of America
350 West Kensington, Suite 117
Mount Prospect, IL 60056
708-670-1780
847-670-1785 (fax)

Voice Information Associates
ASR News
14 Glen Road South
Lexington, MA 02173
617-862-8185
617-863-8790 (fax)

Voysys Corporation
48634 Milmont Drive
Fremont, CA 94538
510-252-1100
510-252-1101 (fax)
www.voysys.com

WTS Bureau Systems, Inc.
2170 Lone Star Drive
Dallas, TX 75212
214-353-5000
214-353-5025 (fax)

Xircom Systems Division
5 Manor Parkway
Salem, NH 03079
603-898-1800
603-894-4545 (fax)
www.xircom.com

INDEX

A

ACDs
Web Callback 4 – 4
What They Do 4 – 2
ADSI Information Sources 3 – 2
AEB Signaling on D/41E Boards Under MS–DOS 6 – 2
Agent Scripting 4 – 39
Airplane Phone Warnings 9 – 1
Alarm Subsystems 8 – 2
AMI WinBios Configuration Problems 6 – 3
ANI Considerations 4 – 7
Answering Consistency 3 – 2
Application Generator
Buying Tips 8 – 18
Simple IVR 3 – 5
UNIX 8 – 19
Application
Development 3 – 3
Discovery Input 1 – 8
Education 1 – 10
Menu Design 3 – 7
Objectives; IVR System 1 – 7

Planning; Ideas 1 – 5
Planning; Questions 1 – 8
Planning; Time 1 – 2
Planning; Tips 1 – 2
Requirements List 1 – 4
Scheduling 3 – 71
Specification 3 – 3
Tools; Choices 8 – 20
Wrong Reasons 1 – 10
Area Code Look–Ups 3 – 9
ASR
Application Design Tips 3 – 10
Arguments 1 – 12
Benefits 8 – 21
Challenges 8 – 24
Considerations 8 – 28
Evaluation 8 – 30
Integrator Guidelines 8 – 28
Opportunities 8 – 21
Vocabulary Break–Down 3 – 10
Auto Attendant
Advanced Features 4 – 10
Critical Questions 3 – 11
Considerations 1 – 13
Security 9 – 2

B

B8ZS Line Encoding Detection 6 – 4
Backup With XCOPY 3 – 11
Block Calls To 900, 976, 970 Etc. 9 – 2
BRI ISDN Installation Tips 7 – 2

Buying Application Tools 8 – 6
Buying Industrial PCs 5 – 2

C

Cables; Bad 7 – 5
Caller ID
  Buying Tips 8 – 32
  Centrex 2 – 2
  LD Abuse 9 – 2
  Planning 1 – 18
  Problems 3 – 12
  Purchase 8 – 32
  Upgrade 1 – 14
  Why Use It 1 – 14
Caller Input Quality 3 – 12
Channel Buffer Size 3 – 15
Consultants
  Honesty 8 – 38
  When To Use 1 – 20
Conventional Memory 3 – 16
Cool Your Phone System 7 – 5
Credit Card Software For IVR 3 – 17, 1 – 21
CSU Companies 2 – 25
Customer Relationship 4 – 11

D

D/41E–SC; Non–SCbus 6 – 6
Debit Calling Service
  Buying Tips 8 – 39
  Planning Tips 1 – 46
Dedicated 800 Lines 2 – 2
Dialback Security 9 – 3
Dialing International Numbers 3 – 19
Dial Tone 6 – 7
DID Mistake 7 – 5
DID Or DNIS Digits 2 – 3
Digit Assignment 1 – 50
DISA
  Maintenance Ports 9 – 3
  VM Fraud 9 – 3
Disaster Prevention 1 – 50
DMI
  Description 4 – 12
  Help Desk 4 – 18
DNIS And DID Digit Passing 2 – 3
Document Imaging 3 – 19
DOS Comm. Software 6 – 7
DSVD Setup 4 – 19
DSVD Technical Support 6 – 7
DTI/1xx T–1 card Installation Tips 7 – 6
DTMF Dialing Levels 6 – 10
DTMF Generators – Acoustically Coupled 6 – 11

E

Ergonomics
- Call Center 4 – 23
- Workplace Creation 4 – 24

Error Handling 3 – 21
European Tip & Ring 7 – 9
Expanded Memory Use 3 – 22
Extended Memory In DOS 3 – 23

F

Failure Planning; LAN 1 – 54
Fax
- Apps Gen 8 – 41
- Archive 7 – 9
- Blaster Features 3 – 24
- Boards And Speed 9 – 4
- Broadcasting Guidelines 1 – 42
- Hidden Cost 9 – 4
- Machine 1 – 41
- Modem Processors 8 – 45
- Planning Tips 1 – 22
- Purchasing 8 – 43
- Reception Truths 1 – 26
- Service Bureau Considerations 1 – 43
- Service Bureau Questions 8 – 51
- Statements 1 – 42

Transmission Quality 1 – 40
Fax-On-Demand
Benefits 1 – 31
Document Changes 3 – 25
Installation Warnings 7 – 9
Integration With WWW 1 – 36
Service Bureau Selection 8 – 52
Services 2 – 4
Software Features 8 – 57
Success Tips 1 – 35
System Parts 6 – 13
System Pricing 8 – 60
System Sizing; Pricing 8 – 62
Via Fax Server 8 – 57
Web – Planning 1 – 38
Who Needs It 8 – 43
Fax Server
Compatibility Issues 7 – 11
Integration Operations 3 – 26
IRQ Problems 7 – 10
LAN Advantages 1 – 25
Pricing 8 – 45
Productivity 1 – 29
Retries 8 – 46
Testing 7 – 12
Vendor Questions 8 – 46
Vendor Segmentation 8 – 50
Why To Use Them 1 – 24
File Handles 3 – 26
Flash ROM Updates 6 – 14
Forms; Save In Text 3 – 32
Forms-Based Tools 3 – 28
Fraud Buster Top Tips 9 – 4

G

Getting Started Successfully 8 – 63
Grounding Problem 7 – 13

H

Hard Disks; Fast 6 – 14
Headset And
Headset Considerations 4 – 26
Headsets; Wireless Kind 8 – 67
Help Desk Support
  Calls 4 – 27
  Roles 4 – 33
Home Agent Planning 4 – 35

I

Installation
  Bills 2 – 5
  On Time 2 – 6
Integration of Data, Voice and Fax 1 – 45

International
Call–Back 1 – 46
Service Planning 1 – 54
Interrupt Dueling 6 – 15
Intranet Planning Tips 1 – 56
ISDN
BRI Lines; Ordering 2 – 13
BRI vs. PRI 2 – 11
BRI vs. SW56 2 – 8
DOS Driver 3 – 32
FAQ 2 – 9
PRI Versus T-1 2 – 18,20
Real 2 – 6
Standards 2 – 8
Why To Use It 2 – 7
IVR
Buying Reasons 8 – 68
Cost Justification 8 – 71
Logic Flow Security 9 – 12
Programming Tips 3 – 32
Selling Approach 8 – 69

J

Japanese Disconnect Supervision 3 – 34

K

Key System Alert 7 – 14

L

LAN
   Cards; Priority 6 – 15
   Segmentation 7 – 14
   Survival 3 – 35
List Management 3 – 35
Load Testing IVR Apps 7 – 20
Logic Flow Considerations 3 – 38
Loop Current Reversal 2 – 3
Loop Versus Ground Start 2 – 21

M

Mail Merge 3 – 40
Memory
   Conflicts 6 – 15
   Saving; How To 3 – 45
   Terms 3 – 43

Meridian 1 PBX T-1 7 – 20
Middleware And CTI 1 – 57
Mission Critical
PC Benefits 5 – 10
PC Platforms 5 – 9
Mistakes; Large Systems 7 – 21
Modem
Call Waiting 6 – 15
Hang-Up 3 – 46

N

NEC Mark II PBX Integration 6 – 16
Norstar
Busy Signal 6 – 16
Caller ID 6 – 17

O

On "Hold" Buying 1 – 60
On-Line Strategies 4 – 37
Operating System
Selection Tips 3 – 46
Fax Servers 3 – 55
Fault Tolerant Tips 8 – 72
Outbound Calling 3 – 56

P

Pager Tone Detection 3 – 57
PBX
  Disconnect Tones 7 – 24
  Fraud 9 – 12
  D/42-SX; SL Boards 3 – 58
  Buying Tips 8 – 73
PC
  Buses; Fans; Slots 5 – 15
  Component Selection 8 – 75
  Switching Configuration 5 – 16
PEB Configuration; HD 6 – 18
Peer To Peer vs. Internet 1 – 61
Power
  Failure Transfer 7 – 26
  Outlets 7 – 27
  Problem Solvers 7 – 27
  Protection On AC; Phone 7 – 26
  Protection; ISDN Apps 7 – 31
Predictive Dialer
  Answering Machine Detection 8 – 78
  Buying Tips 8 – 79
Professional Services Fees 8 – 83
Prompts
  Human Interface 3 – 63
  Recording Tips 3 – 64
  Scripts and Recording Tips 3 – 79
  Silence Is Golden 3 – 59
  Succinct 3 – 60
  Touch – Not Push 3 – 62
  Zero Messages 3 – 59

R

R2MF High Rate Signaling 6 – 18
Reboot Switch For Master Computer 6 – 19
Reboot Your Phone System 7 – 32
Reseller Or Distributor CT Tips 1 – 64
Rotary Pulse Conversion Snags 3 – 70
Routing Tables On Your PBX 7 – 32

S

SC2000 Internal Registers 6 – 21
Screen Pop Test 4 – 38
Script; Development 3 – 74
Selling
  Application Ideas 8 – 84
  Customers; Knowing Them 8 – 91
  Fax On Demand 8 – 84
  Fax Servers 8 – 90
  Simulators For Demos 8 – 96
Serial Port Overruns 6 – 21
Shared Memory – Enabling ISA (Pentium PCs) 6 – 27
Shared RAM Voice Card Exclude 6 – 28
Shoulder Surfing Fraud 9 – 13
Simulators For Testing 7 – 32
Software Interrupt 6D 6 – 28
SOHO–Perfect PC Phone 8 – 96
Staffing Levels For Call Centers 4 – 40

Switching System Selection 8 – 99
Switchroom Security 9 – 13
Synchronous Callback 3 – 80

T

T–1
Carrier Lines; Ordering 2 – 21
Line Installation 2 – 25
Local 2 – 26
Testing Gear 7 – 32
Lock–Out 7 – 34
Troubleshooting Companies 7 – 33

Talk–Off And How To Beat It 3 – 81
Telephone Service Changes 2 – 4
Telex, Fax And Directory Scams 9 – 14
Test
Everywhere 7 – 34
Lifecycle 7 – 36
Platform Economics 7 – 35
Text–To–Speech Evaluation 8 – 101
Thunderstorms 7 – 40
Toll Abuse Responsibility 9 – 14
Toll Restriction 9 – 14
Touch-Tone Charges – Don't Pay 2 – 26
Traffic
Thumbnail Sketch 1 – 68
Tips 2 – 26
Port Sizes 1 – 67
Troubleshooting
PC Boards 7 – 41

T-1 7 – 40
Trunks 7 – 42
TSAPI 3 – 81
Type-Ahead 3 – 82

U

Universal Messaging 8 – 102
UNIX
SRL Parameters 3 – 84
System Kernal Tuning 3 – 83
Upper Memory
D40DRV; TSRs 3 – 85
DOS Device Drivers 3 – 87
QEMM–386 3 – 87
Used Equipment 8 – 105

V

V&H Routing 3 – 88
Video Conferencing
Buying Tips 8 – 107
Cheap System 8 – 108
Virtual Nets Cheap 2 – 28
Voice
Authentication 9 – 14
Card Purchase 8 – 33
Files; RAM Disk 3 – 88

LAN Impact 1 – 70
Mail System; Simple 8 – 109
Processing Menus 3 – 89
Production Outsourcing 8 – 109
Storage 3 – 89
Talent Agency 8 – 111
Warning Messages 3 – 90
Won't Stac 3 – 89

W

Watchdog For PCs 3 – 90
Water Alert 7 – 42
Windows
3.1 And Windows NT Server 3 – 94
95 And EMM386 3 – 92
INI Files 3 – 94
Memory 3 – 94
NT Choice 3 – 96
NT Compiling; Visual C++ 3 – 95
Performance Tips 3 – 92
Programming Preamble 3 – 98
Wireless Phone Considerations 4 – 41, 7 – 13
Workforce Management With Skill Based Routing 4 – 43